NOTES ON THERMODYNAMICS
Hot Oolong Cools

NOTES ON THERMODYNAMICS
Hot Oolong Cools

Hway Chuan Kang

National University of Singapore, Singapore

NEW JERSEY · LONDON · SINGAPORE · BEIJING · SHANGHAI · HONG KONG · TAIPEI · CHENNAI · TOKYO

Published by

World Scientific Publishing Co. Pte. Ltd.

5 Toh Tuck Link, Singapore 596224

USA office: 27 Warren Street, Suite 401-402, Hackensack, NJ 07601

UK office: 57 Shelton Street, Covent Garden, London WC2H 9HE

British Library Cataloguing-in-Publication Data
A catalogue record for this book is available from the British Library.

NOTES ON THERMODYNAMICS
Hot Oolong Cools

ISBN 978-981-12-9037-4 (hardcover)
ISBN 978-981-12-9158-6 (paperback)
ISBN 978-981-12-9038-1 (ebook for institutions)
ISBN 978-981-12-9039-8 (ebook for individuals)

For any available supplementary material, please visit
https://www.worldscientific.com/worldscibooks/10.1142/13763#t=suppl

Typeset by Stallion Press
Email: enquiries@stallionpress.com

Printed in Singapore

Preface

This set of notes, compiled in the past two years from teaching a well-received half-semester course on Thermodynamics, is a supplement to our lectures and tutorials and a guide to more comprehensive texts. We motivate the Laws of Thermodynamics by showing that these really come from empirical observations. We find this instructive because the scientific method may not be obvious in many Thermodynamics texts. For example, starting with everyday experiences, such as a cup of hot tea cooling spontaneously, we use the heat engine as a thinking tool to arrive at the Second Law and the notion of Entropy. We do not hurry through the Laws or merely state them and then simply proceed with applications. All the fundamentals are covered and explained. Our discussions are illustrated with applications to examples, a number of which are hopefully resonant with the broad concerns and interests of students who take STEM classes today. Amongst other examples, we look at the thermodynamics of hydrogen and gasoline engines, hydrogen fuel cells versus the explosive combustion of hydrogen, how efficiently organisms and Spiderman utilize energy, the fizzing of a can of soda and decompression sickness, how atmospheric carbon dioxide affects ocean pH and, worryingly, dissolves the calcium carbonate shells of marine animals. Also, what might happen if you inadvertently fall into a salt lake. We find in-class demonstrations connected to these examples to be interesting to students pedagogically, and thus leverage on these in the notes.

One theme that runs through the book is the efficiency of any thermodynamic machinery, including organisms. However, aside from its immense practical relevance, Thermodynamics is certainly not just a pragmatic subject. It contains deep insight into the passage of Time. In our discussions of the Second Law, we scratch the surface of this, emphasizing that all the processes we observe in our Universe are irreversible. The sentiment is old, but it is probably even more urgent today for all well-educated students to understand the Second Law as much as any pre-eminent work of literature in order, I paraphrase, to progress beyond our Neolithic ancestors. Thermodynamics, apparently, suffers from a bad pedagogical reputation; undeservedly so. In these notes we strove for an informal, readable style but without compromising the rigour. We present Thermodynamics as the macroscopic approach to understanding Nature where heat is involved, drawing upon the microscopic picture of matter only at those points where that helps to intuitively clarify the macroscopic ideas. We introduce the idea of microstates in order to discuss the entropy of mixing, connecting to both Boltzmann and Gibbs' formulations of entropy from the microscopic viewpoint. This is then used, for example, to motivate the formulation of the chemical potential in terms of the activities in non-ideal systems. The goal is to help students appreciate the physical meaning of Thermodynamics and apply its principles at the elementary level.

Our course is mainly for chemistry students who have previously taken general chemistry (and a half-semester of quantum mechanics, which we do not really draw upon). However, non-majors, including non-STEM majors, have reportedly enjoyed and benefitted from the course. These notes are compiled with that broad audience in mind. The mathematics draws only upon elementary algebra and basic notions of high school calculus. Thus, it might serve well as an accessible guide for anyone trying to learn Thermodynamics on their own. These notes contain sufficient material for a complete semester-long course. The chapters are arranged in a traditional manner. The initial chapter provides the background for later discussions. Then the Laws of Thermodynamics in Chapters 2 to 6, and the general framework for equilibrium thermodynamics in

Chapters 7 and 8. This is followed by Chapter 9 which discusses approximate model substances that we use in applications in Chapters 11 to 13. Chapter 10 fleshes out the thermodynamics of the van der Waals gas as a specific important model substance, illustrating phase change.

Contents

List of Examples

Heat and Internal Energy

1.1 Heat, Mechanics, Thermodynamics

Any piece of macroscopic matter consists of a large number of atoms in constant random motion relative to each other, the speeds and the displacements of individual atoms in this random motion being a function of the temperature. This microscopic random motion is heat. Relatable examples of this random motion are the vibrations of the atoms in a piece of solid, and the translation, rotation and vibration of nitrogen and oxygen molecules in the air, which you directly sense as the temperature of a hot iron or the air in your room. Thermodynamics is the science of how heat affects the properties of macroscopic matter. Although heat comes from motion, the scope of Thermodynamics is more than that in Mechanics.

The simple harmonic oscillation of a pendulum, illustrated in Figure 1.1, is a basic example discussed in Mechanics. In this illustrative mechanical system, a mass attached to a string is set into motion to swing back and forth under the influence of gravity. As it oscillates, the energy of the pendulum is interconverted between its kinetic energy of motion and its gravitational potential energy. In an idealised pendulum in Mechanics, there is no friction and its oscillatory motion goes on ad infinitum. A task in Mechanics might be to figure out how the maximum speed of the mass is related to its oscillation amplitude θ. In a real pendulum, as the mass swings, it experiences friction, which (slightly) warms up the air and the mass. An example more typical of a Thermodynamics course would be a pendulum that is set oscillating in an insulated, closed box of air illustrated in Figure 1.2. In this example, the random velocities of the air molecules increase as the pendulum oscillation decreases; the temperature of the air in the box increases from T_1 to T_3 as illustrated.

A typical question in Thermodynamics is: how hot does the mass and surrounding air get when the system finally gets to equilibrium when all the initial mechanical energy is converted to heat. Such interconversion of different forms of energy is the concern of Thermodynamics.

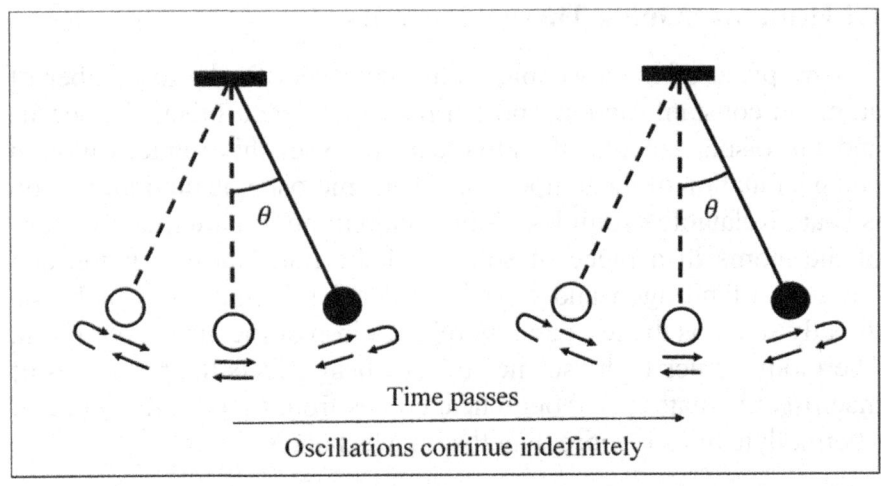

Time passes

Oscillations continue indefinitely

Figure 1.1 A model pendulum in Mechanics. The oscillations continue indefinitely and the maximum displacement θ does not diminish with time. Energy is conserved.

Although heat is a manifestation of the random motion of atoms, the methods in thermodynamics are quite different from those of Mechanics which deal with the bulk motion of bodies. In any macroscopic sample of matter, the number of atoms is enormous. In a one litre box of air, there are approximately 2.5×10^{22} molecules. In seeking to understand the swinging of the pendulum, it would be futile to even attempt to follow the trajectories of that large number of particles colliding with each other, with the walls of the box, and with the oscillating mass. In addition, there are the atoms in the oscillating mass. These share a common bulk velocity, but each of these atoms also has an individual vibrational motion that is random relative to the others. The kinetic energy of random motion of the air molecules and the atoms in the mass constitutes the heat energy in the enclosed box; this is not readily accessible by the methods of Mechanics due to the large number of particles involved. For Thermodynamics, on the other hand, it is a single variable – temperature – that quantifies this random motion of the molecules in the system.

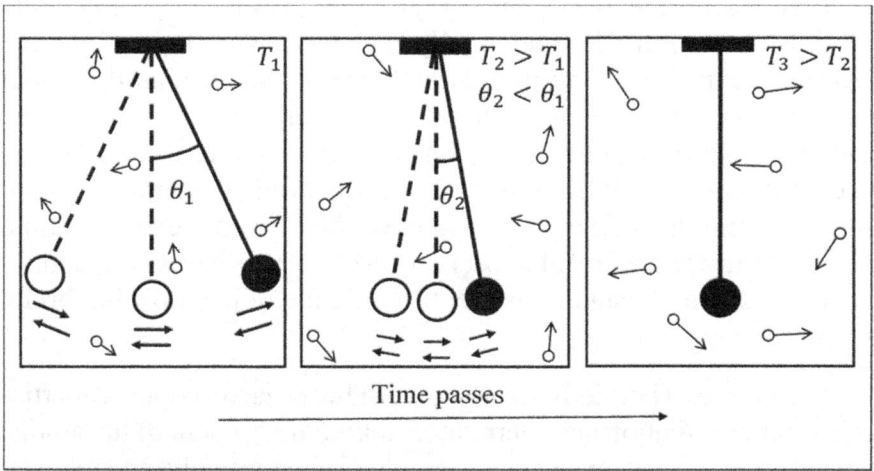

Figure 1.2 A model pendulum in Thermodynamics. The pendulum oscillates inside a box of gas molecules represented by the small circles. As time passes, the oscillation amplitude θ decreases to zero, while the average speed of the air molecules increases. Energy is again conserved. In Thermodynamics, we pay attention to the temperature of the air, which increases as the pendulum slows down.

Aside from the impracticality of keeping track of the trajectories, the fundamental laws of mechanics are not sufficient. Temperature is a necessary fundamental quantity in Thermodynamics. On the other hand, the fundamental equations of motion in neither classical mechanics nor quantum mechanics depend upon temperature, so that heat plays no role. In Thermodynamics, the familiar law of conservation of mechanical energy needs to be extended to include heat interactions. These are absent in problems in mechanics which do not involve a macroscopic object heating up or cooling down. The equations of motion in mechanics are symmetric with respect to time-reversal. This means that if you change the sign of the time variable in the equations of motion, you have the same fundamental laws of mechanics. This time symmetry is clearly not the case for macroscopic systems. These evolve irreversibly in time to settle into states of equilibrium; hence the need to study Thermodynamics. The oscillations of the pendulum in Thermodynamics will gradually diminish in amplitude, and finally the mass will stop oscillating, with its temperature and the temperature of the air slightly higher than at the start of the oscillations. The reverse phenomenon, of an initially stationary pendulum spontaneously starting to oscillate by accumulating energy from its surroundings while the temperature of the surrounding air decreases, will never be observed even though it does not violate the law of conservation of energy. Thermodynamics enables us to understand this fact through one of the most important laws of Nature.

The laws of Thermodynamics govern how macroscopic properties of matter or radiation are interrelated, taking no account of the atomic constitution of matter. Some examples of commonly encountered macroscopic properties are temperature, pressure, density, entropy, heat capacity, compressibility, expansion coefficient, surface tension, refractive index, viscosity, and solubility. For any substance, Thermodynamics gives the general framework to relate these properties to each other. Two such macroscopic properties of a substance are its equation of state and its molar heat capacity. The former tells us how the pressure and density of the substance depend upon temperature, while the latter tells us how much heat we need to raise the temperature of one mole of the substance by one degree.

Thus, a specific task we apply Thermodynamics to is to figure out the general relationship between the equation of state and the molar heat capacity of any substance. It is the encompassing empirical success and the economy of its laws in describing these general relationships that have led Einstein to regard Thermodynamics as "the sole theory of universal content" that "will never be overthrown". The success of the approach taken in Thermodynamics arises from the fact that any macroscopic system consists of a large number of interacting atoms in random motion. Hence, a reasonable approach to understand these systems is to average over these random motions to understand their effects on macroscopic properties. That is the subject of Statistical Mechanics, which provides the connection between Mechanics and Thermodynamics. In these notes we will only draw upon any statistical mechanical considerations when these are elementary and provide some additional clarity for learning Thermodynamics.

1.2 System, surroundings, universe

We define some basic terms used in thermodynamic analysis.

System: this is the part of the Universe we focus on in examining heat, work, matter exchange, or any other thermodynamic change such as a change from liquid state to vapour state. For example, the water vapour in a container is the system when we are considering the thermodynamic changes to the water vapour.

The boundary of the container is a surface only in the mathematical sense. It is ascribed properties such as being thermally insulating or conducting, flexible or rigid, permeable or semi-permeable or impermeable. If the boundary allows heat to be exchanged between the system and its surroundings, it is a diathermal boundary. On the other hand, a thermally insulating boundary is referred to as adiathermal or adiabatic. If the boundary is flexible, mechanical work can be done as the volume of system changes against an external pressure. With a permeable boundary, or a valve, material may enter or leave the system. These properties of the boundary are convenient to invoke in thermodynamic analysis to control the exchange of heat, work and matter, between the system and its surroundings. Because thermodynamics is concerned with macroscopic properties, thermodynamic systems consist of macroscopic amounts of matter.

Surroundings: this is the part of the Universe that the system is exchanging heat, work or matter with. For example, we may be supplying heat to our container of water vapour. The water vapour is our system, the source of heat is the surroundings.

Universe: the system and all the possibly different parts of the surroundings that are needed in doing a thermodynamic analysis. It does not necessarily mean the physical Universe. Hence,

$$universe = system + surroundings.$$

Isolated system: a system that cannot exchange energy and cannot exchange matter with any other object. An isolated system does not interact with its surroundings. An example is a sealed, insulated and rigid container of water vapour; it cannot gain/lose any energy, and it cannot gain/lose any matter. Hence, for an isolated system, the internal energy is a constant.

Closed system: a system that can exchange energy but cannot exchange matter with its surroundings. For example, a container of water vapour which is sealed so that it cannot gain/lose water molecules or any other matter. However, its container wall may be heat conducting so that it can gain/lose heat. Or its container wall may be flexible so that the container can be expanded or compressed by an external force so that the system can gain/lose energy through mechanical work. Or the container may be fitted with a piston on which acts an external pressure. Matter cannot be gained/lost by the system, but the water vapour may expand and push the piston against an external pressure, thus exchanging energy through mechanical work.

Open system: a system that can exchange energy and matter with its surroundings. For example, water vapour in a container with porous, heat conducting or flexible walls.

We next review three basic thermodynamic properties that are probably already familiar to you. This review provides the starting platform for our further discussions.

1.3 The *PVT* equation of state

A familiar example of an equation of state is that for the ideal gas

$$PV = nRT$$

where P is pressure, V is volume, T is temperature, n is number of moles of gas, and R is a constant number known as the gas constant. The ideal gas equation of state describes a gas with

- zero intermolecular interaction
- atoms/molecules that are approximated as point particles with zero volume.

The ideal gas equation of state gives a reasonably good description of real gases when the temperature is high and the pressure is low. At high temperatures the random kinetic energy of motion of gas molecules dominates the potential energy of intermolecular interaction. At low pressures, the volume of gas molecules is negligible compared to the volume occupied by the gas.

Another example of an equation of state which we will encounter in later chapters is the van der Waals equation:

$$\left(P + \frac{an^2}{V^2} \right)(V - nb) = nRT.$$

The constants a and b are known as van der Waals coefficients. The former accounts for the strength of the attractive intermolecular interaction, while the latter accounts for the strength of the repulsive intermolecular interaction. Although it is not quantitatively accurate, especially at high pressures, the van der Waals equation of state is important because it was the first model to provide physical insight into the condensation of a gas. It maintains the form of the ideal gas equation of state. P in the ideal gas equation of state is replaced by $\left(P + \frac{an^2}{V^2} \right)$, while V is replaced by $(V - nb)$.

Within the bulk of the gas, the molecular neighbourhood of each gas molecule is isotropic on average so that the attractive molecular interactions average to zero. However, molecules of the gas that are close to the walls of its container have a net attractive interaction into the bulk of the gas because that is the direction where there are other gas molecules. The rate of collision of molecules with the walls of the container is proportional to the density of the gas. For each of these colliding molecules the number of attractive interactions pulling it toward the bulk of the gas is also proportional to the density of the gas. Hence, the correction term is proportional to the square of the density of the gas giving the factor of $\frac{n^2}{V^2}$. The attractive interaction lowers the pressure from what it would have been if there were no interaction. Thus, the term to describe the pressure of this gas within the form of the ideal gas equation is $\left(P + \frac{an^2}{V^2} \right)$.

An effect of the repulsive intermolecular interaction is to create an excluded volume around each molecule in the gas, because each molecule repels other molecules that approach it too closely. Thus, the coefficient b can be effectively considered as setting a non-zero volume for each molecule. Hence, subtracting this so-called excluded volume from the total volume occupied by the gas gives us $(V - nb)$. Putting these two corrections into the ideal gas equation of state results in the van der Waals equation of state.

An equation of state that is obtained from more fundamental/systematic arguments is the virial equation of state which is as follows:

$$PV = nZRT$$

$$Z = 1 + \frac{B(T)}{V} + \frac{C(T)}{V^2} + \ldots$$

where Z is known as the compressibility factor and is a function of the temperature. The coefficients $B(T)$, $C(T)$, etc, in this series are known as the virial coefficients. The first term in the virial equation of state captures the ideal gas dependence of the pressure upon temperature and volume. The subsequent terms provide corrections that depend upon successively higher powers of the density through the $1/V^m$ factors. The compressibility factor is

$$Z(T) \equiv \frac{PV}{nRT},$$

so that measurements of pressure, volume and temperature directly gives the value for $Z(T)$. Using the methods of statistical mechanics, the virial coefficients can be obtained theoretically given the intermolecular interaction potential.

At sufficiently high pressures for any gas, the molecules are close enough such that their intermolecular interactions are predominantly repulsive. You would intuitively expect that the pressure is higher than in an ideal gas at the same temperature and pressure. This results in a compressibility factor that is larger than unity. Conversely, at lower pressures, when the average distance between gas molecules is larger, the average intermolecular interaction is attractive, and the pressure is lower than in an ideal gas. Hence, compressibility factors for most gases at "normal" pressures and temperatures are less than unity. At very low pressures, all gases tend towards ideal gas behaviour. Thus, as the pressure tends to zero for any gas, we have

$$\lim_{P \to 0} Z(T) = 1.$$

This limiting behaviour of all gases as the pressure tends towards zero is important in defining the ideal gas temperature scale.

§ *Example 1.1 The van der Waals gas*

The van der Waals parameters for carbon dioxide are $a_{CO_2} = 3.658 \times 10^{-1}$ N m^4 mol^{-2} and $b_{CO_2} = 4.29 \times 10^{-5}$ m^3 mol^{-1}. Calculate the molar volume v of carbon dioxide at a temperature of 273.15 K and a pressure of 5 bar. Calculate its compressibility factor Z_{CO_2} at this temperature and pressure. Do the same for helium for which the van der Waals parameters are $a_{He} = 3.455 \times 10^{-3}$ N m^4 mol^{-2} and $b_{He} = 2.38 \times 10^{-5}$ m^3 mol^{-1}.

$$1 \text{ bar} = 10^5 \text{ N m}^{-2}$$

For one mole of the van der Waals gas,

$$\left(5 \times 10^5 + \frac{a}{v^2}\right)(v - b) = 273.15R.$$

We can solve this cubic equation for v, getting the solution

$$v = 4.421 \times 10^{-3} \text{ m}^3 \text{ mol}^{-1} = 4.421 \text{ dm}^3 \text{ mol}^{-1}.$$

Thus, under these conditions, the compressibility factor of carbon dioxide is

$$Z_{CO_2} = \frac{Pv}{RT} = 0.97336.$$

Since the compressibility factor is less than one, we can conclude that the intermolecular attractive interactions dominate the repulsive interactions.

Similarly, the molar volume of helium estimated from the van der Waals parameters is

$$v = 0.004564 \text{ m}^3 \text{ mol}^{-1},$$

giving a compressibility factor of

$$Z_{He} = \frac{Pv}{RT} = 1.00494.$$

For helium, the repulsive intermolecular interactions dominate.   §

1.4 Internal energy

Each atom in any piece of matter has

- kinetic energy due to its random thermal motion,
- potential energy due to its interaction with its neighbouring atoms/molecules.

Generally, these two components make up what is called the internal energy of the object because these energies derive from physical phenomena internal to it. For example, when the temperature of the object changes, both of these components of internal energy can change substantially.

Internal energy does not include kinetic energy due to bulk motion of the object. For example, the kinetic energy due to bulk flow of a liquid is not part of its internal energy. Internal energy also does not include the interaction with an external field. For example, a solid object at the surface of the Earth interacts with the Earth's gravitational field so that raising or lowering its height changes its total energy. The total energy of a system is equal to the sum of its internal energy, kinetic energy of bulk motion and the potential energy from interaction with an external field. Since a potential energy due to interaction with an external field does not solely "belong" to the *object* nor to the external field, we need some clarification on when it is convenient not to include this in the internal energy of a *system*.

Any object of mass m at the surface of the Earth has a gravitational potential energy equal to mgh, where h is the height of the object relative to some arbitrary reference altitude. Consider a gold bar sitting on your desk. We can quite conveniently consider its internal energy to be just the sum of the random kinetic energy due to the thermal vibrations of the gold atoms, plus the potential energy that is due to the interatomic interactions between them. Both of these components of energy are significantly dependent upon the temperature of the gold bar. As the temperature rises, the vibrational kinetic energy increases significantly, for example. The gravitational potential energy, even though it arises from the interaction of each gold atom with the Earth's gravitational field, can be conveniently excluded from the internal energy. This is because the average altitude of the gold bar is just the altitude of your desk, and not dependent upon its temperature; mgh for the gold remains a constant regardless of its temperature, and can thus be conveniently excluded from its internal energy. Of course, the total energy – the internal energy plus the gravitational potential energy – obeys the law of conservation of energy.

What about a column of water vapour in the atmosphere? The molecules again have both kinetic energy arising from the random thermal motion of the molecules and intermolecular potential energy arising from their interactions with each other. Now, when the temperature of the column of water vapour is changed, the average altitude of the water molecules will change because the gas column can expand upward when warmed. Thus, the potential energy due to the interaction of each water molecule with the Earth's gravitational field is dependent upon the temperature of the water vapour column.   The thermodynamic system can now be conveniently considered to be the water vapour column plus the gravitational field, rather than just the water vapour. The internal energy of this system is therefore the sum of

- the kinetic energy of random thermal motion of the water molecules,
- the potential energy of intermolecular interactions between them,
- the gravitational potential energy from the interaction of each water molecule in the Earth's gravitational potential

1.5 Internal energy of ideal gases

It is useful to briefly describe how the kinetic theory of gases gives us an estimate of how much internal energy there is in a gas. In an ideal gas the molecules are non-interacting, so that the energy of each molecule is simply the kinetic energy due to its motion. From classical mechanics, this is just $\frac{1}{2}mv^2$, where m is the mass of a molecule and v its speed. Thus, the internal energy of an ideal gas is simply the average of this over all the molecules in the gas. A basic idea in statistical mechanics is that, in a thermally equilibrated population of molecules at temperature T, the probability of molecules with energy ϵ is

$$p(\epsilon) \propto exp\left(-\frac{\epsilon}{k_B T}\right).$$

This is called the Boltzmann factor for energy ϵ at temperature T. The constant k_B is the Boltzmann constant. Therefore, at temperature T, the fraction of ideal gas molecules with speed v is proportional to the Boltzmann factor

$$exp\left(-\frac{mv^2}{2k_B T}\right)$$

where we are considering a monatomic ideal gas such as argon; its internal energy is just its translational kinetic energy, that is ϵ is just $\frac{1}{2}mv^2$. An argon atom with speed v can have its velocity vector (v_x, v_y, v_z) in any direction with

$$v_x^2 + v_y^2 + v_z^2 = v^2,$$

where v_x, v_y and v_z are the velocity components along the x, y and z directions. This means that the fraction of atoms with speed between v and $v + dv$ is proportional to the volume of the spherical annulus of radius v and thickness dv centred at the origin of velocity space. Putting this together with the Boltzmann factor, the fraction of argon atoms with speed between v and $v + dv$ is equal to

$$C4\pi v^2 \, exp\left(-\frac{mv^2}{2k_B T}\right) dv$$

where the constant C is a normalisation factor so that the integral of these fractions over all possible speeds from zero to infinity is equal to one:

$$\int_0^\infty C4\pi v^2 \, exp\left(-\frac{mv^2}{2k_B T}\right) dv = 1.$$

This gives

$$C = \left(\frac{m}{2\pi k_B T}\right)^{\frac{3}{2}}.$$

Therefore, the probability distribution of speeds in a monatomic ideal gas where the atoms have mass m is

$$p(v) = 4\pi \left(\frac{m}{2\pi k_B T}\right)^{\frac{3}{2}} v^2 exp\left(-\frac{mv^2}{2k_B T}\right).$$

This central result of the kinetic theory of gases is known as the Maxwell-Boltzmann distribution of speeds in an ideal gas. From this we calculate the mean speed

$$v_{mean} = \int_0^{\infty} v p(v) \, dv$$

$$= 4\pi \left(\frac{m}{2\pi k_B T}\right)^{\frac{3}{2}} \int_0^{\infty} v^3 exp\left(-\frac{mv^2}{2k_B T}\right) dv$$

$$= 4\pi \left(\frac{m}{2\pi k_B T}\right)^{\frac{3}{2}} \left(\frac{\sqrt{2}k_B T}{m}\right)^2 = \left(\frac{8k_B T}{\pi m}\right)^{\frac{1}{2}}.$$

Using the Maxwell-Boltzmann distribution, we can also easily calculate the average kinetic energy per atom:

$$\left\langle \frac{1}{2} mv^2 \right\rangle = \int_0^{\infty} \left(\frac{1}{2} mv^2\right) p(v) dv = \frac{3}{2} k_B T.$$

Thus, for an ideal gas, molar kinetic energy is

$$\frac{3}{2} N_A k_B T = \frac{3}{2} RT$$

where N_A is the Avogadro number. The relationship between the Boltzmann constant k_B and the gas constant R, namely $R = N_A k_B$, allows us to match the results of Statistical Mechanics and Thermodynamics, connecting the microscopic to the macroscopic.

§   *Example 1.2 How fast are the air molecules in your room?*

Calculate the root-mean-square velocity v_{rms} of an oxygen molecule at room temperature and pressure. Approximate the oxygen molecule as a hard sphere with a radius of 1.78×10^{-10} m.

Estimate the volume swept out by an oxygen molecule in one second given the root-mean-square velocity you have calculated above. Calculate the number of molecules per unit volume in air at room temperature and pressure. Then estimate the number of collisions per second for an oxygen molecule in the air.

Do you think it is likely that an oxygen molecule in your room moves from one end of the room to the other within 1/100 s.

We have

$$v_{rms} = \left(\frac{3k_BT}{m}\right)^{1/2}$$

$$= 482 \text{ m s}^{-1}.$$

At this speed, the oxygen molecule will traverse 3 m in about 6×10^{-3} s if it does not undergo any collisions.

The collision volume $\mathcal{V}$ of hard sphere molecules of radius r travelling at speed v_{rms} is

$$\mathcal{V} = \pi(2r)^2 v_{rms} = 1.92 \times 10^{-16} \text{ m}^3 \text{ s}^{-1}.$$

The air in your room is at atmospheric pressure. At this pressure, using the ideal gas equation of state, the number of molecules per unit volume is estimated to be

$$N = \frac{P}{k_BT} = 2.46 \times 10^{25} \text{ m}^{-3}.$$

Hence, the collision rate for an oxygen molecule is $\mathcal{V}N$, which equals about 4.72×10^9 collisions per second.

Each time the oxygen molecule collides, its direction of travel is randomly changed. Thus, it is unlikely that an oxygen molecule can traverse a straight path of 3 m from one end of the room to the other in 1/100 s. §

The collision rate above should more accurately be calculated using the mean *relative speed* v_{rel} because molecules are not stationary relative to each other. Thus, the volume swept out is given by

$$\mathcal{V} = \pi(2r)^2 v_{rel}.$$

Because the velocities of any two molecules are generally not along the same direction, the mean relative speed is related to the mean *speed* by

$$v_{rel} = \sqrt{2}v_{mean}.$$

The mean speed for oxygen molecules at room temperature is 444 m s^{-1}. Therefore, the number of collisions per second for a molecule is, more accurately,

$$\mathcal{V}N = \sqrt{2}(2r)^2 \left(\frac{8k_B T}{m}\right)^{\frac{1}{2}} \frac{P}{k_B T} = 6.05 \times 10^9.$$

We have so far only considered the translational kinetic energy of the ideal gas. A monatomic gas only has translational kinetic energy, so its average internal energy per mole is $\frac{3}{2}RT$. The considerations above can also be exactly applied to the translational energy of any polyatomic molecule. Hence, the contribution of the translational kinetic energy to the internal energy per mole of any gas is $\frac{3}{2}RT$, regardless of whether it is monatomic or polyatomic. How is this rationalised using a microscopic picture of a gas? Each atom in a monatomic gas can move with independent velocities v_x, v_y and v_z along the three Cartesian directions. It can be mathematically demonstrated using the Boltzmann distribution that each of these three translational degrees of freedom contributes $\frac{1}{2}RT$ of kinetic energy per mole. This is called the principle of equipartition of energy: in general, each degree of freedom of motion contributes to the internal energy an average amount of energy equal to $\frac{1}{2}RT$. This is true only if molecular motion obeys the laws of classical mechanical. Unfortunately, because an accurate description of molecular motion is quantum mechanical, this is only approximate.

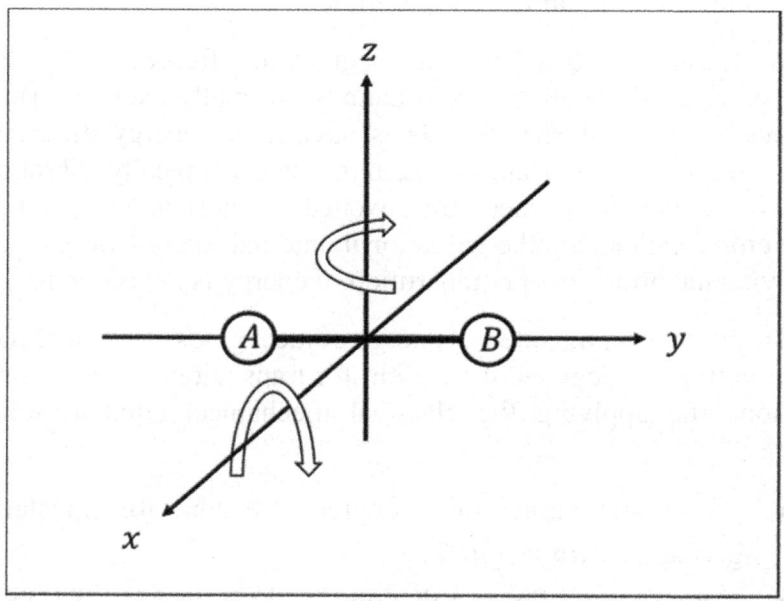

Figure 1.3 The two possible rotational modes of a diatomic molecule AB. Arrows show molecular rotations about the x and z axes. Because the quantum state of a diatomic is cylindrically symmetric about its axis, there is no rotational mode about the y-axis.

In addition to translational kinetic energy, a diatomic molecule also has rotational kinetic energy. The molecule rotates about two principal axes. This is illustrated in Figure 1.3 for rotation about the x and z axes. There is no rotation about the y axis because the molecular structure of the diatomic is cylindrically symmetric about its bond axis. Hence, quantum mechanically, there can be no rotational excitation about that axis giving diatomic molecules two degrees of freedom for rotational motion. The same considerations apply for any linear polyatomic molecule such as CO_2. Together with the translational degrees of freedom, the internal energy of diatomics and linear polyatomics is thus $\frac{5}{2}RT$ per mole. For a non-linear polyatomic there is an additional third degree of freedom for rotation. Therefore, the internal energy of non-linear polyatomics at temperature T is equal to $3RT$ per mole.

What about the vibrational degrees of freedom? At usual temperatures, these are not substantially thermally excited. This is because vibrational quantum states have larger energy differences than rotational and translational quantum states. Typically, vibrational states are not fully thermally excited at normal temperatures. Therefore, estimating the vibrational internal energy of gases by applying the principle of equipartition of energy is not accurate.

To summarise, the internal energy of ideal gases can be estimated by counting the degrees of freedom for translational and rotational motion, and applying the classical mechanical equipartition of energy:

- monatomic gases: three degrees freedom for translation, giving $U = nu = \frac{3}{2}nRT$,
- diatomics and linear polyatomics: three degrees of freedom for translation and two degrees of freedom for rotation, giving $U = nu = \frac{5}{2}nRT$,
- non-linear polyatomics: three degrees of freedom for translation and three degrees of freedom for rotation, giving $U = nu = 3nRT$.

We will use the letter U for the internal energy of an object, and u for its molar internal energy. Later in this chapter we will examine how accurate these estimates of internal energy are for various gases, and what we can learn from the deviations from kinetic theory predictions.

1.6 Heat capacity

The molar heat capacity C of a substance is defined as the ratio of the heat supplied ΔQ to the change in the temperature ΔT of one mole of the substance. In the limit of an infinitesimal change in temperature:

$$C \equiv \lim_{\Delta T \to 0} \frac{\Delta Q}{\Delta T}.$$

If we heat a substance at constant volume, all the energy supplied as heat goes into its internal energy. We write this as follows:

$$C_V \equiv \lim_{\Delta T \to 0} \left(\frac{\Delta Q}{\Delta T}\right)_V$$

with the subscripts V indicating that this is for heating at constant volume. Thus, the heat needed to warm up one mole of a substance by ΔT at constant volume is

$$(\Delta Q)_V = C_V \Delta T.$$

C_V is the constant volume molar heat capacity of the substance.

Similarly, when we heat one mole of the same substance at constant pressure, we have

$$C_P \equiv \lim_{\Delta T \to 0} \left(\frac{\Delta Q}{\Delta T}\right)_P,$$

so that in general, the heat needed to warm up a gas by ΔT at constant pressure is

$$(\Delta Q)_P = C_P \Delta T.$$

C_P is the constant pressure molar heat capacity of the substance. The heat capacities C_V and C_P are not equal. When any object expands and pushes against the external pressure while being heated, in addition to increasing the internal energy of object, heat must also be supplied to do work for volume expansion. That is,

$$(\Delta Q)_P > (\Delta Q)_V.$$

Thus, the constant pressure heat capacity is larger than the constant volume heat capacity.

$$C_P > C_V,$$

the difference being due to the mechanical work done to expand against the external pressure. Now, mechanical work ΔW is the product of force F and distance moved by the force Δx, that is

$$\Delta W = F\Delta x.$$

Consider a volume change ΔV where the cross-sectional area is A. Hence, the change in the volume is given by

$$\Delta V = A\Delta x,$$

where Δx is the distance moved by the force. Also from mechanics, we have pressure equal to force per unit area. Therefore, the work that is done by a system is equal to

$$\Delta W = F\Delta x = \frac{P}{A} A\Delta x = P\Delta V,$$

when it undergoes a volume change ΔV. The expanding system has to do work pushing against the external pressure in its expansion. This extra energy that the gas needs in order to do mechanical work on the surroundings is part of the energy that needs to be supplied as heat to the gas in order to raise its temperature by ΔT at constant pressure. With this expression for the work done by a system on the surroundings, the difference between the heat capacities C_P and C_V is

$$C_P\Delta T = C_V\Delta T + P\Delta V$$

where the volume change is what accompanies heating at constant pressure. Since no volume expansion work is done, constant volume processes are known as isochoric processes. Similarly, constant pressure processes are known as isobaric processes. The isochoric heat capacity is always less than the isobaric heat capacity.

This relationship between the heat capacities is particularly easy to quantify for an ideal gas given the equation of state $PV = nRT$. For one mole of an ideal gas constant pressure, the change in the volume with temperature is given by

$$P\Delta V = R\Delta T.$$

Hence, we have

$$C_P = C_V + R.$$

In the previous section we have obtained kinetic theory predictions of the internal energy of ideal gases. For a monatomic ideal gas, for example a box of argon, the internal energy is $\frac{3}{2}RT$. When a substance is heated up at constant volume, the change is only in its internal energy since there is no expansion work needed. Hence, for argon, the constant volume heat capacity is

$$C_V = \frac{3}{2}R.$$

Adding to this the contribution from the work done pushing back against the external pressure of the surroundings due to volume expansion, we have the constant pressure heat capacity of argon equal to

$$C_P = \frac{5}{2}R$$

In a similar way we can get the following estimates of heat capacities for diatomic and polyatomic ideal gases applying the kinetic theory of gases:

- ideal monatomic gases:

$$C_V = \frac{3}{2}R \qquad\qquad C_P = \frac{5}{2}R$$

- ideal diatomics and linear polyatomics:

$$C_V = \frac{5}{2}R \qquad\qquad C_P = \frac{7}{2}R$$

- ideal non-linear polyatomics:

$$C_V = 3R \qquad\qquad C_P = 4R$$

Notice that the constant pressure and constant volume heat capacities of an ideal gas are constants. In particular, these quantities are independent of the temperature of the ideal gas.

We have discussed the equation of state, internal energy and heat capacity of ideal gases so that we can use the ideal gas as a simple model substance to begin to illustrate the principles of thermodynamics. In a nutshell, thermodynamics gives the universal laws for relating the different temperature-dependent macroscopic properties of a substance. Indeed, in arriving at $C_P = C_V + R$ we made use of the result that the work done by an expanding ideal gas, $P\Delta V$, is equal to $nR\Delta T$. Thus, there is a relationship between the PVT behaviour of a substance and its heat capacities. This is a simple example that does not require much thermodynamics. We will encounter other examples that better illustrate the power of thermodynamics. For example, we will see later that starting from $PV = nRT$, we can deduce that the internal energy of an ideal gas is dependent only upon its temperature, and not on its pressure and volume.

§   *Example 1.3 Relationship between internal energy and equation of state*

Show that for a gas that obeys the ideal gas equation of state its internal energy is independent of volume.

In Chapter 7, using the First and Second Laws of Thermodynamics, we will find that for all substances

$$\left(\frac{\partial U}{\partial V}\right)_T = T\left(\frac{\partial P}{\partial T}\right)_V - P,$$

where the subscripts denote the variables that are held constant in each of the two partial derivatives. This is a relationship that holds in general between the dependence of the internal energy upon its volume at constant pressure and the dependence of its pressure upon temperature at constant volume. We know that the equation of state of an ideal gas is $PV = nRT$, so that we can use the above relationship to determine how the internal energy U depends upon the volume of the system at constant temperature. This is done as follows.

$$\left(\frac{\partial P}{\partial T}\right)_V = \frac{nR}{V}.$$

Hence, on the right-hand side

$$T\left(\frac{\partial P}{\partial T}\right)_V - P = T\left(\frac{nR}{V}\right) - \frac{nRT}{V} = 0.$$

Therefore,

$$\left(\frac{\partial U}{\partial V}\right)_T = 0.$$

Conclusion: given just the equation of state of the ideal gas, we can deduce that its internal energy is independent of volume at constant temperature. §

The importance of Thermodynamics is that it allows you to establish such general relationships between macroscopic properties, rather than merely providing numerical values for any one of the thermodynamic variables of a system under various conditions, although this latter is undoubtedly of great utility. In any case, it is useful in learning thermodynamics to examine model substances for which we can estimate concrete properties of real systems. In the following three examples we do this by considering the heat capacity of ideal gases to get a sense of why the properties of real gases can deviate from ideal gas behaviour.

§   *Example 1.4 Heat capacity of monatomic gases*

The constant pressure heat capacity of some monatomic gases are tabulated here for a temperature of 298.15 K and 1 bar, the so-called standard ambient temperature and pressure. What conclusion can be drawn from this?

Gas	C_P in $JK^{-1}mol^{-1}$
He	20.786
Ne	20.786
Ar	20.786
Kr	20.786
H (atomic hydrogen)	20.79

The kinetic theory prediction for the constant pressure heat capacity of monatomic gases is $\frac{5}{2}R \approx 20.785\ JK^{-1}mol^{-1}$. The data in the Table shows that the classical mechanical approximation for the internal energy of monatomic gases is rather accurate.   §

Another way of stating this is that at 298.15 K, the translational quantum states are fully excited. From your quantum mechanics class you would have learned that the typical difference in translational energy levels of gases is much smaller than the order of magnitude of thermal energy, which is RT per mole of atoms. From quantum mechanics the energies of the ground state and the first excited state of a particle-in-a-3D-box are $\dfrac{3h^2}{8mL^2}$ and $\dfrac{6h^2}{8mL^2}$, where h is Planck's constant, m is the mass of the atom, and L is the length of the box. We apply this particle-in-a-box model to a gas of helium atoms. If we have a quite normal-sized box of 1 m^3 of helium atoms, we obtain an energy difference of about 2.5×10^{-41} J between the ground state and the first excited state. Even with the underestimate of L this is rather many orders of magnitude *smaller* than the thermal energy of 4.12×10^{-21} J at a temperature of 298.15 K; it is more than tricky to do translational spectroscopy. Translational excitation of all gases are fully thermally excited at any "normal" temperature,

§   *Example 1.5 Heat capacity of some diatomic gases*

The constant pressure heat capacities of the gases oxygen, nitrogen and hydrogen chloride are tabulated below for standard temperature and pressure. Comment on the components of the internal energy in these gases.

Gas	C_P in $JK^{-1}mol^{-1}$
O_2	29.355
N_2	29.125
HCl	29.12

Because the constant pressure heat capacity is rather close to $\frac{7}{2}R \approx$ 29.099 $JK^{-1}mol^{-1}$, this data suggests that there are 5 degrees of freedom, which is as expected since these are diatomic molecules. As it turns out, from statistical mechanics, oxygen, nitrogen and hydrogen chloride rotational quantum states are substantially excited for temperatures above approximately 2.07, 2.86 and 15.2 K, respectively. These are referred to as the rotational excitation temperatures. For most gases, these rotational excitation temperatures are low compared to room temperature. Amongst the most common gases, H_2 has a notably high rotational excitation temperature of 85.4 K because of its light atoms; this is still considerably below room temperature. Hence, at room temperature the rotational motion of most gases are expected to be fully thermally excited, and the classical mechanical estimate of the rotational contribution to heat capacity is rather good.   §

§   *Example 1.6 Heat capacity of some other diatomic gases*

The constant pressure heat capacities of fluorine, chlorine and bromine at a temperature of 298 K and a pressure of one atmosphere are tabulated below. Compare these to the heat capacities of the diatomic gases in Example 1.5.

Gas	C_P in $JK^{-1}mol^{-1}$
F_2	31.30
Cl_2	33.91
Br_2	36.02

These are diatomic molecules. Thus, there are two rotational modes, leading to the classical mechanical estimate for the constant pressure heat capacity of $\frac{7}{2}R$ per mole. We have seen that rotational motion is generally fully excited at room temperature.

The measured heat capacities for these diatomics are larger than what we might expect for translational and rotational excitation. This data indicates that halogen molecules are vibrationally excited at ambient temperature and pressure since their constant pressure molar heat capacities exceed $\frac{7}{2}R$ substantially. §

A vibrational mode has both potential energy and kinetic energy. Thus, extending the principle of equipartition of energy, we count two degrees of freedom for vibrational motion, giving $C_V = \frac{7}{2}R$ and $C_P = \frac{9}{2}R$ per mole. At the same temperature, the vibrational motion for bromine is more thermally excited than for fluorine. From quantum mechanics we know that the separation between successive vibrational energy levels is hf, where f is the vibrational frequency of the chemical bond in the diatomic treated as a harmonic oscillator. The vibrational frequency of F_2 is 7.1×10^{13} s^{-1} while that for Br_2 is 9.7×10^{12} s^{-1}, so that the vibrational energy levels are much closer in bromine than in fluorine. For comparison, the vibrational frequency of HCl (Example 1.5) is 8.6×10^{13} s^{-1}. This is higher than that for fluorine; vibrational motion is less thermally excited in hydrogen chloride than in fluorine.

We now take a brief detour on how to count the number of vibrational modes in molecules. In a diatomic with one chemical bond, it is clear that there is one vibrational mode. How about polyatomic molecules? In general, a molecule with N atoms has $3N$

degrees of freedom in its motion because each atom can move in three directions. In any linear molecule, there are three translational and two rotational modes, while in a non-linear molecule there are three translational and three rotational modes. Hence, the number of vibrational mode is

$$3N - 5$$

for linear molecules, and

$$3N - 6$$

for non-linear molecules. An additional point to note is that the translational and rotational motion of a molecule are "free". That is, they do not take place in a potential. Hence, these modes contain only kinetic energy. However, the vibrations of a molecule involve the stretching of bond lengths and the bending and torsion of bond angles which occur in the potential wells of the chemical bonds. There is thus both kinetic and potential energy components associated with each vibrational mode. Hence, applying the principle of equipartition of energy, each fully thermally excited vibrational mode contributes $2 \times \frac{1}{2}RT = RT$ to the internal energy of a molecule. For a diatomic this contribution is RT, leading to $C_V = \frac{7}{2}R$.

§   *Example 1.7 Heat capacity of carbon dioxide – temperature dependence*

The dependence of the constant pressure heat capacity upon temperature for carbon dioxide is tabulated here for a pressure of one atmosphere. Comment on the temperature dependence.

Temperature K	C_P in $JK^{-1}mol^{-1}$
243.16	35.120
273.16	36.352
323.16	38.465

Carbon dioxide is a linear molecule. Hence, translation and rotational motions contribute $\frac{7}{2}R$ per mole. But this is only approximately 29.099 JK^{-1}mol^{-1}. Thus, we see a clear vibrational contribution to the heat capacity of CO_2 which increases with temperature. Carbon dioxide is a linear molecule with three atoms, so that $3N - 5 = 4$; it thus has four vibrational modes, giving $C_V = 5\left(\frac{R}{2}\right) + 4R = \frac{13}{2}R$. Assuming that the ideal gas equation of state is accurate for the temperature and pressure under consideration, $C_P = \frac{15}{2}R$. Thus, the constant pressure heat capacity of carbon dioxide converges to 62.35 JK^{-1}mol^{-1} at sufficiently high temperatures to fully excite its vibrational modes. This data shows that the vibrational modes are not yet fully thermally excited at 323.16 K. Incidentally, three of the four vibrational modes are infrared active, and are responsible for the greenhouse effect of CO_2. §

Heat is the central actor in Thermodynamics. This chapter introduces some thermal properties of ideal gases so that we have a starting stage upon which to discuss the laws of Thermodynamics. There are only four of these laws, but they completely frame all the macroscopic equilibrium properties of matter that arise from heat interactions.

Zeroth and First Laws

2.1 Thermodynamic state, intensive and extensive variables

Consider a thermodynamic system consisting of n moles of water. To be specific, say the temperature is 298.15 K and the pressure is 1 bar so that the system is one homogeneous phase, which happens to be the liquid phase at this temperature and pressure. The actual phase, so long as it is homogeneous, is not important to the discussion here. The internal energy of the system is equal to U and its volume equal to V. If we have a second system at the same temperature and pressure, but containing $2n$ moles of water, we say that the size of the second system is twice that of the first. Let us examine a macroscopic portion in the bulk of the first system consisting of m moles of water, and similarly a portion of the second system also consisting of m moles of water. We will not find a difference in any macroscopic physical properties – for example, temperature, pressure, internal energy, volume, heat capacity, mass – between these subsets of the two systems; we say that the two systems and the two subsets of these systems are in the same thermodynamic state.

Properties such as the internal energy and the volume are referred to as extensive properties because these are proportional to the size of the system given the same thermodynamic state. The second system will have an internal energy equal to $2U$ and a volume equal to $2V$ because it contains twice the number of moles in the first system. Similarly, each of the two subsets consisting of m moles have internal energy equal to $\frac{m}{n}U$ and volume equal to $\frac{m}{n}V$.

On the other hand, properties such as temperature and pressure, which are independent of system size, are referred to as intensive properties. The temperatures and pressures of subsets of the two systems above are the same as those of both systems, namely, 298.15 K and 1 bar pressure. Properties such as viscosity, refractive index and compressibility are additional examples of intensive quantities. Some intensive properties are trivially derived from extensive ones. For example, the molar internal energy $u = U/n$ is an intensive quantity; it is simply the internal energy per mole of the system, and therefore not dependent upon system size. Similarly, the molar volume $v = V/n$. The values of any intensive property for two systems are the same as long as these systems are in the same thermodynamic state.

2.2 Equilibrium

A system is in equilibrium when none of its macroscopic properties such as n, P, V, T, ... is changing with time. All its intensive properties have settled down to their respective constant values. We then say that it is in a state of thermodynamic equilibrium. We already know from the previous section that all intensive properties of a system are determined once the thermodynamic state of the system is specified. But what about the inverse question of how many intensive properties need to be specified in order to determine the state of a system?

Consider a simple system consisting of an ideal gas. The equation of state written in terms of the molar volume v is

$$Pv = RT.$$

Specifying the values of any two of the variables P, v and T, fully determines the third. All other intensive properties of the gas – compressibility, density, molar heat capacities – are also completely determined once any two of P, v and T are given. If we need an extensive physical property, for example the mass or the volume of the system, we will additionally need to know the system size n.

Thus, for a homogeneous system consisting of one chemical species, only two intensive properties need to be specified in order to fully determine the thermodynamic state of the system. For now, we motivate this by our simple consideration of the ideal gas equation of state. We will discuss in a later chapter the number of intensive variables needed to establish thermodynamic equilibrium in more general terms. We will also discuss the equilibrium condition in systems consisting of multiple chemical species and/or multiple phases.

For a heterogenous system with more than one phase, it is necessary to know the size of the system in order to completely specify the state of the system. An example of a heterogenous system is an isolated system consisting of liquid water and water vapour in equilibrium with each other. All the intensive properties of each phase are fully determined if the pressure and temperature of the system are given. However, to completely specify all the physical properties, such as how much of each phase is present or the total volume of such a system, we need an additional physical quantity such as the total number of moles of water present. We will discuss such systems when we consider phase equilibrium.

2.3 What is Temperature?

The Zeroth Law of Thermodynamics states:

If a system A is in thermal equilibrium with system B, and if B is in thermal equilibrium with system C, then A is also in thermal equilibrium with C.

It is a commonplace empirical observation that there are situations in which heat is exchanged between systems and, on the other hand, there are also situations in which zero net heat exchange occurs. We say that the former systems are not in thermal equilibrium while the latter systems are. The Zeroth Law of Themodynamics is grounded on these rather simple empirical observations that lead us to the concept of temperature that is not quite just a qualitative assessment of "hotness" on the basis of sensation.

Consider two gases that are in separate containers connected through a piston so that mechanical energy can be exchanged between them. At equilibrium, the pressure of the two gases must be equal so that the forces on the piston are mechanically balanced. However, mechanical equilibrium is only a subset of thermodynamic equilibrium for the two gases if their connecting wall allows heat to be exchanged. Thus, to generally characterise thermodynamic equilibrium, aside from equating the pressures of the two gases, we need to equate a thermodynamic variable θ which measures the "hotness" of each gas. Say we have two systems A and B, possibly consisting of different substances, in equilibrium. As we have seen in the previous section, the thermodynamic state of a substance is completely specified if two intensive properties are given. The value of θ for each system has to be a function of the thermodynamic state of the system, so that thermodynamic equilibrium means

$$\theta_A = f_A(\alpha_A, \beta_A)$$
$$\theta_B = g_B(\gamma_B, \delta_B)$$
$$\theta_A = \theta_B$$

where α, β, γ and δ are intensive properties; the pair of properties α and β do not have to be the same as γ and δ, and f_A and g_B are functions that could depend on the specific substances constituting the systems A and B, respectively.

If we have another system C which is also in equilibrium with A, then we have

$$\theta_A = \theta_C.$$

Then the Third Law tells us that

$$\theta_B = \theta_C.$$

Thus, we can consider system A to be a thermometer and the value of θ_A to be temperature. All systems that when placed in thermal equilibrium with A results in the same value of θ are at the same temperature. The question is what might be a suitable system A to use as a thermometer? And how do we define the function f_A that determines θ in terms of the intensive properties α and β. That is, how do we define θ so that it is a measurable quantity that is preferably not dependent upon the particular substance A. A rough analogy might be to ask how we make rulers such that the length of one centimeter is not dependent upon the property of the wood or metal or plastic used to make the ruler. The behaviour of gases in the limit of zero pressure gives us an answer.

Consider the schematic setup in Figure 2.1 where one mole of any gas is placed in thermal equilibrium with an object at some fixed degree of "hotness". Using the piston, we can adjust the pressure and the volume while keeping the gas in thermal equilibrium with the object. For each value of pressure, we calculate the product of the pressure and volume

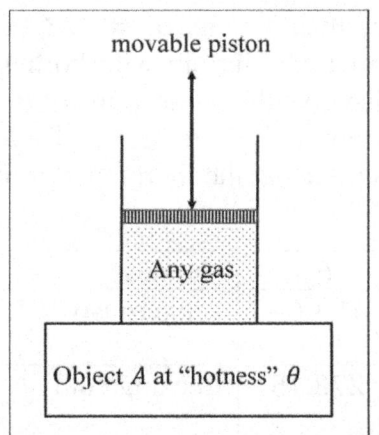

$$\theta = \frac{PV}{R}$$

with pressure P measured in units of Nm^{-2}, volume V measured in units of m^3 and the constant R equal to 8.314.

Figure 2.1 A movable piston is used to adjust the pressure and volume of the gas which is placed in thermal contact with an object A at fixed "hotness" θ.

For an object at any particular "hotness", we gradually reduce the pressure to sufficiently low values so that θ for the gas approaches a constant limiting value:

$$\lim_{P \to 0} \frac{PV}{R} = \theta_0.$$

We know that this constant limit is approached because the ideal gas equation of state is true for any gas in the limit of zero pressure. This is essentially just Boyle's law. This same value θ_0 is obtained regardless of the chemical identity of the gas. Also, the same value is obtained for any other object that is at thermal equilibrium with object A. Hence, the value of θ_0 characterises the set of objects that are in thermal equilibrium with each other; it is an objective measure of the degree of hotness.

To be concrete, let's consider numerical values. And a few systems which have constant "hotness" that might be used to thermally equilibrate the gas with. These are listed in the Table below. By their nature these systems maintain constant "hotness". Indeed, these can be used as natural thermostats. If system V is pure water boiling at a pressure of 1 bar, then we get $\theta_0 = 373.15$. Using any other object that is in thermal equilibrium with boiling pure water at the same pressure will also give this same value of θ_0. If we measure θ_0 for a few constant "hotness" systems, the Table below shows the values we would obtain using the thermometer in Figure 2.1

	system	θ_0	"hotness"
I	$CO_2(solid) + CO_2(gas)$ at 1 bar	194.64	dangerously cold
II	Ice + water at 1 bar	273.15	freezing cold
III	$H_2O(solid) + H_2O(liquid) + H_2O(gas)$	273.16	freezing cold
IV	boiling ethanol at 1 bar	351.52	very hot
V	Boiling water at 1 bar	373.15	dangerously hot

In the last column we qualitatively describe the degree of "hotness" of each of these systems. Our gas setup is clearly a thermometer, albeit a rather clumsy one because many PV measurements have to be made in order to find the $P \to 0$ limit giving the value of θ_0 for each temperature. The measurements are independent of the gas used; different gases used in the piston will reproduce the same value of θ_0 for the same object used as the constant "hotness" system. Any set of objects that are in thermodynamic equilibrium with each other will give the same value of θ_0. As a corollary, any two objects that have the same value of θ_0 will be found to be in equilibrium with each other, and conversely objects which do not have the same value of θ_0 will be observed not to be in equilibrium with each other.

The set of numbers in the column for θ_0 is thus a temperature scale. Although the thermometer uses the properties of a gas, the scale is independent of the actual gas used; it is a universally applicable scale. This is what you probably already know as the ideal gas temperature scale. The choice of setting R to be equal to 8.314 is arbitrary; a different value used for R would result in a different but equivalent temperature scale. You are familiar with R as the gas constant. The modern convention is to set the temperature for system *III* to be exactly equal to 273.16, denoted simply as 273.16 K, where K stands for Kelvin. This means that the arbitrary choice is made of dividing the temperature difference between zero Kelvin and the temperature of system *III* into exactly 273.16 units. System *III* is when solid water, liquid water and water vapour are equilibrium. This is known as the triple point of water. For this equilibrium to be maintained, the degree of "hotness" and the pressure is well-defined. This is selected to set the Kelvin scale because it is easier to set up the system at triple point, than to use the combination of the boiling and melting points of water. The latter was previously used to define the Kelvin temperature scale, which then made the arbitrary choice of dividing the temperature difference between boiling water and melting ice into exactly 100 units, as in the older Celsius scale.

We have only mentioned zero Kelvin. We will discuss this in Chapter 6. Here, we only state that the temperature of zero Kelvin cannot be reached in any real system. It is the unreachable lower limit of "hotness" for any equilibrium system.

Another empirical "hotness" scale that you are undoubtedly familiar with can be constructed by using the expansion of liquid mercury in a capillary tube, i.e. a mercury thermometer. Setting the temperature of boiling water and melting ice at one atmosphere to be 100°C and 0°C, respectively. We then measure how hot an object is by measuring the length of the mercury column in the capillary tube. We evenly divide this length difference between 100°C and 0°C into 100 intervals, giving us the Celsius scale. This is satisfactory for practical purposes, but not too satisfactory in a fundamental way. In the limit of zero pressure, all gases obey Boyle's law $PV = nR\theta_0$ as a law of Nature. And only systems with the same θ_0 are in thermal equilibrium. The product of pressure and volume is exactly linear with respect to the ideal gas temperature for this fundamental reason which is independent of the process of defining the temperature scale. On the other hand, using the mercury thermometer to define a temperature scale means that the linearity between the expansion of mercury and the resultant temperature scale is assumed. As it turns out, the expansion of mercury with temperature is quite closely, but not exactly, linear with respect to the Kelvin temperature scale. It is one of the better substances to use in a thermometer for many "normal" applications because its expansion is close to linear, but it is not satisfactory in a fundamental way.

2.4 The Joule experiment

We have seen in Chapter 1 that the difference between Mechanics and Thermodynamics is heat. We have also discussed in Section 1.4 the concept of the internal energy of a thermodynamic system. The Joule experiment, performed with the apparatus illustrated in Figure 2.2 investigates how we change the internal energy of a system. A fixed amount of water is placed in a rigid container with adiathermal walls that are also impermeable to matter. This is a closed system; the amount of material in it is constant. Since the container walls are rigid, no mechanical work can be done through volume expansion or contraction. The container walls are adiathermal so heat does not flow through it.

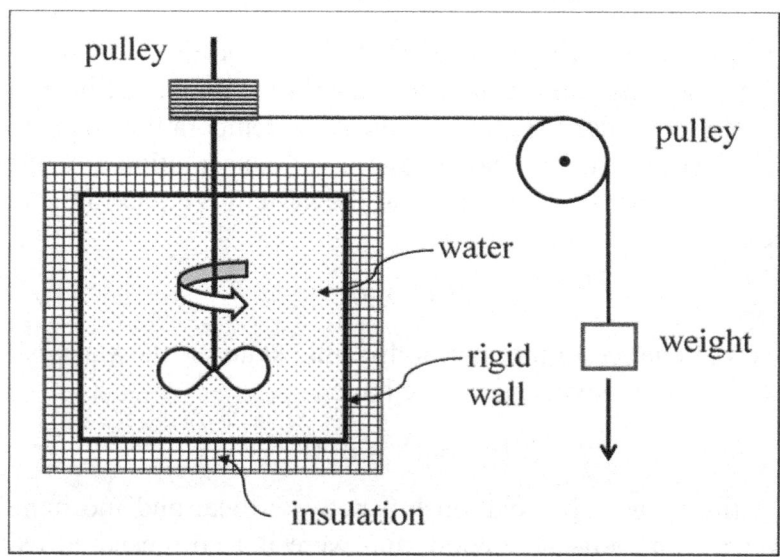

Figure 2.2 Joule experiment. Weight falls, turning a paddle wheel which delivers energy to the water contained in a box with rigid, adiathermal walls.

The system consists only of the water inside the container. The weight and the paddle wheel are not part of the system but rather only a means of performing mechanical work on the system. The paddle wheel is made of adiathermal material so it does not exchange heat with the water it is in contact with. As the weight falls and the paddle wheel rotates inside the container, work is done pushing the water into motion. When the weight has fallen to its final position and the paddle wheel has come to rest, a certain amount of mechanical work would have been done on the system $\Delta_0 W$. Because of the viscous friction in the water, this work is converted into internal energy of the water. Finally, after the bulk motion of the water has ceased, the increase in the internal energy would be ΔU so that by the conservation of energy,

$$\Delta U = \Delta_0 W.$$

This increase in the internal energy is tracked by a corresponding change in the temperature of the water $\Delta T = T_f - T_i$.

In a second experiment with a slightly modified apparatus to allow heat flow into the system, we supply an amount of heat $\Delta_0 Q$ to the system in order to reproduce the same temperature change ΔT as in the first experiment above. Again, by conservation of energy, the increase in the internal energy of the system is equal to the heat supplied,

$$\Delta U = \Delta_0 Q.$$

Because the change in the thermodynamic state of the water is the same as in the first experiment, we have

$$\Delta U = \Delta_0 Q = \Delta_0 W.$$

At a time when the relationship between heat and mechanical energy was not well-understood, and were indeed measured using different units, Joule arrived at the "mechanical equivalent of heat" by doing such experiments. That is, he demonstrated that heat and work are simply different forms of energy and provided an "exchange rate" to convert between them.   This is so well understood today that we do not even pause to ponder this, but in the 1800's this insight was big conceptual progress.

2.5 The First Law

We imagine performing additional experiments of this type where now both heat and work are supplied to the system in each experiment. The amounts of work ΔW and heat ΔQ are adjusted so that in each experiment the net amount of work and heat is given by

$$\Delta W + \Delta Q = \Delta_0 Q = \Delta_0 W.$$

We would then find that the temperature rise ΔT is the same as in the first two experiments, indicating that the same change of thermodynamic state has occurred for the water. With conservation of energy this is accompanied by the same change in the internal energy

$$\Delta U = \Delta W + \Delta Q = \Delta_0 Q = \Delta_0 W.$$

We thus state that

the change in the internal energy of a system is equal to the amount of work done on the system plus the amount of heat supplied to the system: $\Delta U = \Delta Q + \Delta W$.

This is the quantitative statement of the First Law of Thermodynamics. It is essentially the conservation of energy with the following highlights:

- heat is a form of energy
- the internal energy is the measure of the total energy of the system
- the change in the internal energy of a system is exactly equal to the work done on it and the heat supplied to it; there is no free lunch.

It is conceptually simple and rather straightforward to put numerical quantities into. But there are a few important points to take note.

Sign convention. In this equation ΔW is the work done on the system and ΔQ is the heat supplied to the system. If in any process that work is actually performed by the system on its surroundings then ΔW is a negative quantity. Similarly, if heat is lost by the system to its surroundings, then ΔQ is negative. It is sometimes convenient to change these sign conventions. For example, in some situations, such as when the system is an engine, it might be "intuitive" to have a sign convention that takes ΔW to be the work done *by the system* on the surroundings. Then, the equation for the internal energy change would be

$$\Delta U = \Delta Q - \Delta W.$$

Similarly, it could be convenient to sometimes use the sign convention that ΔQ is the heat lost by the system. It is best to think of which way the heat and work exchange is going in any specific case to get the correct signs for these quantities, rather than to merely operate using a fixed sign convention.

Internal energy U is a <u>state function</u> of the system. Any change ΔU depends only upon the change in the state of the system, for example going from its initial state at T_i to its final state at T_f, and not upon the actual path taken to effect that change. In the first Joule experiment described above, the change in ΔU is due only to work done on the system. In the second experiment, the same internal energy change is due only to heat supplied to the system. In the other additional experiments, the same change ΔU is reached by various combinations of ΔW and ΔQ. Regardless of the actual path taken to change the system from a specific initial state to a specific final state, the change ΔU is the same. This property of any state function is critical to understanding thermodynamics.

In contrast, work ΔW and heat ΔQ are not state functions. The amounts of work done on the system and heat supplied to it are different in the different experiments above even though the initial and final states are the same in all these experiments. Thus, we refer to work and heat as <u>path functions</u> because they depend on the path take to effect any change of state. For a finite change, we have

$$\Delta U = \Delta W + \Delta Q,$$

while for an infinitesimal change, it is convenient to write

$$dU = dW + dQ.$$

At this point we have to be rather careful about the meanings of the differentials dU, dW and dQ. If a thermodynamic quantity x is a state function, then it is completely determined once the state of the system is defined. A few examples of these state functions would be internal energy U, pressure P, volume V, temperature T, which are familiar. Another, perhaps less familiar, example might be the isothermal compressibility κ_T; it does not matter which specific state function. If the state of a system changes from state i to state f, then the corresponding change in a state function x is simply

$$\Delta x = x_f - x_i$$

where x_i and x_f, are the values of x at the initial and final states, respectively. We frequently need to calculate Δx, through an integral

$$\Delta x = \int_{initial\ state}^{final\ state} dx$$

where any path through thermodynamic states can be taken to reach state f from state i; that is, the value of Δx is independent of the path taken. Mathematically, we say that dx is an exact differential.

The work and heat exchanged between a system and its surroundings, ΔW and ΔQ, are not the properties of a thermodynamic state but rather are dependent on the specific path taken to go from one state to another. Thus, we are not able to simply write quantities W_f and W_i to denote the amount of work in the system at final and initial states. We can only talk about the work done on a system as it undergoes a specific process transforming it from one state to another. Therefore, the following are meaningless equations:

$$\Delta W = W_f - W_i$$
$$\Delta Q = Q_f - Q_i.$$

And, similarly, the following integrals are not meaningful in thermodynamics:

$$\int_{W_i}^{W_f} dW \text{ or } \int_{Q_i}^{Q_f} dQ.$$

In thermodynamics, as we have noted above, ΔW and ΔQ are path functions. Connecting to mathematics, we say that the differentials dW and dQ are not exact differentials. A notation that is fairly widely used to denote this is to write these inexact differentials as $đW$ and $đQ$, with a strikethrough in the letter d. We will not use this notation, but instead always note that work and heat are path specific.

How then are we to calculate ΔW and ΔQ? We outline a specific example here expanding upon it in a subsequent section. Consider an ideal gas that expands from an initial state with volume V_i to a final state with volume V_f. We additionally specify that the path taken in this expansion is such that at every point along the path, the volume and pressure are related by $PV = nRT$ with the temperature constant. For this reversible process, we are able to write

$$\Delta W = \int_{initial\ state}^{final\ state} dW$$

$$= \int_{V_i}^{V_f} P dV$$

$$= \int_{V_i}^{V_f} \frac{nRT}{V} dV$$

$$= RT \ln \frac{V_f}{V_i}.$$

In the first integral, we need to recognise that dW is not an exact differential, so we cannot simply integrate this to obtain $\Delta W = W_f - W_i$. Notice that in this integral, the limits are simply stated as the initial and final states. It is not meaningful to write these limits as W_f and W_i denoting the amount of work in the system at final and initial states. In the second line the work done by the gas in any infinitesimal step of a volume expansion is equal to PdV, and importantly, we are now able to write specific thermodynamic quantities V_i and V_f for the limits of the integral. In order for this line of the equation to be valid, it is important that the quantities P and V are well defined for the entire path taken to go from V_i and V_f. We have intentionally specified above that the path taken in this expansion is such that at every point along the path, the volume and pressure are related by $PV = nRT$. Thus, our calculation of ΔW here is valid only for this particular path. Indeed, if a path for volume expansion is taken such that the pressure and volume are not well-defined along the path, then it may not be possible to directly calculate ΔW in the way we have done here. The thinking behind the third line is straightforward. We have used the ideal gas equation of state to replace P with V in the integrand depending only on the variable V.

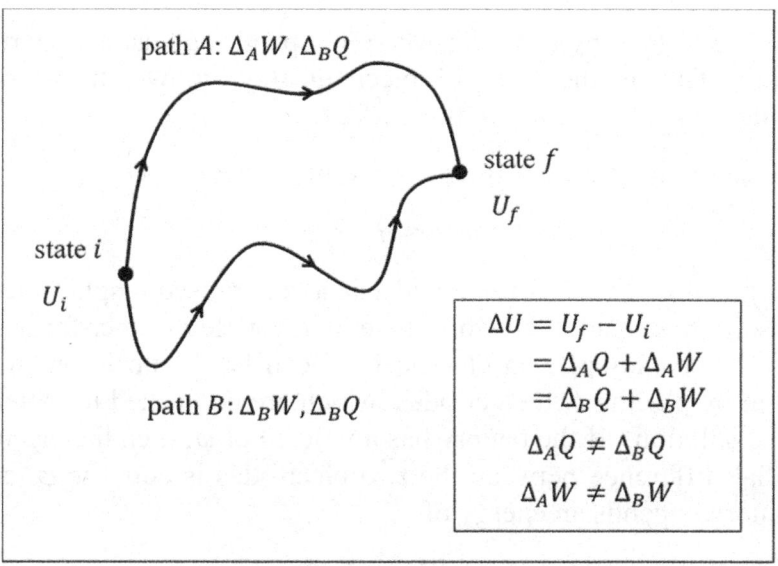

Figure 2.3 First Law of Thermodynamics. A and B are any two arbitrary processes transforming the system from state i to state f. The internal energy is a state function. Work and heat are path functions.

We re-examine the calculation of ΔW and ΔQ in the next section. Figure 2.3 summarises pictorially what we have learned from the First Law.

2.6 Other forms of work

In Mechanics we learn is that work done is equal to the force exerted on an object times the distance the object is moved by the force. If a force F moves an object over a distance dL, then the work done is equal to

$$dW = FdL.$$

As we have seen in Section 1.5 this leads to

$$dW = PdV$$

for the work done by a system when it expands against an external pressure. This is the form of mechanical work which we will encounter most frequently in these notes.

We can write the work interaction quite generally as

$$dW = fdx$$

where f is a generalised force and x is a generalised displacement. It is useful to consider the work done by a couple of other forms of energy. For example, an electric field can be set up in an ionic solution by placing two electrodes, which are connected to battery, into the solution. If the battery has a voltage of φ, then the electric potential difference between the two electrodes is equal to φ, and the battery expends an energy of

$$dW = \varphi d\mathcal{Q}$$

for an amount of charge $d\mathcal{Q}$ that is transferred from the electrode at high potential to the electrode at lower potential. This is the energy that the battery puts into the system when an ionic solution is electrolysed. Thus, the electric potential difference between the electrodes is the generalised driving force in this case, and the charge transferred is the corresponding generalised displacement. For a finite amount of charge transferred ΔQ, the work done is

$$\Delta W_{elec} = \int_{Q_i}^{Q_f} \varphi dQ = \varphi \Delta Q.$$

With φ measured in volts and Q measured in coulombs, then the electric current I is measured in amperes. If this current is driven by a battery of constant voltage φ through a heating element for a duration of time Δt, then the total electrical work done by the battery over this duration of time is

$$\Delta W_{elec} = \varphi \Delta Q = \varphi I \Delta t.$$

If the battery itself is considered as the system, then

$$\Delta W_{elec} = -\varphi \Delta Q$$

because electrical work by the battery would be energy lost by the system. In the Joule experiment, we can first convert the gravitational potential energy of the falling weight to electrical energy, storing this electrical energy in a battery. Then using this battery to power a heating element in the water, this would be the electrical work delivered from the battery to increase the internal energy of the water.

The molecules at the surface of a liquid have a net attractive interaction into the bulk of the liquid. Hence, molecules at a liquid surface have a higher energy than molecules in the bulk. Extra energy is therefore needed to increase the area of a liquid surface or a liquid film. The surface tension is equal to this extra energy per unit area. Therefore, in order to increase the area of, for example a soap film, work needs to be done against the surface tension, giving

$$dW = \gamma dA$$

with γ the surface tension and dA the increase in surface area.

2.7 Reversible and irreversible processes

We first describe the reversible expansion of an ideal gas from an initial volume V_i to a final volume V_f at constant temperature. How this can be achieved is illustrated in Figure 2.4 below.

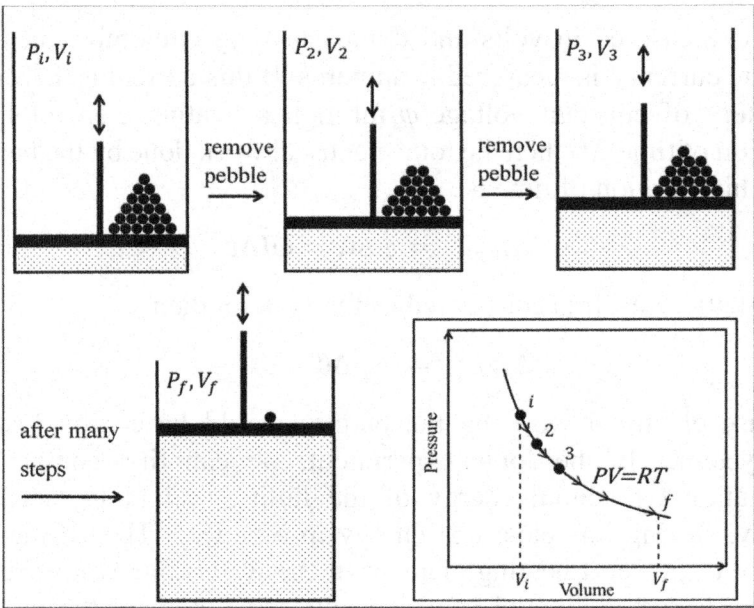

Figure 2.4 Pebbles are removed one by one. In the limit when the pebble size goes to zero, we have a reversible expansion. The solid line gives the pressure dependence upon volume during the process.

We have an ideal gas in a piston with diathermal walls that allow heat exchange with a heat reservoir so that the gas can be at a constant temperature T. Because the temperature stays the same throughout the process, it is known as an isothermal process. The force on the piston is due to the pile of little pebbles on it. To increase the gas volume reversibly, we remove one pebble at a time, let the system equilibrate, and then remove the next pebble. A reversible process is obtained in the limit where each pebble is infinitesimally small, so that in each step the increase in volume is an infinitesimal dV. Hence, the process can be reversed at any point by adding, rather than removing, an infinitesimally small pebble. In this limit, the system is at thermodynamic equilibrium at all times during the process because it is always only infinitesimally perturbed away from equilibrium. All macroscopic properties are completely defined at each step along such a process. Since a reversible process takes infinitesimally small steps, any finite change requires an infinite number of steps. A reversible process is the limit when a real process is carried out infinitely slowly. Reversible processes are therefore really only idealisations. But they are essential for analysing real systems because at every point in a reversible process all thermodynamic properties of a system are completely defined. Why is this useful?

Because thermodynamic properties are well-defined along a reversible process, we can always calculate any path function such as ΔW and ΔQ for a reversible process. For example, when the volume of the system is changed reversibly from V_i to V_f the pressure and temperature at every point along the process have well-defined values depending on the reversible process used. Since the mechanical work done by an expanding system for an infinitesimal volume change is dV, the work done by the system in a reversible process is

$$\Delta W = \int_{V_i}^{V_f} P dV.$$

In the same way, the heat gained by the system when it transforms from an initial state with temperature T_i to T_f in a reversible process is

$$\Delta Q = \int_{T_i}^{T_f} C dT$$

where C is the heat capacity, which can be C_P or C_V, respectively, depending upon whether the process is at constant pressure or constant volume. Hence, both ΔW and ΔQ can be readily calculated in any specific reversible process. This is pointedly not the case for

an irreversible process. In general, it is not possible to calculate
these path functions for an irreversible path because the
thermodynamic properties of a system are not defined in an
irreversible process.

An ideal gas can expand irreversibly in an infinite number of
ways from initial volume V_i to V_f while keeping the temperature the
same for the initial and final states. We illustrated one such
irreversible process in Figure 2.5.

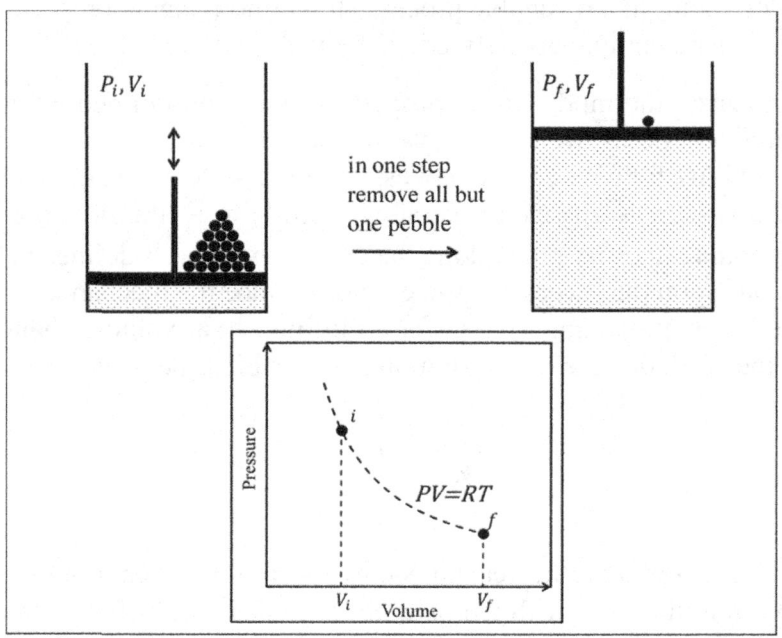

*Figure 2.5 Many pebbles are removed in one step giving an
irreversible expansion. In general, only the initial and final states
are well-defined in irreversible processes.*

In contrast to the process illustrated in Figure 2.4, all the pebbles
except for one are removed in a single step. The states of the system
during an irreversible process are generally not in thermodynamic
equilibrium so that the process is not representable by a continuous
curve as in Figure 2.4 on the *PV* diagram. For example, the

temperature and pressure of the system are not likely uniform throughout the system during the process. Thus, the thermodynamic properties of a system are only generally only well-defined at the initial and final states which are specified to be at thermodynamic equilibrium. In Figure 2.5, we have drawn a dashed line that gives the pressure dependence upon volume at constant temperature for an ideal gas. This is drawn only to indicate that the initial and final states are thermodynamic equilibrium states at the same temperature. This dashed line does not represent the states of the system during this irreversible process.

Since it is not possible to calculate ΔW and ΔQ for an irreversible process, the change in the internal energy in irreversible processes cannot be calculated using $\Delta U = \Delta W + \Delta Q$. However, U is a state function, and ΔU depends only on initial and final states. Thus, to calculate ΔU for an irreversible path, we first figure out a reversible path that connects the initial and final states of the given irreversible path. Then, we calculate ΔW and ΔQ for that reversible path. This gives ΔU for the reversible path, but then that is equal to the internal energy change for the irreversible path because U is a state function.

2.8 Properties of the internal energy

The First Law is completely encapsulated by

$$dU = dQ + dW$$

written with the sign convention that dQ and dW are the heat *absorbed by* the system and the work *done on* the system, respectively. For systems where the only type of work interaction is through its volume change, we have

$$dU = dQ - PdV.$$

For these systems, the change in the internal energy at constant volume is

$$(dU)_V = (dQ)_V$$

or for a finite change,

$$(\Delta U)_V = (\Delta Q)_V.$$

We have used the subscript V to emphasise that this applies only for constant volume processes. We denote the internal energy change for isochoric processes as follows:

$$\left(\frac{\partial U}{\partial T}\right)_V = \lim_{\Delta T \to 0} \left(\frac{\Delta Q}{\Delta T}\right)_V.$$

It is worthwhile to examine our notation. On the left-hand side, we have used the partial derivative notation. In Section 2.2 we have seen that the thermodynamic state of a (single-phase) system is dependent upon only two variables. Thus, the internal energy of the system is a function of not one but two variables, for example V and T. The partial derivative notation in this equation is thus necessary to denote the change of the internal energy with respect to T, while keeping V constant. On the right-hand side, we have written the change of Q with respect to T in the limit of ΔT going to zero. The subscript reminds us that the process we are considering is at constant volume. We have specifically not used the same partial derivative notation as on the left-hand side of the equation because Q is a path function. Writing the right-hand side as $\left(\frac{\partial Q}{\partial T}\right)_V$ would mean that Q is a state function, $Q = Q(V, T)$, which is not correct. From the definition of the constant volume heat capacity

$$C_V \equiv \lim_{\Delta T \to 0} \left(\frac{\Delta Q}{\Delta T}\right)_V,$$

we obtain

$$\left(\frac{\partial U}{\partial T}\right)_V = C_V$$

for systems in which the only form of work is through volume expansion or compression.

We further explore the internal energy as a function of the variables V and T. For any function of two variables, $U = U(V, T)$, the change in internal energy dU is related to the changes in volume and temperature dV and dT by:

$$dU = \left(\frac{\partial U}{\partial T}\right)_V dT + \left(\frac{\partial U}{\partial V}\right)_T dV.$$

We have already seen that the coefficient of dT is equal to C_V. What about the coefficient of dV, namely $\left(\frac{\partial U}{\partial V}\right)_T$? The First Law requires that, in general

$$dU = dQ - PdV.$$

If we simply equate the corresponding terms, we might conclude that

$$dQ = \left(\frac{\partial U}{\partial T}\right)_V dT = C_V dT$$

$$-PdV = \left(\frac{\partial U}{\partial V}\right)_T dV.$$

These two above equations are not correct, even though it seems quite reasonable to think of dQ as a heat capacity multiplied by dT. As we have emphasised earlier, dQ is equal to $C_V dT$ only for constant volume processes. The heat capacity of a system depends upon the condition under which the temperature is changed; it is not generally true that dQ is equal to $C_V dT$. And $\left(\frac{\partial U}{\partial V}\right)_T$ is equal to a quantity known as the internal pressure π_T, which is *not* equal to pressure P. That is,

$$\left(\frac{\partial U}{\partial V}\right)_T = \pi_T \neq P.$$

In any general process, dQ depends upon the change in temperature dT and also the change in volume dV. We will disentangle these two contributions after we have discussed the Second Law of Thermodynamics, which will shed further light on the nature of heat.

2.9 The internal energy of an ideal gas

Let us examine the First Law for an ideal gas $dU = dQ - PdV$. From the principle of equipartition of energy, we know that the internal energy of an ideal gas is

$$dU = C_V dT$$

which is a function only of the temperature. Even though you see the subscript V on the right-hand side, this equation is always true as long as the system is an ideal gas, regardless of whether the process is at constant volume or not. Thus, in the case of a monatomic ideal gas, we immediately have a constant volume heat capacity of

$$C_V = \left(\frac{\partial U}{\partial T}\right)_V = \frac{3}{2} nR.$$

Since the internal energy of an ideal gas is independent of the volume, we may write, in general,

$$\left(\frac{\partial U}{\partial T}\right)_V = \left(\frac{dU}{dT}\right) = C_V.$$

The First Law for ideal gases is then given by

$$C_V dT = dQ - PdV.$$

If a process is isothermal, then the left-hand side is zero, so that

$$dQ = PdV.$$

In an isothermal expansion of a gas, heat is needed even though the internal energy remains constant. This is because the gas expands against an external pressure, so that energy needs to be supplied to the gas to perform this expansion work.

When a gas is expanded or compressed in a container with insulated (adiathermal/adiabatic) walls, no heat interaction with the surroundings occurs, and

$$dQ = 0.$$

Such processes are known as adiabatic processes. For an ideal gas undergoing an adiabatic process, we thus have

$$C_V\,dT = -P\,dV.$$

In an expansion, either isothermal or adiabatic, where work is done by the gas in pushing against an external pressure, $dW = -P\,dV$ is a negative quantity. Since the heat capacity is a positive quantity, this means that an ideal gas expanding adiabatically cools down. We have already seen that the internal energy of a gas is a function of only its temperature. Consequently, the internal energy of an adiabatically expanding gas decreases. While an expanding isothermal gas draws heat energy from its surroundings to perform the expansion work, in an adiabatic expansion, the heat energy needed to perform the expansion work is drawn from the thermal kinetic energy of the gas. Thus, it cools down.

Finally, we consider an isochoric process in an ideal gas, that is a process in which $dV = 0$. Applying the First Law to an isochoric process gives

$$C_V\,dT = dQ$$

so that we simply recover the result we have obtained above for the heat capacity of a constant volume expansion for any system. Since C_V for an ideal gas is a constant, we have for any finite change in temperature in an isochoric process,

$$\left(\frac{\Delta Q}{\Delta T}\right)_V = C_V.$$

Note that this relationship is valid only when the process is isochoric; hence the need for the subscript V on the left-hand side. Indeed, this relationship is true for isochoric processes in any system, not only for ideal gases, as long as the only form of work done by the system is volume expansion/compression work.

In contrast, we note that for an ideal gas, no matter what the process is, it is always true that

$$\frac{\Delta U}{\Delta T} = C_V.$$

This is valid even for processes which are not isochoric. This is because the ideal gas internal energy is a linear function only of its temperature, with the proportionality constant being equal to C_V. This relationship is not valid for real gases. For a real gas, the internal energy is generally a function of both temperature and volume, so that $dU = C_V dT$ does not completely describe how its internal energy changes. The dependence of the internal energy of a real gas on T and V, or equivalently on T and P, can be traced to the non-zero intermolecular interactions. Molecules in real gases interact.

Applications of the First Law

We discuss some thermodynamics ideas that follow from the First Law. We begin with some typical "problems" making use of the ideal gas as a simple model substance to make the discussion concrete. Then we will introduce the concept of Enthalpy and use that to discuss thermochemistry applications, the throttling of real gases and intermolecular interactions.

3.1 Using the First Law

3.1.1 Reversible isothermal expansion

§ *Example 3.1 Ideal gas reversible isothermal expansion*

One mole of an ideal gas at temperature T expands reversibly and isothermally from an initial volume of V_i to a final volume of V_f. This process is illustrated in Figure 3.1.

Calculate the work done on the gas ΔW and the heat absorbed by the gas ΔQ.

Defining $dS \equiv dQ/T$, calculate the change in S for the process.

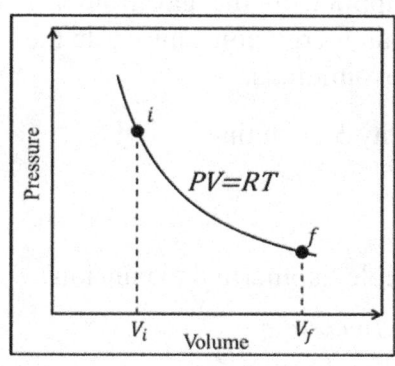

Figure 3.1 Pressure vs volume for an ideal gas expanding isothermally

The internal energy of an ideal gas depends only upon its temperature. For an isothermal expansion the temperature stays constant, and therefore the internal energy of the gas stays constant. With $dU = 0$, we have from the First Law

$$dW = -dQ$$

so that we have $\Delta W = -\Delta Q$. For an ideal gas expanding reversibly, the work done *on* the gas is

$$\Delta W = -\int_{V_i}^{V_f} P \, dV$$

$$= -\int_{V_i}^{V_f} \frac{RT}{V} \, dV$$

where we have used the ideal gas equation of state $PV = nRT$ in the second line. Integrating this, the work done on the gas is equal to

$$\Delta W = -RT \ln \left(\frac{V_f}{V_i} \right).$$

The gas expands, so that $V_f > V_i$. Therefore, ΔW is a negative quantity: work is actually done *by* the gas in expanding against the external pressure. This is a reversible expansion so that the external pressure is equal to the pressure of the gas at all times.

Since $\Delta Q = -\Delta W$, we have

$$\Delta Q = RT \ln \left(\frac{V_f}{V_i} \right).$$

This is a positive quantity: heat is supplied to the gas from the surroundings in order to keep its internal energy constant while the gas performs expansion work on the surroundings.

Looking ahead, let us define a quantity S such that

$$dS \equiv \frac{dQ}{T}.$$

We want to calculate ΔS for this reversible, isothermal expansion.

$$\Delta S = \int_{initial\,state}^{final\,state} dS = \int_{initial\,state}^{final\,state} \frac{dQ}{T}$$

$$= \frac{1}{T}\Delta Q$$

$$= R \ln\left(\frac{V_f}{V_i}\right).$$

You should notice see that we have not made any other assumptions on the quantity S aside from its definition. In particular, we have not assumed that it is a state function. In the second line, we have moved $\frac{1}{T}$ out of the integration because the temperature is constant. We will revisit this calculation of ΔS after we have discussed the Second Law in Chapter 5. §

3.1.2 Reversible cyclic process

§ *Example 3.2 Ideal gas reversible cyclic process*

One mole of an ideal gas undergoes the cyclic process illustrated in Figure 3.2. AB is a reversible isobaric expansion, BC is a reversible isochoric process, and CA is a reversible isothermal compression. CA is the reverse of the process we looked at in Example 3.1. The pressures and volumes of the states A, B and C are labelled in the figure.

Calculate ΔW, ΔQ, ΔU and ΔS for each of the processes AB, BC, CA and for the entire cyclic process in terms of P_i, V_i and V_f. Define $dS \equiv dQ/T$ as in Example 3.1

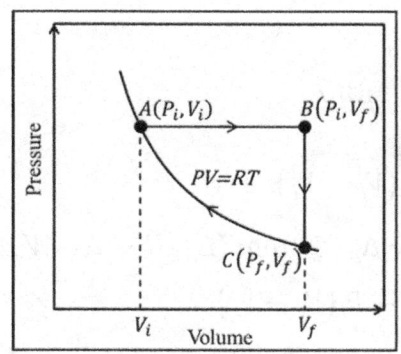

Figure 3.2 A cyclic process for an ideal gas

Along *AB* the pressure is constant, so applying the ideal gas equation of state

$$\frac{P_i V_i}{T_A} = \frac{P_i V_f}{T_B}$$

Thus,

$$T_B = (V_f/V_i)T_A$$

Thus, $T_B > T_A$. In order to increase the volume of an ideal gas while keeping its pressure constant, the temperature has to increase. This reversible process can be performed by successively placing the gas into thermal contact with reservoirs at gradually increasing temperatures while maintaining the same pressure throughout, perhaps by using a piston. Because the temperature increases, the internal energy of the gas also increases by

$$\Delta_{AB} U = \int_{T_A}^{T_B} C_V \, dT$$

$$= C_V (T_B - T_A)$$

$$= C_V \frac{P_i}{R} (V_f - V_i).$$

The first two lines are clear because the change in the internal energy of an ideal gas is equal to the change in its temperature multiplied by C_V, which is a constant for an ideal gas. We write the third line only to express the result in terms of P_i, V_i and V_f by simply using the ideal gas equation of state.

Similarly, the work done on the gas along the path AB can be calculated by integrating $-P dV$ from state A to state B because each state along the path is at thermodynamic equilibrium, and thus the pressure is well-defined,

$$\Delta_{AB} W = - \int_{V_i}^{V_f} P \, dV$$

$$= -P_i (V_f - V_i).$$

Then we use the First Law to calculate $\Delta_{AB} Q$ from $(\Delta_{AB} U - \Delta_{AB} W)$

$$\Delta_{AB} Q = C_V (T_B - T_A) + P_i (V_f - V_i).$$

Examine this to see that the heat supplied to the gas is simply equal to the energy needed to raise the temperature of the gas plus the energy to do the volume expansion work against the constant external pressure. From our above expression for $\Delta_{AB}U$,

$$(T_B - T_A) = \frac{P_i}{R}(V_f - V_i),$$

so that

$$\Delta_{AB}Q = C_V(T_B - T_A) + R(T_B - T_A)$$
$$= C_P(T_B - T_A),$$

since $C_P = C_V + R$ for ideal gases from Section 1.5. AB is a constant pressure process, so we could have obtained this result immediately since the heat gained in a constant pressure process is simply equal to the constant pressure heat capacity multiplied by the change in temperature.

BC is an isochoric process. Hence, we immediately have

$$\Delta_{BC}W = 0.$$

The temperature for state C is the same as that for state A, so that the change in the internal energy is

$$\Delta_{BC}U = C_V(T_C - T_B) = C_V(T_A - T_B)$$
$$= -\Delta_{AB}U$$
$$= C_V\frac{P_i}{R}(V_i - V_f),$$

which is a negative quantity; the internal energy decreases from B to C. From the First Law the heat supplied to the gas along BC is thus equal to $\Delta_{BC}U$.

CA is the reverse of the process in Example 3.1. Thus, from the results in Example 3.1, we have

$$\Delta_{CA}W = RT \ln\left(\frac{V_f}{V_i}\right)$$

$$\Delta_{CA}Q = -RT\ln\left(\frac{V_f}{V_i}\right)$$

$$\Delta_{CA}U = 0.$$

Again, looking ahead, we calculate the quantity ΔS for each of the steps taken in this cyclic process, given that $dS \equiv dQ/T$.

$$\Delta_{AB}S = \int_{state\,A}^{state\,B} \frac{dQ}{T}$$

$$= \int_{state\,A}^{state\,B} \frac{dU}{T} - \int_{state\,A}^{state\,B} \frac{dW}{T}$$

$$= \int_{T_A}^{T_B} \frac{C_V}{T}dT + \int_{V_A}^{V_B} \frac{P}{T}dV$$

$$= \int_{T_A}^{T_B} \frac{C_V}{T}dT + \int_{V_A}^{V_B} \frac{R}{V}dV$$

$$= C_V \ln\left(\frac{T_B}{T_A}\right) + R\ln\left(\frac{V_B}{V_A}\right)$$

$$= C_P \ln\left(\frac{V_B}{V_A}\right).$$

It is useful to reread Section 1.5 if the steps in the second and third lines are not entirely clear. In the last step we have used the ideal gas equation of state to equate T_B/T_A to V_B/V_A since AB is at constant pressure. In the same way, we get

$$\Delta_{BC}S = C_V \ln\left(\frac{T_C}{T_B}\right) = C_V \ln\left(\frac{T_A}{T_B}\right) = -C_V \ln\left(\frac{V_B}{V_A}\right)$$

$$\Delta_{CA}S = R\ln\left(\frac{V_A}{V_B}\right) = -R\ln\left(\frac{V_B}{V_A}\right).$$

We add up these step changes to calculate ΔS for the cyclic process:

$$\Delta S = \Delta_{AB}S + \Delta_{BC}S + \Delta_{CA}S = 0.$$

It appears that the change in the quantity S in this cyclic process is equal to zero; S might be a state function. We will come back to this example in Chapter 5. §

3.1.3 Irreversible adiabatic expansion

§   *Example 3.3 Ideal gas irreversible adiabatic expansion*

One mole of an ideal gas expands adiabatically from V_i to V_f against zero external pressure, as illustrated in Figure 3.3. The expansion is against zero pressure, so it is irreversible. Calculate the change in the internal energy of the gas ΔU. Again, we define $dS = dQ/T$. Given that the process is irreversible, are we able to calculate ΔS?

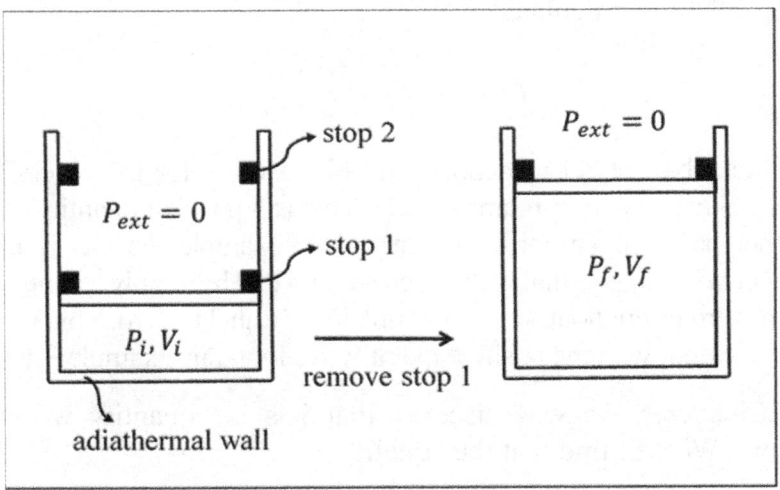

Figure 3.3 Irreversible expansion of an ideal gas in box with adiathermal walls. Removal of Stop 1, allows the gas to expand quickly, irreversibly, up to the volume when the piston is held in place using Stop 2. The process occurs spontaneously.

Since the expansion is against zero external pressure, no volume expansion work is done. Therefore, ΔW is zero. The process is also adiabatic, so ΔQ is also zero. Therefore, using the First Law ΔU must be zero; the internal energy does not change in this process. Because the gas is an ideal gas, its internal energy is a function only of the temperature. Since there is no change in U, there is also no change in the temperature.

Thus, this process takes the gas from the state (V_i, T_i) to (V_f, T_i). These are the same initial and final states in the reversible isothermal expansion in Example 3.1. ΔU is the same for both these processes because U is a state function, that is, ΔU depends only upon the initial and final states and not on the path that is taken. On the other hand, ΔW and ΔQ are path functions. In this example, both are zero, but in Example 3.1, they are non-zero.

What about ΔS? Using $dS \equiv dQ/T$ along the process in Example 3.1, we calculated

$$\Delta_{3.1} S = R \ln \left(\frac{V_f}{V_i} \right)$$

using the subscript 3.1 to denote that this result is for the reversible isothermal expansion in Example 3.1. This is a positive quantity. On the other hand, if we consider that in this example the gas is in a container with adiathermal walls, then we find that by simply setting dQ equal to zero throughout, we would obtain ΔS equal to zero. This is not at all consistent with the positive quantity we found in Example 3.1.

In Chapter 5 we will discover that S is the quantity we call Entropy. We will find that the equality

$$dS = \frac{dQ}{T}$$

only applies for reversible processes. For any irreversible process such as the expansion here, it turns out that

$$\Delta S > 0,$$

consistent with the calculation in Example 3.1. This is an important conceptual point we will understand in Chapter 5. §

3.1.4 The ideal gas adiabat

§   *Example 3.4 Ideal gas reversible adiabatic expansion*

One mole of an ideal gas expands reversibly and adiabatically from V_i to V_f against an external pressure that is always adjusted to equal the pressure of the gas. This is illustrated in Figure 3.4. The initial state of the gas is given by V_i and T_i. The final state of the gas is given by V_f and T_f.

Find out how the final volume and temperature depend upon the initial volume and temperature by applying the First Law.

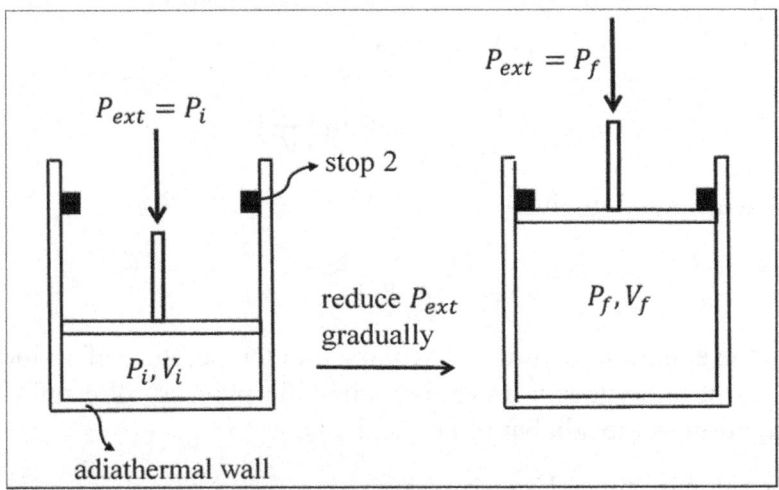

Figure 3.4 In contrast to Example 3.3, the external pressure on the piston is now gradually reduced from the initial value of P_i to the final value of P_f so that the expansion is reversible.

The First Law of Thermodynamics states that

$$dU = dQ + dW$$

For an adiabatic process, dQ is equal to zero, and

$$dU = dW$$

which for an ideal gas gives

$$C_V dT = -P dV$$

as we seen in Section 2.9. There are three variables in this equation: P, V and T. However, only two variables are independent because they are related by the ideal gas equation of state. In particular, we substitute $\frac{RT}{V}$ for P. Hence, in terms of V and T, we have

$$C_V \frac{dT}{T} = -R \frac{dV}{V},$$

separating the variables on each side of the equation. Integrating this from (V_i, T_i) to (V_f, T_f), we obtain

$$C_V \ln\left(\frac{T_f}{T_i}\right) = -R \ln\left(\frac{V_f}{V_i}\right).$$

This is rearranged to give

$$V_i T_i^{\frac{C_V}{R}} = V_f T_f^{\frac{C_V}{R}}.$$

This is the equation relating the volume and temperature of an ideal gas when it undergoes a reversible adiabatic process. We refer to this equation as the adiabat of an ideal gas.

Since this is an ideal gas, we have

$$\frac{T_f}{T_i} = \frac{P_f V_f}{P_i V_i}.$$

We use this to replace the temperature in terms of volume and pressure in the equation of the ideal gas adiabat giving:

$$P_i V_i^{\frac{C_V + R}{C_V}} = P_f V_f^{\frac{C_V + R}{C_V}}.$$

Similarly, in terms of the temperature and pressure, the equation of the ideal gas adiabat is:

$$P_i^{\frac{C_V}{R}} T_i^{\frac{C_V+R}{C_V}} = P_f^{\frac{C_V}{R}} T_f^{\frac{C_V+R}{C_V}}.$$

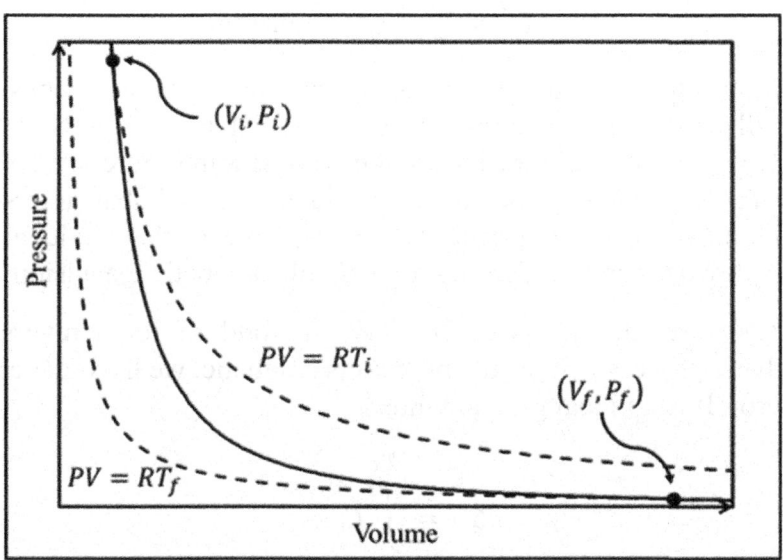

Figure 3.5 The ideal gas adiabat. The dashed lines are ideal gas isotherms at T_i and T_f. The solid line is the adiabat that passes through the states (V_i, P_i) and (V_f, P_f). The equation of the adiabat is $PV^\gamma = constant = P_i V_i^\gamma = P_f V_f^\gamma$.

We have seen that the constant pressure heat capacity of an ideal gas is

$$C_P = C_V + R.$$

Defining

$$\gamma = \frac{C_P}{C_V}.$$

We can rewrite these equations for the ideal gas adiabat as

$$T_i V_i^{\gamma-1} = T_f V_f^{\gamma-1} = constant,$$

$$P_i V_i^{\gamma} = P_f V_f^{\gamma} = constant,$$

$$P_i^{1-\gamma} T_i^{\gamma} = P_f^{1-\gamma} T_f^{\gamma} = constant.$$

The constant for each of these equations is not a universal constant, but rather is a constant along a specific adiabat defined by the values of P_i, V_i, T_i. We illustrate the ideal gas adiabat in Figure 3.5. Notice that as the volume increases the ideal gas adiabat intersects isotherms of decreasing temperature. This means that the ideal gas cools down when it undergoes a reversible adiabatic expansion. §

What are the values of ΔU, ΔW, ΔQ and ΔS for a reversible adiabatic process? Since the process is adiabatic, we have ΔQ equal to zero. It is an ideal gas, and thus

$$\Delta U = \int_{T_i}^{T_f} C_V \, dT$$
$$= C_V (T_f - T_i)$$

in terms of the initial and final temperatures. Since T_f is lower than T_i, the internal energy decreases. Using the equation of the adiabat, this can be expressed in terms of the pressure or volume change of the process. ΔW is equal to ΔU from the First Law and is thus also negative; work is done by the gas on the surroundings. Because the process is adiabatic, the energy needed for this expansion work done by the gas is drawn from the internal energy of the ideal gas. Thus the random thermal motion of the gas is reduced and the gas cools. What about ΔS? Because this process is reversible, we are able to use the definition $dS = dQ/T$. Since it is adiabatic, we immediately have ΔS equal to zero. We will see in Chapter 5 that this is true in general for any system, consisting of any substance, undergoing a reversible adiabatic process.

3.2 Enthalpy

The origins of Thermodynamics had much to do with measurements of heat exchange between systems and their surroundings. If the surroundings are ambient, and thus more or less at constant pressure, then the heat exchange measured is really the quantity we today call enthalpy. For example, when we burn charcoal in a *hibachi* (or in a steam-engine) a palpable amount of heat is released. That heat released is the enthalpy that the surroundings has gained from the combustion system. We start with defining enthalpy in terms of the thermodynamic quantities that we have already encountered. Internal energy, pressure, volume are state functions. The enthalpy is defined, in general, as

$$H \equiv U + PV,$$

which makes it a state function because it is defined completely in terms of state functions.

Then an infinitesimal change in the enthalpy is

$$dH = dU + PdV + VdP$$
$$= (dQ - PdV) + PdV + VdP$$
$$= dQ + VdP.$$

In the second line, we have used the First Law for a system in which work dW is solely due to volume expansion.

For systems in which there are other kinds of work, there will be an additional term in the expression for dH. Just to flesh this out, for example we have a system that has volume expansion and length extension work, perhaps like a rubber band. Then

$$dW = -PdV + FdL$$

taking the volume and the length as independent variables. Thus, the enthalpy change would be given by

$$dH = dQ + FdL + VdP$$

with the additional non-*PV* work contribution. We will not consider these cases as they do not add to the basic concepts we cover in these notes.

Going back to systems with only *PV* work, we have

$$dH = dQ + V dP$$

so that at constant pressure, we have

$$dH = dQ.$$

Hence, the change in the enthalpy of a system is equal to the change in its heat, as long as the system is undergoing a constant pressure process. In particular, we can write

$$\left(\frac{\partial H}{\partial T}\right)_P = \lim_{\Delta T \to 0} \left(\frac{\Delta Q}{\Delta T}\right)_P.$$

This is in parallel to the equation, discussed in Section 2.8, for internal energy change at constant volume

$$\left(\frac{\partial U}{\partial T}\right)_V = \lim_{\Delta T \to 0} \left(\frac{\Delta Q}{\Delta T}\right)_V.$$

Thus, we have

$$\left(\frac{\partial U}{\partial T}\right)_V = C_V$$

$$\left(\frac{\partial H}{\partial T}\right)_P = C_P.$$

Physically, for a constant volume process, the change in heat ΔQ is equal to the change in the internal energy. Similarly, for a constant pressure process, ΔQ is equal to the change in the quantity we have defined as enthalpy. This definition is particularly useful in Chemistry because many chemical reactions are carried out in open reaction vessels and thus occur at constant ambient pressure. We thus have

$$\Delta U = \int_{T_i}^{T_f} C_V \, dT = C_V \left(T_f - T_i \right)$$

$$\Delta H = \int_{T_i}^{T_f} C_P \, dT = C_P \left(T_f - T_i \right),$$

the first line for constant volume processes and the second line for constant pressure processes, applicable to all substances. This second equation is useful for calculations of enthalpies of reactions at arbitrary temperatures, starting from tabulated reference or standard state enthalpies. This, really simply an application of the First Law, is the basis of thermochemistry.

3.3 Thermochemistry

Enthalpy is a state function. Thus, like internal energy change, ΔH depends only upon the initial and the final states in a process; it is independent of the path taken to transform the system from the initial to the final state. Therefore, even though heat is a path function, for constant pressure processes, ΔQ depends only upon the initial and final states. Indeed, even if the process is not performed at constant pressure throughout, but the initial and final states are at the same pressure, we will have ΔH equal to the ΔQ measured in a constant pressure process. With $\Delta U = \Delta Q$ at constant volume and $\Delta H = \Delta Q$ at constant pressure, changes in the internal energy and enthalpy of a substance can be readily measured through calorimetry. Note that these measurements do not yield the absolute values of internal energy or enthalpy, but rather values relative to reference states at standard pressure and temperature.

Two common types of calorimeters are the bomb-calorimeter and the constant-pressure calorimeter. The former is essentially a closed container which is thermally insulated and has rigid walls so it is an isolated system. The amount of heat released by a reaction, for example, is then indicated by measuring the temperature change of the calorimeter, the heat capacity of which is known from prior calibration. In a bomb-calorimeter, the reaction system is at constant volume so that the heat measurement yields the change in the internal energy. In a constant-pressure calorimeter, the process occurs at constant pressure so that the heat measurement yields the change in enthalpy. From such measurements, tables such as the enthalpy of formation of compounds at standard states can be compiled. These are then used to "read off" the enthalpy of any reaction between tabulated compounds even if no direct measurements of the specific reaction are available.

For each element, the most stable state at 298.15 K and 1 bar is used as the standard state. If the most stable state is a solid, its most stable allotrope is selected. For example, graphite is the most stable state for carbon at 298.15 K and 1 bar. For hydrogen, the most stable state at 298.15 K and 1 bar is H_2 gas, so that is the standard state.

Consider the reaction between graphite and hydrogen gas

$$C(graphite) + 2H_2(gas) \rightarrow CH_4(gas).$$

At standard temperature and pressure, the enthalpy change of this reaction is

$$\Delta_{rxn}H^{\emptyset} = H^{\emptyset}[CH_4(gas)] - H^{\emptyset}[C(graphite)] - 2H^{\emptyset}[H_2(gas)]$$

with the standard state at 298.15 K and 1 bar denoted by the superscript $\emptyset$. The enthalpy of each element at its standard state is set to zero,

$$H^{\emptyset}[C(graphite)] = 0$$

$$H^{\emptyset}[H_2(gas)] = 0$$

so that the standard enthalpy of reaction is also the enthalpy of formation of methane from its elements:

$$\Delta_{rxn}H^{\emptyset} = H^{\emptyset}[CH_4(gas)] - H^{\emptyset}[C(graphite)] - 2H^{\emptyset}[H_2(gas)]$$

$$= \Delta_f H^{\emptyset}[CH_4(gas)] = -74.81 \text{ kJ mol}^{-1}$$

with the subscript f denoting "formation". Since the enthalpy of reaction is negative, heat is released by the reacting system to its surroundings. This is an exothermic reaction.

Similarly, the combustion of carbon in oxygen to form carbon dioxide,

$$C(graphite) + O_2(gas) \rightarrow CO_2(gas)$$

and the combustion of hydrogen gas to form liquid water,

$$H_2(gas) + \frac{1}{2} O_2(gas) \rightarrow H_2O(liq)$$

have standard enthalpies of reaction equal to, respectively,

$$\Delta_{rxn} H^{\emptyset} = H^{\emptyset}[CO_2(gas)] = \Delta_f H^{\emptyset}[CO_2(gas)]$$

$$= -393.51 \text{ kJ mol}^{-1}$$

$$\Delta_{rxn} H^{\emptyset} = H^{\emptyset}[H_2O(liq)] = \Delta_f H^{\emptyset}[H_2O(liq)]$$

$$= -285.83 \text{ kJ mol}^{-1}$$

since we have conventionally also set $H^{\emptyset}[O_2(gas)]$ to zero. Both of these reactions are exothermic. From calorimetry measurements of these reactions, we are thus able to determine the standard enthalpy of formation for $CH_4(gas)$, $CO_2(gas)$ and $H_2O(liq)$. The method can be applied to any compound.

Now, we consider the combustion of methane gas in oxygen to form carbon dioxide and water:

$$CH_4(gas) + 2O_2(gas) \rightarrow CO_2(gas) + 2H_2O(liq).$$

Because enthalpy is a state function, the standard enthalpy of reaction is simply equal to the standard enthalpies of the products minus the standard enthalpies of the reactants, so that

$$\Delta_{rxn}H^{\emptyset} = \{H^{\emptyset}[CO_2(gas)] + 2H^{\emptyset}[H_2O(liq)]\}$$
$$-\{H^{\emptyset}[CH_4(gas)] + 2H^{\emptyset}[O_2(gas)]\}$$
$$= \Delta_f H^{\emptyset}[CO_2(gas)] + 2\Delta_f H^{\emptyset}[H_2O(liq)]$$
$$-\Delta_f H^{\emptyset}[CH_4(gas)]$$
$$= -890.36 \text{ kJ mol}^{-1}$$

using what we have found above for the enthalpies of formation of methane, carbon dioxide and liquid water. From this you can see that the enthalpy of combustion of one mole of methane is equal to the enthalpies of the following step-wise processes:

1. disassemble one mole of methane into hydrogen gas and graphite; enthalpy change is equal to $-\Delta_f H^0[CH_4(gas)]$
2. reassemble the graphite and oxygen gas into one mole of carbon dioxide; enthalpy change is equal to $\Delta_f H^0[CO_2(gas)]$
3. reassemble the hydrogen gas and oxygen gas into two moles of water; enthalpy change is equal to $2\Delta_f H^0[H_2O(liq)]$.

These steps are illustrated here:

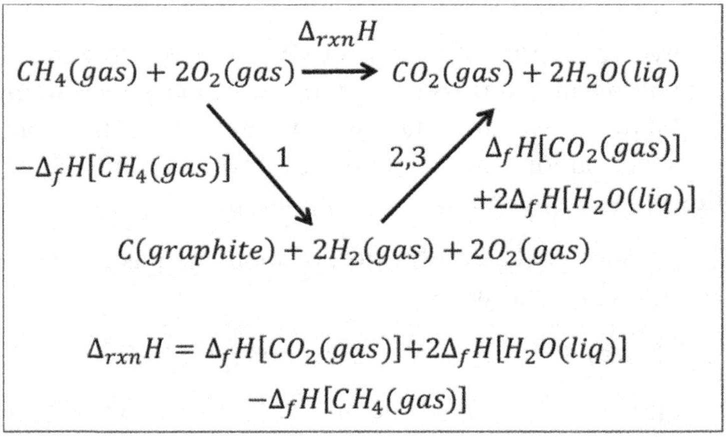

Figure 3.6 Hess's Law

The enthalpy change in a chemical reaction from the initial reactant state to the final product state is independent of the reaction path taken to go from reactants to products. The enthalpy change of the direct methane combustion path $\Delta_{rxn}H$, is equal to the sum of the enthalpy changes of the steps 1, 2 and 3 above, as illustrated in Figure 3.6. This is Hess's Law. Thus, without actually carrying out the reaction for the combustion of methane, we can obtain its enthalpy of reaction from the calorimetric data for the reactions forming methane, carbon dioxide and water. This is why tables of standard enthalpies of formation for various compounds are rather convenient for calculating the enthalpies of reactions.

We will generally need enthalpies of reaction at temperatures and pressures other than standard states. To extend this discussion, let us say that we want to calculate the enthalpy of reaction at 400 K and 1 bar, instead of at the standard condition of 298.15 K and 1 bar. At 400 K and 1 bar, water is a gas rather than a liquid. To apply Hess's Law, we consider the transformations illustrated in Figure 3.7. The enthalpies of reaction are $\Delta_{rxn}H^{\emptyset}$ and $\Delta_{rxn}H^{1}$ at standard temperature and at 400 K, respectively, for the combustion of one mole of methane at 1 bar. The following are the enthalpy changes for each step illustrated in Figure 3.7:

Step a: $\Delta_a H$ is the enthalpy change when one mole of methane and two moles of oxygen are cooled from 400 K to 298.15 K at a pressure of 1 bar. Since $\left(\frac{\partial H}{\partial T}\right)_P = C_P$, we have

$$\Delta_a H = \int_{400}^{298.15} \{C_p[CH_4(gas)] + 2C_p[O_2(gas)]\}\, dT.$$

Generally, the heat capacities C_P can be functions of temperature and pressure.

Step b: $\Delta_b H$ is the enthalpy change when we raise the temperature of one mole of carbon dioxide and two moles of liquid water from

298.15 K to 373.15 K. The latter temperature is, of course, the boiling point of water at 1 bar. Similarly, to $\Delta_a H$, we have

$$\Delta_b H = \int_{298.15}^{373.15} \{C_p[CO_2(gas)] + 2C_p[H_2O(liq)]\} \, dT.$$

Step c: $\Delta_c H$ is the enthalpy change when two moles of liquid water is vaporised at 373.15 K and 1 bar. The phase change from liquid water to water vapour takes place at constant pressure, so

$$\Delta_c H = 2\Delta_{vap} H[H_2O, 1 \text{ bar}].$$

Step d: $\Delta_d H$ is the enthalpy change when one mole of carbon dioxide and two moles of water vapour is heated from 373.15 K to 400 K, keeping the pressure at 1 bar.

$$\Delta_c H = \int_{373.15}^{400} \{C_p[CO_2(gas)] + 2C_p[H_2O(gas)]\} \, dT$$

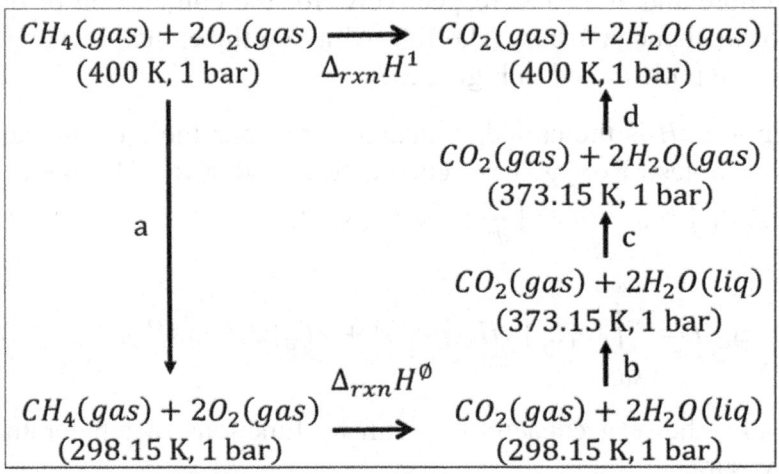

Figure 3.7 Temperature and phase change in Hess's Law.

From Hess's Law, we have

$$\Delta_{rxn}H^1 = \Delta_a H + \Delta_{rxn}H^{\emptyset} + \Delta_b H + \Delta_c H + \Delta_d H$$

so that the enthalpy of reaction at 400 K and one bar can be calculated from the standard enthalpy of reaction. Hence, from tabulations of enthalpies of formation at standard states, enthalpies of reaction at other temperatures can be calculated fairly straightforwardly using heat capacities and enthalpies of phase change. The effect of pressure on reaction enthalpy will be discussed later.

3.4 Enthalpy of an ideal gas

The enthalpy is defined in terms of internal energy by

$$H = U + PV.$$

Thus, the change in enthalpy is

$$dH = dU + d(PV).$$

For an ideal gas $dU = C_V dT$ and $d(PV) = RdT$ per mole. Hence, the enthalpy change of an ideal gas is simply given by

$$dH = (C_V + R)dT$$

$$= C_P dT.$$

Therefore, the enthalpy of an ideal gas is a function only of its temperature. From Section 1.5 on the heat capacities of ideal gases, we thus have

$$U = \frac{3}{2}RT$$

$$H = \frac{5}{2}RT$$

per mole of a monatomic ideal gas. Similarly, for diatomic and polyatomic ideal gases according to our discussion of the principle of equipartition of energy. The internal energy and the enthalpy of ideal gases depend only upon the temperature.

3.5 Enthalpy of an isolated system

Frequently chemical reactions are carried in isolated systems. For example in a bomb-calorimeter which is thermally insulated and at constant volume. In such a system, $\Delta Q = 0$ and $\Delta V = 0$. Hence, by the First Law, $\Delta U = 0$. What about ΔH?

$$\Delta H = \Delta U + \Delta(PV)$$

$$= V\left(P_f - P_i\right)$$

since the volume is a constant. Hence, unlike the internal energy, the enthalpy of an isolated system is generally not a constant.

3.6 Throttling a gas: general considerations

We have already discussed some examples of gas expansions: reversible isothermal expansion; reversible adiabatic expansion; irreversible adiabatic expansion against zero external pressure. In the first two cases, the pressure of the gas is equal to the external pressure while in the third, the external pressure is zero. Equipped with the concept of enthalpy, we can now discuss the irreversible adiabatic expansion of a gas against an external pressure that is less than the gas pressure, but not equal to zero. This process can be performed by expanding the gas from high pressure to low pressure through a valve (essentially a narrow opening) or through a porous plug (essentially, many narrow openings).

These processes are called throttling, and are schematically shown in Figure 3.8, for expansion through a porous plug. These are very common processes in cryogenics. They occur, for example, in your refrigerator and air-conditioners where the refrigerant is cooled down by expansion through a valve from a high to a low pressure. The gas expands adiabatically through a porous plug with an upstream pressure P_i and volume V_i to a downstream pressure P_f and volume

V_f. We neglect the change in the kinetic energy due to bulk motion. We take the flow of the gas to be horizontal so that we also neglect the change in gravitational potential energy of the gas.

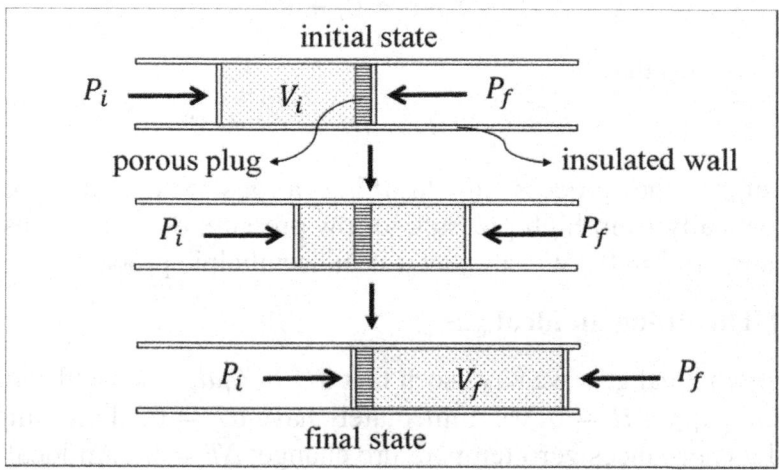

Figure 3.8 Throttling a gas through a porous plug.

From the First Law,

$$\Delta U = \Delta Q + \Delta W.$$

The process is adiabatic, so that

$$\Delta U = \Delta W.$$

What is the work done on the gas equal to? On the upstream side of the porous plug, work is done on the gas by the pressure P_i pushing the volume V_i of gas through the porous plug. On the downstream side, work is done by the gas against a pressure P_f pushing the volume V_f of gas. Hence, the net work done on the gas is equal to

$$\Delta W = P_i V_i - P_f V_f.$$

But since ΔU is equal to ΔW, we have

$$\Delta U = U_f - U_i$$

$$= P_i V_i - P_f V_f.$$

We rearrange the terms to get

$$U_f + P_f V_f = U_i + P_i V_i$$

which means that

$$H_f = H_i.$$

Therefore, the process of throttling a gas where it expands adiabatically from high pressure to low pressure occurs at constant enthalpy, $\Delta H = 0$. We say that it is an isenthalpic process.

3.6.1 Throttling an ideal gas

For an ideal gas, we have seen that $dH = C_P dT$. Since throttling is isenthalpic $\Delta H = 0$, we immediately have $dT = 0$. Throttling an ideal gas produces zero temperature change, $\Delta T = 0$. An ideal gas cannot be used as the coolant in a refrigerator.

Because the internal energy of an ideal gas depends only upon its temperature, we also have $\Delta U = 0$.

3.6.2 Throttling a real gas

To consider the throttling of a real gas, we will get a little bit ahead of ourselves by considering the enthalpy as a function of temperature and pressure, namely

$$H = H(T, P).$$

With this, the change in enthalpy is dependent upon both the change in temperature and the change in pressure:

$$dH = \left(\frac{\partial H}{\partial T}\right)_P dT + \left(\frac{\partial H}{\partial P}\right)_T dP.$$

We have seen that the enthalpy of an ideal gas depends only upon its temperature, this property being traced back to its internal energy being a function only of its temperature. This is because the internal energy of an ideal gas consists solely of the kinetic energy of random motion of its molecules since intermolecular interaction is zero in an ideal gas. Hence, the contribution of $\left(\frac{\partial H}{\partial P}\right)_T dP$ to dH is zero for an ideal gas. This, however, is not so for a real gas with non-zero intermolecular interactions.

As we have seen, throttling is an isenthalpic process so that dH is equal to zero. Therefore, we have

$$0 = \left(\frac{\partial H}{\partial T}\right)_P dT + \left(\frac{\partial H}{\partial P}\right)_T dP.$$

The structure of this equation is a general one for a mathematical relationship between three variables, here H, T and P. The zero on the left-hand side means we are considering just those situations in which enthalpy is a constant. This equation being true for throttling allows us to write

$$\left(\frac{\partial H}{\partial T}\right)_P dT = -\left(\frac{\partial H}{\partial P}\right)_T dP$$

at constant H. Taking the derivative on both sides with respect to pressure at constant H, we have

$$\left(\frac{\partial H}{\partial T}\right)_P \left(\frac{\partial T}{\partial P}\right)_H = -\left(\frac{\partial H}{\partial P}\right)_T.$$

Note the subscript H on the second factor on the left-hand side for the partial derivative of temperature with respect to pressure in an isenthalpic process. Since, in general

$$\left(\frac{\partial H}{\partial P}\right)_T = \frac{1}{\left(\frac{\partial P}{\partial H}\right)_T},$$

we obtain

$$\left(\frac{\partial H}{\partial T}\right)_P \left(\frac{\partial T}{\partial P}\right)_H \left(\frac{\partial P}{\partial H}\right)_T = -1.$$

Although we have obtained this by considering specifically the enthalpy as a function of temperature and pressure, you should note that we have not used any thermodynamics in getting to this result. This relationship for a product of three partial derivatives is simply a mathematical result that holds for any three variables that are mutually related. For example, if we have variables x, y and z related by

$$z = z(x, y),$$

then

$$\left(\frac{\partial z}{\partial x}\right)_y \left(\frac{\partial x}{\partial y}\right)_z \left(\frac{\partial y}{\partial z}\right)_x = -1.$$

This is the triple product rule also known as the cyclic chain rule or Euler's chain rule.

In relation to the throttling process for a real gas with intermolecular interactions, it allows us to calculate how temperature changes with pressure when a gas is expanded isenthalpically. This temperature change upon throttling is called the Joule-Thomson effect. From the triple product rule, we get

$$\left(\frac{\partial T}{\partial P}\right)_H = -\frac{\left(\frac{\partial H}{\partial P}\right)_T}{\left(\frac{\partial H}{\partial T}\right)_P}$$

$$= -\frac{\left(\frac{\partial H}{\partial P}\right)_T}{C_P}.$$

The quantity $\left(\frac{\partial T}{\partial P}\right)_H$ is called the Joule-Thomson coefficient because the investigation of how temperature changes with pressure when a gas is throttled was first done by Joule and Thomson. It is denoted by μ_{JT},

$$\mu_{JT} \equiv \left(\frac{\partial T}{\partial P}\right)_H = -\frac{\left(\frac{\partial H}{\partial P}\right)_T}{C_P}.$$

As was stated in Chapter 1, Thermodynamics is about relating the different macroscopic properties of a substance to each other. It would be nice to be able to obtain the Joule-Thomson coefficient of a substance from its equation of state. We explore this here. We have

$$dH = \left(\frac{\partial H}{\partial T}\right)_P dT + \left(\frac{\partial H}{\partial P}\right)_T dP$$

$$= C_P dT + \left(\frac{\partial H}{\partial P}\right)_T dP.$$

We have also seen in Section 3.2, that

$$dH = dQ + V dP.$$

These two different expressions for dH look enticingly alike, but we have to *resist* the temptation to equate the corresponding terms to get

$$C_P dT = dQ$$

$$\left(\frac{\partial H}{\partial P}\right)_T dP = V dP.$$

These two equations here are, in general, not true! Indeed, $\left(\frac{\partial H}{\partial P}\right)_T$ contains two contributions, one of which is V and another of which when added to $C_P dT$ gives dQ for a reversible process. We are not able to explicitly obtain this second term for now because we have not yet discussed the Second Law. We will return to this in Section 7.7 where we find out that we can obtain the Joule-Thomson coefficient of a gas from its equation of state, and calculate it for a

van der Waals gas in Section 7.9. For now, we will be content with
a qualitative understanding of

$$\mu_{JT} = -\frac{\left(\frac{\partial H}{\partial P}\right)_T}{C_P}.$$

In an ideal gas $\left(\frac{\partial H}{\partial P}\right)_T$ is zero, so that μ_{JT} is also zero. In a real gas,
the Joule-Thomson coefficient can be either positive or negative
depending upon the sign of $\left(\frac{\partial H}{\partial P}\right)_T$. For most real gases at room
temperatures, it is positive. Notable exceptions are hydrogen and
helium. Since most gases have a positive μ_{JT} at room temperature,
a Joule-Thomson expansion of most gases around room temperature
produces a cooling effect. This cooling effect of the expansion of a
refrigerant gas in your refrigerator or air-conditioner is the physical
basis for how these devices function.

3.7 Intermolecular interactions and μ_{JT}

What is the connection to intermolecular interactions? We sketch
in Figure 3.9 the typical shape of the dependence of the intermolecular
interaction potential upon the distance between a pair of molecules in
a gas. The minimum energy position is when the two molecules are
at a distance of r_{eq} apart. For most gases at "normal" or room
temperature (and pressure), the average separation between pairs of
molecules is close to the bottom of the intermolecular potential well
and slightly larger than r_{eq}, so that attractive interactions dominate the
repulsive interactions. When a gas expands, the average separation
increases as illustrated by the solid arrow in Figure 3.9. This increases
the internal potential energy of the gas. Under adiabatic conditions in
a throttling expansion, this increase in internal potential energy
has to be drawn from the kinetic energy component of the internal
energy, so that the random molecular speeds in the expanding
gas decreases. Thus, most gases cool down when throttled. We have
seen in Example 1.1 that in carbon dioxide at a temperature of
273.15 K and a pressure of 5 bar, the van der Waals approximation

tells us that its compressibility at these conditions is less than one and that the attractive interactions dominate the repulsive interaction. Throttling of carbon dioxide from 5 bar to atmospheric pressure cools down the gas.

For gases such as hydrogen and helium at room temperature, the average separation is close to but smaller than r_{eq}, with repulsion dominating their intermolecular interactions. Thus, when these gases are throttled, their internal potential energy decreases, as illustrated by the dashed arrow in Figure 3.9. Hydrogen and helium warm up when expanded at room temperature. This is consistent with the approximation in Example 1.1 which gives a compressibility of greater than one for helium; the repulsive interactions dominate the attractive interaction.

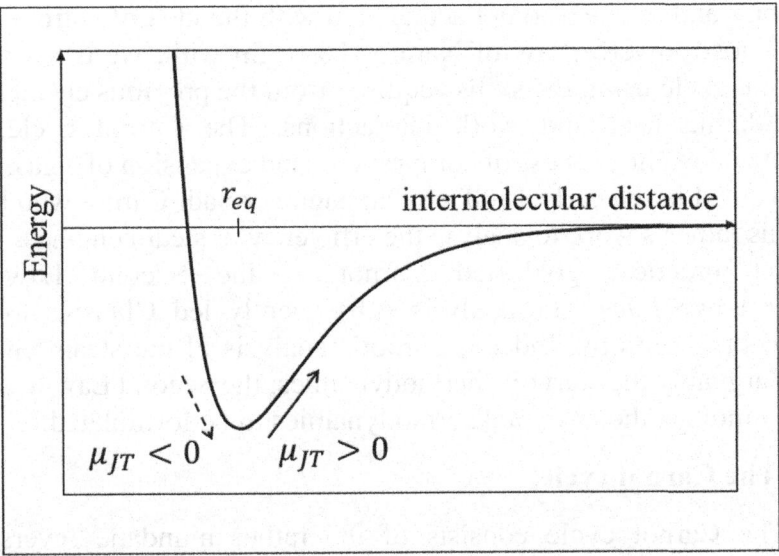

Figure 3.9 The variation of intermolecular potential energy with distance between two molecules. The internal energy of real gases consists of the kinetic energy of random thermal motion plus this intermolecular potential energy.

4

Heat Engines and the Carnot Cycle

If all human interactions were simply driven by a relentless natural increase of, say, net happiness, and if we knew how to calculate the amount of happiness in any situation, we would be able to predict the course of any human interaction. Such an encompassing law of human interaction would be on par with the Second Law of Thermodynamics for understanding Nature. And happiness would just be the analogue in human affairs of the quantity we call entropy in describing Nature. We will take this chapter and the next to get acquainted with the idea of entropy and this most powerful law of Nature. To begin with, we discuss the Carnot cycle using the skills acquired from the previous chapters in calculating heat and work interactions. The Carnot cycle, an idealized cyclic process of compression and expansion of a gas, was conceived in the early 1800's by an engineer Sadi Carnot who built on his father's work to analyse the efficiency of steam engines. That rather practical goal led Carnot to the Second Law of Thermodynamics. His analysis subsequently led Clausius to the concept of entropy. Indeed, Carnot's analysis of the steam engine was arguably the start of Thermodynamics; the Second Law was the first amongst the laws of Thermodynamics to be formulated.

4.1 The Carnot cycle

The Carnot cycle consists of the rather mundane reversible isothermal and adiabatic processes illustrated in the *PV* diagram below:

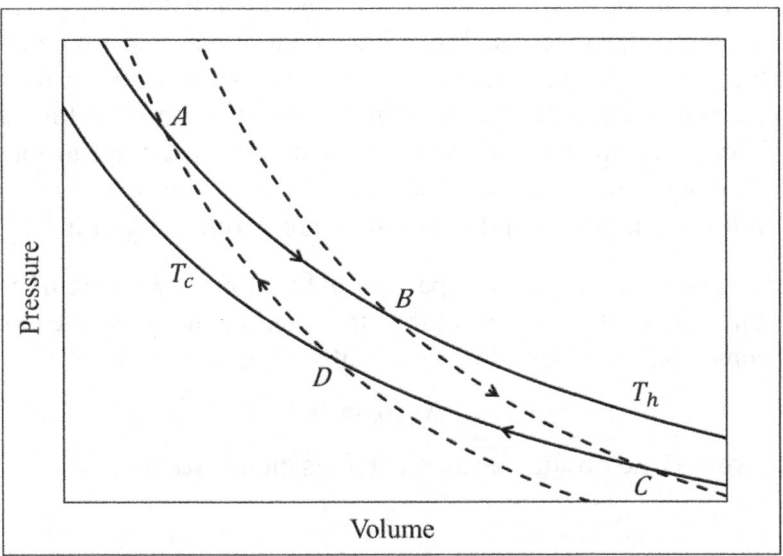

Figure 4.1 The ideal gas Carnot cycle on a PV diagram. Solid lines are isotherms at temperatures T_h and T_c. Dashed lines are adiabats. The cycle is $A \rightarrow B \rightarrow C \rightarrow D \rightarrow A$.

AB is a reversible isothermal expansion from V_A to V_B at temperature T_h.

BC is a reversible adiabatic expansion from (V_B, T_h) to (V_C, T_c). Along BC the temperature falls from T_h to a lower temperature T_c.

CD is a reversible isothermal compression from V_C to V_D at temperature T_c.

DA is a reversible adiabatic compression from (V_D, T_c) to (V_A, T_h). Along DA the temperature rises from T_c back to the higher temperature T_h.

The system starts from state A and ends up back at state A after each cycle. We analyse the Carnot cycle with an ideal gas as the working substance, i.e. the stuff that is getting expanded and compressed in the cycle obeys the equation of state $PV = nRT$.

However, as we will find, the central result is not dependent on what the working substance is. Indeed, the central result is not dependent on the details of the expansion or compression, whether these are isothermal or adiabatic or any combinations of these, as long as the cycle consists entirely of reversible steps. We use the Carnot cycle in this discussion of the Second Law because it helps us to focus on the roles of the hotter and colder heat reservoirs at T_h and T_c.

Reversible isothermal expansion AB: Along AB the temperature remains constant. Because of this, the internal energy of the working substance, which is an ideal gas, is also constant:

$$\Delta_{AB} U = 0.$$

The work done on the gas as it expands along AB is

$$\Delta_{AB} W = \int_A^B dW$$

$$= - \int_{V_A}^{V_B} P\, dV$$

$$= - \int_{V_A}^{V_B} \frac{nRT_h}{V}\, dV$$

$$= -nRT_h \ln\frac{V_B}{V_A}.$$

Because it is expanding reversibly, the external pressure acting on the gas along AB is equal to the equilibrium pressure of the gas, $P = \frac{nRT_h}{V}$. Using the First Law, we have

$$\Delta_{AB} Q = \Delta_{AB} U - \Delta_{AB} W$$

$$= nRT_h \ln\frac{V_B}{V_A}.$$

Net work is done *by* the gas as it expands from A to B against the external pressure; $V_B > V_A$. Hence, $\Delta_{AB}W$ is a negative quantity. Thus, $\Delta_{AB}Q$ is a positive quantity. Net heat is absorbed by the gas from the heat reservoir to keep the gas isothermal at T_h as it expands.

Reversible adiabatic expansion BC: No heat is lost or gained by the gas as it expands along BC because this is an adiabat. Therefore:

$$\Delta_{BC}Q = 0.$$

The temperature of the gas changes from T_h to T_c. Because the gas is ideal, its internal energy depends only upon temperature so that the change in the internal energy of the gas is equal to

$$\Delta_{BC}U = \int_{T_h}^{T_c} nC_V \, dT$$

$$= nC_V(T_c - T_h).$$

Using the First Law, we have

$$\Delta_{BC}W = \Delta_{BC}U - \Delta_{BC}Q$$
$$= nC_V(T_c - T_h).$$

The internal energy of the gas decreases because $T_c < T_h$. $\Delta_{BC}W$ is negative, the expanding gas doing work against the external pressure.

Reversible isothermal compression CD: Along CD the temperature is constant at T_c. Thus, the working substance being an ideal gas, remains at constant internal energy

$$\Delta_{CD}U = 0.$$

Similarly to AB, the work done along CD is equal to

$$\Delta_{CD}W = -\int_{C}^{D} P \, dV$$

$$= -\int_{C}^{D} \frac{nRT_c}{V} \, dV$$

$$= -nRT_c \ln \frac{V_D}{V_C}.$$

Using the First Law, we have

$$\Delta_{CD}Q = \Delta_{CD}U - \Delta_{CD}W$$
$$= nRT_c \ln \frac{V_D}{V_C}.$$

Now, net work is done *on* the gas as it is compressed from C to D by the external pressure; $V_C > V_D$. Hence, $\Delta_{CD}W$ is a positive quantity. Thus, $\Delta_{CD}Q$ is a negative quantity. Net heat is lost by the gas to the heat reservoir to keep the gas isothermal at T_c as it is compressed.

Reversible adiabatic compression DA. No heat is lost or gained by the gas as it is compressed along DA because this is an adiabat. Therefore:

$$\Delta_{DA}Q = 0.$$

The temperature of the gas changes from T_c to T_h. Again, because the gas is ideal, its internal energy depends only upon temperature so that the change in the internal energy of the gas is equal to

$$\Delta_{DA}U = \int_{T_c}^{T_h} nC_V \, dT$$
$$= nC_V(T_h - T_c).$$

Using the First Law, we have

$$\Delta_{DA}W = \Delta_{DA}U - \Delta_{DA}Q$$
$$= nC_V(T_h - T_c).$$

The internal energy of the gas increases because $T_c < T_h$. $\Delta_{DA}W$ is positive, work is done on the gas by compressing.

For the entire Carnot cycle we have the following results for the net work

$$\Delta W = -nRT_h \ln \frac{V_B}{V_A} + nC_V(T_c - T_h) - nRT_h \ln \frac{V_D}{V_C} + nC_V(T_h - T_c)$$

$$= -nRT_h \ln \frac{V_B}{V_A} - nRT_h \ln \frac{V_D}{V_C},$$

and net heat interactions

$$\Delta Q = nRT_h \ln \frac{V_B}{V_A} + nRT_c \ln \frac{V_D}{V_C} = -\Delta W.$$

From the work and heat interactions summed over the four steps of the Carnot cycle, we have

$$\Delta U = \Delta Q + \Delta W = 0,$$

as expected for the change in a state function in a cyclic process.

4.2 Caloric, language, sign convention

We pause a slight bit to remind ourselves of a point made in Chapter 1 on the nature of heat. At the time when Carnot was examining steam engines, heat was thought of as a substance known as caloric. We now know heat is the kinetic energy of random molecular motion. We have tried to be careful in the previous section by using the term "heat interaction" in order not to suggest a mental picture of a substance moving from a heat reservoir to the gas system, or vice versa. However, we also say that heat is "absorbed from the reservoir by the system", possibly making it seem as though heat is a substance, which it is not. We will continue with some of these language lapses, for example stating that heat "flows" from one object to another, simply for convenience, not precision of expression.

The sign convention that we mostly adhere to in this book is that a heat interaction ΔQ has a positive sign if heat flows into the system from its surroundings, and is negative when heat is lost from the system to its surroundings (alas, somewhat consistent with thinking about heat as a substance). However, it is sometimes convenient, and perhaps clearer in the presentation, to break this sign convention by referring to heat as a positive quantity all the time and always indicating the direction it is flowing, much like it is a substance. When we discuss heat engines in the next section we will do just this.

4.3 Heat engines

If we consider the Carnot cycle as the operating process of a heat engine, we see that:

- An amount of heat equal to $Q_h = nRT_h \ln \frac{V_B}{V_A}$ (a positive quantity) is supplied to the system by the heat reservoir at temperature T_h (note the sign convention!).
- An amount of heat equal to $Q_c = nRT_h \ln \frac{V_C}{V_D}$ (a positive quantity) is lost from the system to the heat reservoir at temperature T_c.
- An amount of work equal to the difference $Q_h - Q_c$ is done by the system each time it goes around one Carnot cycle.

In general, the efficiency of any heat engine that is supplied an amount of energy Q_h and does an amount of work equal to $Q_h - Q_c$, is defined as

$$\eta = \frac{work\ done\ by\ system}{energy\ supplied\ to\ system}$$
$$= \frac{Q_h - Q_c}{Q_h}$$
$$= 1 - \frac{Q_c}{Q_h}.$$

From our calculations above, for the specific heat engine operating with the Carnot cycle and using an ideal gas as its working substance, this quantity is equal to

$$\eta = \frac{nRT_h \ln \frac{V_B}{V_A} - nRT_c \ln \frac{V_C}{V_D}}{nRT_h \ln \frac{V_B}{V_A}}$$

$$= 1 - \frac{T_c}{T_h} \frac{\ln \left(\frac{V_C}{V_D}\right)}{\ln \left(\frac{V_B}{V_A}\right)}.$$

The path BC is an adiabat. Therefore, from Example 3.4,

$$T_h V_B^{\gamma-1} = T_c V_C^{\gamma-1}.$$

This is similarly the case for the adiabat DA:

$$T_h V_A^{\gamma-1} = T_c V_D^{\gamma-1}.$$

Combining these two equations, we have for the Carnot cycle in Figure 4.1

$$\frac{V_B}{V_A} = \frac{V_C}{V_D},$$

and thus

$$\ln \left(\frac{V_B}{V_A}\right) = \ln \left(\frac{V_C}{V_D}\right).$$

Hence, the efficiency of the heat engine is given by

$$\eta = 1 - \frac{T_c}{T_h} \frac{\ln \left(\frac{V_C}{V_D}\right)}{\ln \left(\frac{V_B}{V_A}\right)}$$

$$= 1 - \frac{T_c}{T_h}.$$

Comparing this with the general definition for the efficiency of a heat engine, we have

$$\frac{T_c}{T_h} = \frac{Q_c}{Q_h}$$

where, to remind ourselves, Q_h is the quantity of heat *supplied to* the heat engine from the hotter reservoir at T_h, and Q_c is the quantity of heat *lost from* the heat engine to the colder reservoir at T_c; both Q_c and Q_h are positive numbers. Because we will use the heat engine to a large extent in our later discussions we illustrate it in a generic schematic diagram in Figure 4.2 below.

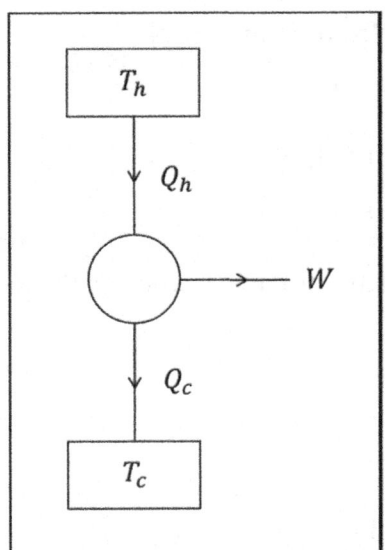

Figure 4.2 Schematic representation of a heat engine. The box T_h is a heat reservoir at the hotter temperature and the box T_c is a heat reservoir at the colder temperature. Q_h is the heat that flows (!) from the hotter reservoir to the engine denoted by the circle. Q_c is the heat lost from the engine to the colder reservoir. W is the work done by the engine. If the heat engine is operating a Carnot cycle with an ideal gas as the working substance we have $T_c/T_h = Q_c/Q_h$.

Since all the steps $A \rightarrow B \rightarrow C \rightarrow D \rightarrow A$ in the Carnot cycle are reversible, we can reverse the entire cycle so that we take the path $A \rightarrow D \rightarrow C \rightarrow B \rightarrow A$. Operating in the forward direction, we have a heat engine that produces work W by extracting some heat Q_h from reservoir T_h and losing some heat Q_c to reservoir T_c. The work obtained is equal to $Q_h - Q_c$. Operating in the reverse direction $A \rightarrow D \rightarrow C \rightarrow B \rightarrow A$, we extract an amount of heat Q_c from the colder reservoir T_c and lose an amount of heat Q_h to the hotter reservoir T_h. With the cycle reversed, this means an amount of work W, again equal to $Q_h - Q_c$, is involved, but now it is work to be supplied to the engine. Note that operating the system in the reversed direction we are pumping heat from a colder object to a hotter object. A moment's reflection will convince you that this is not a spontaneous process; with no other external agency, heat always flows from hot to cold, not the other way around. The operation of the Carnot cycle as a heat pump is illustrated in Figure 4.3. Compare this with the diagram of the heat engine in Figure 4.2, noting particularly the directions of the heat transfers and the work.

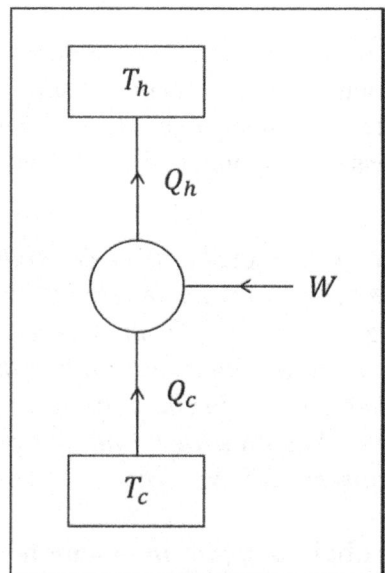

Figure 4.3 Schematic representation of a heat pump. Q_h is now the heat that flows from the engine to the hotter reservoir. Q_c is the heat that flows from the colder reservoir to the engine. W is the work done on the engine in order for it to pump heat from the colder reservoir to the hotter reservoir. As in Figure 4.2, if the heat pump is operating a Carnot cycle with an ideal gas as the working substance we have $T_c/T_h = Q_c/Q_h$.

4.4 Clausius and Kelvin statements: First look at the Second Law

Fundamental laws of Nature are statements arrived at by the application of inductive reasoning to summarise large amounts of empirical observations. In this section we discuss in detail the Clausius and Kelvin statements which are two logically equivalent statements of the Second Law that directly connect to empirical observations. These statements *seem* to apply only to certain types of physical phenomena. However, through some bit of deductive reasoning relying upon using the device of heat engines introduced in the previous section, we will show that the Clausius and Kelvin statements apply to all phenomena once we understand the concept of entropy, which we will really get to only in the next chapter.

The Clausius statement for the Second Law is: "It is impossible to devise a system which, operating in a cyclic process produces no effect other than the transfer of heat from a colder to a hotter body." In this statement, it is important to note:

a) "… operating in a cyclic process …"
b) "… produces no effect other than …"

These points emphasize that at the end of the process, the system and all parts of the universe are back at their respective original states, except that heat has been transferred from a colder object to a hotter one. We visually represent a process that violates the Clausius statement by the following diagram

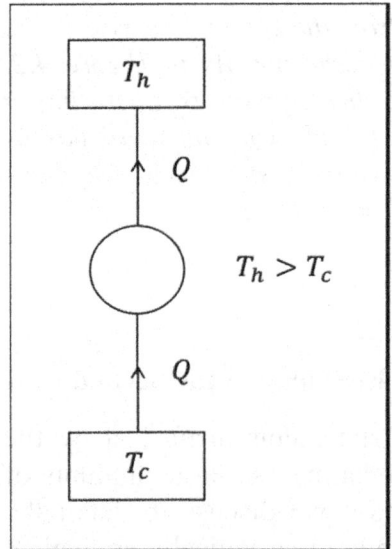

Figure 4.4 A cyclic process that violates the Clausius statement, by transferring heat Q from a colder reservoir to a hotter reservoir with no other change in the state of the universe. Such a process has never been observed in Nature.

An obvious *apparent* violation is an air-conditioner, which in the process of keeping your room cool, pumps heat from the cooler room out into the hotter Singapore outdoors. Air-conditioners do not violate the Clausius statement because electrical energy is drawn from some source (a power plant somewhere) which would have its thermodynamic state changed for each cycle of the air-conditioner compressor. We note that no cup of hot tea in a cool room has ever been observed to spontaneously get even hotter by absorbing heat from the room, leaving the latter even cooler than before, with no change to the rest of the universe. If you ever observe this to really happen, it would be a noteworthy empirical observation that invalidates the Clausius statement.

The Kelvin statement of the Second Law is "It is impossible to devise a system which, operating in a cyclic process produces no effect other than the extraction of heat from a reservoir and completely converting this energy into mechanical work." There are many areas of the world which are hotter than we might prefer them to be, for example, the sea waters around Singapore. The Kelvin statement assures us that it is not possible to:

a) extract heat energy from these areas,
b) completely convert this energy into work,
c) produce/require no other change in any other parts of the universe.

The violation of the Kelvin statement is illustrated in Figure 4.5.

We have encountered Joule's paddle wheel experiment in which a falling weight, through its connection to a pulley system, turns the paddle wheel and warms up the water in an closed box. The Kelvin statement assures us that no one has or will observe the reverse process where the water in the box cools down, losing heat energy which is completely converted to the kinetic energy of the paddle wheel which turns, and lifts the weight back up to its original height, with no other change in the rest of the universe.

Perhaps, closer to your everyday experience, you have observed a bouncing ball gradually losing its kinetic energy as its bounce gradually get lower until it stops bouncing. You have never observed a ball, initially motionless on the floor, spontaneously beginning to bounce higher and higher by absorbing heat from its surroundings and converting this into the mechanical work needed to lift the ball against gravity. We emphasize that the Clausius and Kelvin statements are summaries of empirical observations, not *a priori* postulates. We next show that these two statements are logically equivalent. If the

Clausius statement is not true, then the Kelvin statement is also not true. And vice versa.

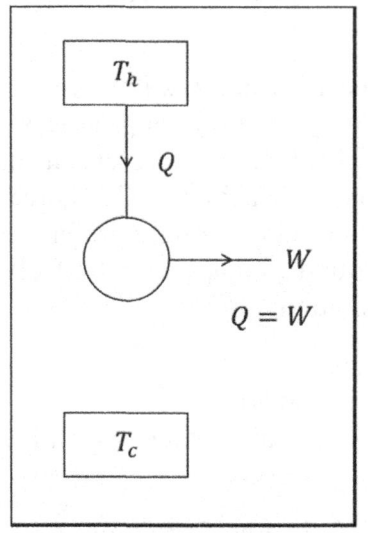

Figure 4.5 A cyclic process that violates the Kelvin statement, by absorbing heat Q from a reservoir and completely converting this energy into work so that $W = Q$, with no other change in the state of the universe. Such a process has never been observed in Nature.

Let us first violate the Kelvin statement. Consider the system illustrated in Figure 4.6. We have a cyclically operating device A which extracts heat Q from a reservoir at temperature T and completely converts it into work W per cycle, with no other change to the universe. Then we can use that work to drive a Carnot heat pump B of an appropriate size to extract heat Q_c from a colder reservoir $T_c < T$, and pump heat $Q_h = Q_c + W = Q_c + Q$ into the reservoir at temperature T, with no other change in the universe. Thus, the overall effect of the system per cycle is to transfer an amount of heat equal to $Q_h - Q = Q_c$ to the hotter reservoir from the cold reservoir, with no other change to the rest of the universe. This is essentially a spontaneous transfer of heat from the cold reservoir to the hot reservoir, violating the Clausius statement. Thus, if the Kelvin statement is not true, then the Clausius statement is not true.

The steam engines in the early days of Thermodynamics were not very efficient machines. This meant that the amount of heat going into an engine from the heat source is only slightly larger than the amount coming out into the heat sink from the engine. The point that we take from Kelvin's statement is that the latter, which always less than the former, is *never* zero, no matter how efficient you make your engine. This is the important point in Kelvin's summary of empirical observations.

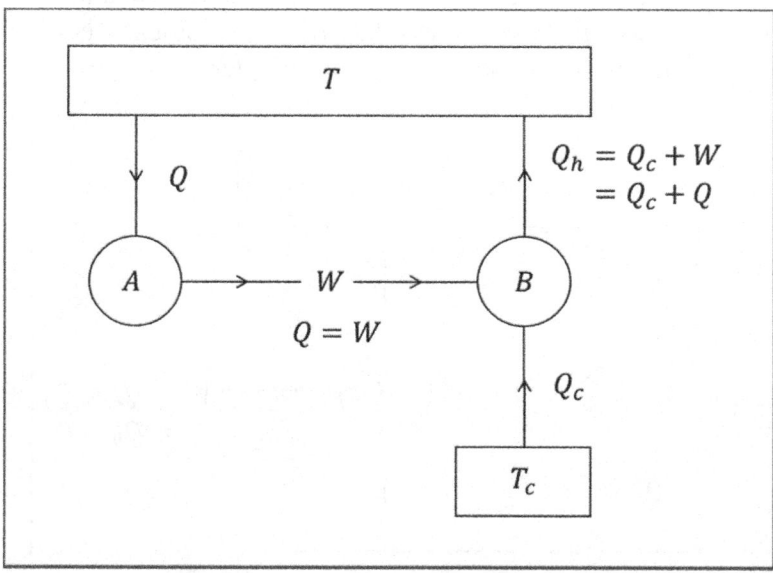

Figure 4.6 We have a cleverly constructed machine A that violates the Kelvin statement by completely converting heat Q into work W, with no other change to the universe. We connect A to a Carnot heat pump B which uses the work W to pump heat $Q_c + Q$ back into the hot reservoir. The net result is violation of the Clausius statement.

Let us now begin with violating the Clausius statement. Using the system illustrated in Figure 4.7, we have a device A that cyclically transfers heat Q from a colder reservoir T_c to a hotter reservoir T_h, without any other change in the universe. We then connect a Carnot engine B of an appropriate size to operate between the same two heat reservoirs. This absorbs heat Q_h from the hotter reservoir and loses heat $Q_c = Q$ to the colder reservoir, while doing work W. Thus, the net heat change is zero for the cold reservoir, while the hot reservoir loses heat $Q_h - Q$, and W is equal to $Q_h - Q_c = Q_h - Q$. Therefore, we essentially have a system that completely converts heat absorbed from the hot reservoir into work, violating the Kelvin statement. If the Clausius statement is not true, then the Kelvin statement is not true. Since the set of all circumstances that violate the Kelvin and Clausius statements are identical, these two statements are logically equivalent. We will use them in later sections of this chapter.

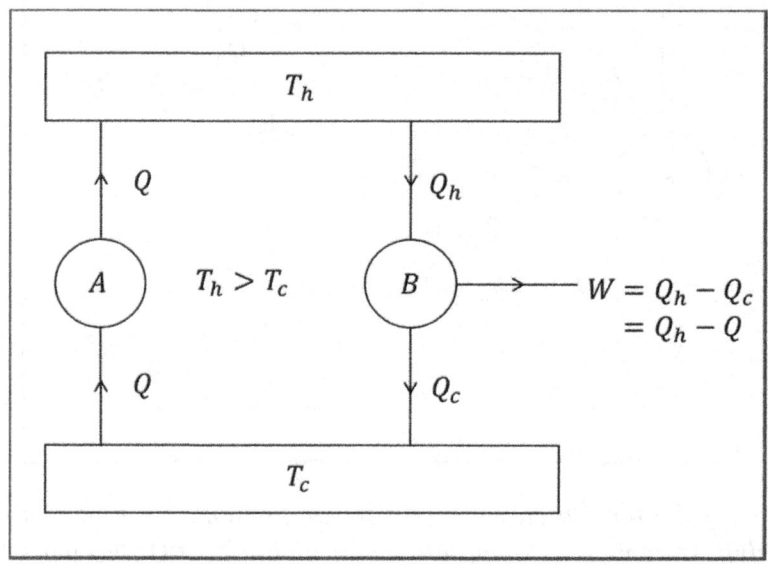

Figure 4.7 We connect a deviously designed machine A that violates the Clausius statement by transferring heat Q from the cold reservoir to the hot reservoir with no other change to the universe. Then a Carnot engine operating between the reservoirs supplies heat $Q_c = Q$ back to the cold reservoir, while producing work W. The net result is violation of the Kelvin statement.

4.5 No engine is more efficient than the Carnot engine!

A ball bouncing on the floor eventually stops bouncing. Similarly, a cup of hot tea in a cool room eventually cools down to the temperature of the room. If you examine any of the microscopic jigglings of atoms in these systems, you will not find anything that violates the fundamental time reversible laws of motion, either classical or quantum. Therefore, the time reverse of these natural spontaneous processes do not violate fundamental mechanics. Even though, in particular, they do not violate the law of conservation of energy, you will never observe the time-reverse of the slowing down of the bouncing ball, or the time-reverse of the cooling down of the cup of hot tea. These two time-reversed processes, which you will never observe, are specific examples of the impossible transformations forbidden by the Kelvin and Clausius statements. Using these precise statements to examine heat and work interactions in general, we will

a) show that there exists an important thermodynamic state function, Entropy,
b) define and calculate entropy for thermodynamic systems, and
c) arrive at a general statement of the Second Law of Thermodynamics in terms of entropy.

A Carnot engine draws energy in the form of heat Q_h from a hotter reservoir at T_h, loses heat to a colder reservoir at T_c, and performs work $W = Q_h - Q_c$. As seen in Section 4.3 above, its efficiency is

$$\eta_{Carnot} = \frac{work\ done\ by\ system}{energy\ supplied\ to\ system}$$
$$= \frac{W}{Q_h} = \frac{Q_h - Q_c}{Q_h}$$
$$= 1 - \frac{Q_c}{Q_h}$$
$$= 1 - \frac{T_c}{T_h}.$$

Is it possible for any heat engine operating between the same temperatures to be more efficient that the Carnot engine? That is, is it possible to find an engine which has a value of $\frac{Q_c}{Q_h}$ that is smaller than $\frac{T_c}{T_h}$?

Let us assume that there exists a super-engine with an efficiency greater than that for a Carnot engine, $\eta_{super} > \eta_{Carnot}$. We will see whether the logical consequences of that assumption are consistent with Nature. Consider connecting the super-engine with a Carnot engine in the configuration illustrated here:

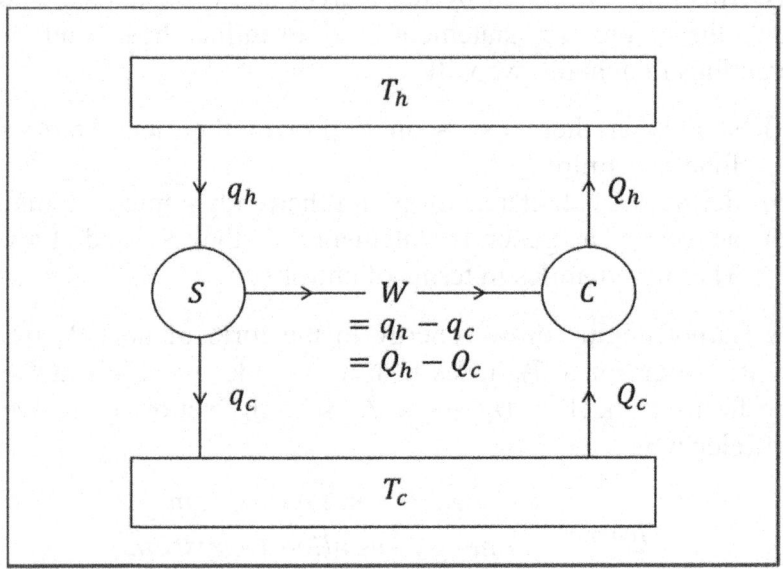

Figure 4.8 S is a super-engine with efficiency higher than that of the Carnot engine C. In this configuration, the work produced by the super-engine is used to drive the Carnot engine running in reverse as a heat pump.

We are running a Carnot cycle in reverse so that we have a Carnot heat pump. The work from the super-engine is equal to

$$W = q_h - q_c$$

and we appropriately size the engines so that this is equal to the work required to run the Carnot engine, which is

$$W = Q_h - Q_c,$$

from which we have

$$q_h - q_c = Q_h - Q_c.$$

If the super-engine is more efficient than the Carnot cycle run as an engine, we have

$$\eta_{super} = \frac{W}{q_h} > \eta_{Carnot} = \frac{W}{Q_h}.$$

Therefore, if it is indeed true that the efficiency of the super-engine is greater than that of a Carnot engine, we will have

$$Q_h > q_h.$$

However, we have already seen that $q_h - q_c = Q_h - Q_c$. Rearranging this, we have

$$Q_h - q_h = Q_c - q_c > 0.$$

Upon examining Figure 4.8, you will see that $Q_h - q_h$ is the net heat lost to the hotter reservoir and $Q_c - q_c$ is the net heat absorbed from the colder reservoir. Both of these quantities are positive from our arguments thus far, and hence we have a situation in which we have a cyclic process that moves heat from a colder reservoir to a hotter reservoir, with no other change to the thermodynamic state of the rest of the universe. But this violates the Clausius statement! Hence, we conclude that $\eta_{super} > \eta_{Carnot}$ is not consistent with any observations of heat and work interactions in Nature. By this reductio ad absurdum argument, we have found that no engine can be more efficient than a Carnot engine. The efficiency of any engine is less than or, at best, equal to that of the Carnot engine.

4.6 All reversible engines are equally efficient

The Carnot cycle is a reversible cycle, and thus the Carnot engine is a reversible engine; when run in the reverse direction, it functions as a heat pump. Not all engines operate with reversible cycles, so not all engines are reversible. If the super-engine in the previous section is reversible, we can switch the roles of the two engines in Figure 4.8. We can then run the super-engine in reverse as a heat pump driven by the work output from the Carnot engine, which now is run in the forward direction. This configuration is illustrated in Figure 4.9 with the generic reversible super-engine, now labelled R. We use this configuration to compare the efficiency of the Carnot engine with, specifically, any other reversible engine.

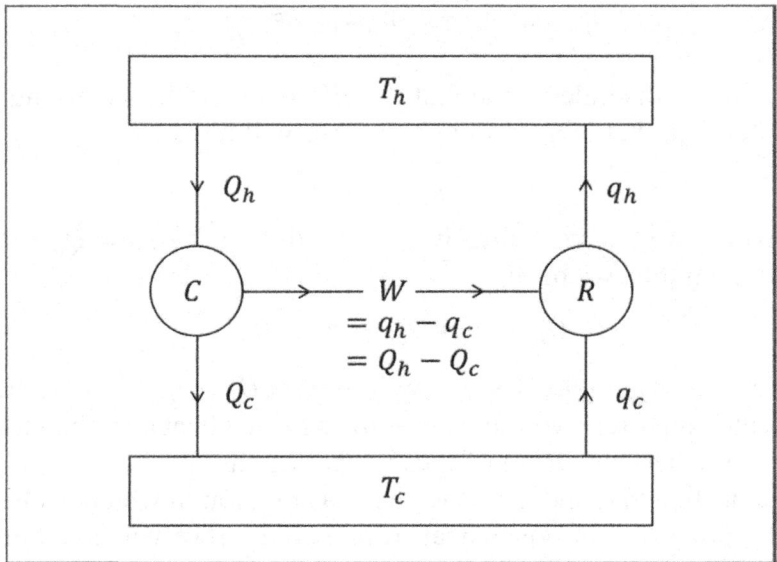

Figure 4.9 In this configuration R is a reversible engine run as a heat pump driven by the work produced by a Carnot engine C.

Using the same considerations as above, the First Law gives us

$$W = q_h - q_c = Q_h - Q_c.$$

Looking at Figure 4.9, we now see that $q_h - Q_h$ and $q_c - Q_c$ are the heat amounts pumped into the hotter reservoir and pumped from the colder reservoir, respectively. And these quantities are equal.

If any generic reversible engine is less efficient than the Carnot engine, $\eta_{Carnot} > \eta_{rev}$, then

$$\frac{W}{Q_h} > \frac{W}{q_h},$$

and

$$Q_h < q_h.$$

This means that $q_h - Q_h = q_c - Q_c$ are both positive quantities. Looking at Figure 4.9, we see that the system operating in a cyclic process transfers heat from the colder reservoir to the hotter reservoir, with no other change in thermodynamic state of the rest of the universe. And this violates the Clausius statement. Thus, the assumption $\eta_{Carnot} > \eta_{rev}$ cannot be valid. We conclude that *no* reversible engine can have an efficiency less than that of the Carnot engine.

This leaves the two possibilities

$$\eta_{Carnot} < \eta_{rev}$$

and

$$\eta_{Carnot} = \eta_{rev}.$$

But, we have seen in the previous section that the Carnot engine efficiency cannot be less than the efficiency of any other engine, ruling out the first possibility. Hence, combining these results, we have the following two conclusions about the efficiencies of all engines:

$$\eta_{Carnot} = \eta_{rev},$$

$$\eta_{Carnot} > \eta_{irrev},$$

where the subscripts "*rev*" and "*irrev*" denote reversible and irreversible engines. Operating between a hot and a cold reservoir, the efficiency of any engine is given by

$$\eta = 1 - \frac{q_c}{q_h},$$

where q_h is the amount of heat supplied to the engine from the hotter reservoir and q_c is the amount of heat lost by the engine to the colder reservoir in the process of performing work $W = q_h - q_c$. From our results, we have the following important general results comparing the heat interactions of the Carnot engine with any other engine:

$$\left(\frac{q_c}{q_h}\right)_{Carnot} = \left(\frac{q_c}{q_h}\right)_{rev}$$
$$\left(\frac{q_c}{q_h}\right)_{Carnot} < \left(\frac{q_c}{q_h}\right)_{irrev}.$$

Working with any two specific reservoir temperatures, all reversible engines deliver work with the same efficiency. These conclusions seem rather mundane, or merely of practical interest rather than fundamental importance. However, they contain the sobering truth that, consistent with all our experience of this Universe, Time flows only in one direction. We will get to that in the next chapter. For now, we discuss a smaller but also rather important truth, namely that there is such an objective, measurable quantity as an absolute temperature that is independent of the means we use to measure it. But first a real-life example of a heat engine, in approximation.

4.7 Hurricanes are heat engines

These powerful storms can be modelled as heat engines with the hot reservoir being the warm ocean and the cold reservoir being the upper atmosphere. Indeed, they can be *approximately* described by the Carnot cycle, as illustrated in a vertical cross-section of a storm in Figure 4.10. At the centre of these storms is a calm, low pressure cylindrical region called the eye indicated in the figure. Because this is at a low pressure of 0.9 bar, strong winds blow air into the centre of the storm from the surrounding high pressure regions typically at 1.03 bar. Together with the Coriolis force due to the rotation of the Earth, these winds spiral around and into the eye of the storm. As the air is blown over the warm ocean from high pressure to low, it expands and picks up water vapour due to evaporation. A typical ocean surface temperature at the eye is 30°C. Thus, heat is absorbed into the atmosphere from the oceans in the isothermal expansion AB. The warm air in the eye is then forced by the wind up into lower pressure and cooling down in the upper atmosphere in an adiabatic expansion, illustrated by the arrow BC. In the upper atmosphere, and far away from the centre of the storm, this humid air loses heat to the cold surroundings at about -70°C, resulting in an isothermal compression CD. The cold air then sinks, undergoing an adiabatic compression DA, feeding back into the high pressure region at A. Clearly, a real storm does not consist of reversible processes as the Carnot cycle does, but nonetheless the approximation is instructive: it is the temperature difference between the warm ocean and the cold upper atmosphere that gives these storms their enormous energy. We can estimate the amount of this energy as follows.

Take the eye of a typical storm to be about 500 m high and 100 km in radius. The pressure in the eye is typically low at about 0.9 bar. From this, we can estimate the number of moles of air in the eye to be

$$n = \frac{PV}{RT} \cong 5.6 \times 10^{14}.$$

With this estimate of the number of moles of air, we can estimate the amount of heat absorbed from the ocean in the isothermal expansion AB in Figure 4.10. From the discussion in Section 4.1, this is

$$\Delta Q = nRT_h \ln\left(\frac{P_A}{P_B}\right)$$
$$\cong 1.9 \times 10^{14} \text{ kJ}$$

per cycle of the heat engine. Using the Carnot engine efficiency, we have

$$\eta = 1 - \frac{T_c}{T_h}$$
$$= 1 - \frac{(273 - 70)}{(273 + 30)} \cong 0.33$$

Thus, modelling a hurricane as a Carnot engine, the amount of energy released is 6.3×10^{13} kJ for each cycle of the engine. This is about the energy released by 3000 atomic bombs. Of course, a storm is not a reversible process, so the efficiency has to be less than that of the Carnot engine. This is an overestimate, but indicates the enormous amount of energy in these storms. It underscores the amount of heat energy that is stored in oceans, which are increasing in temperature due to global warming. Imagine harnessing such an enormous source of energy.

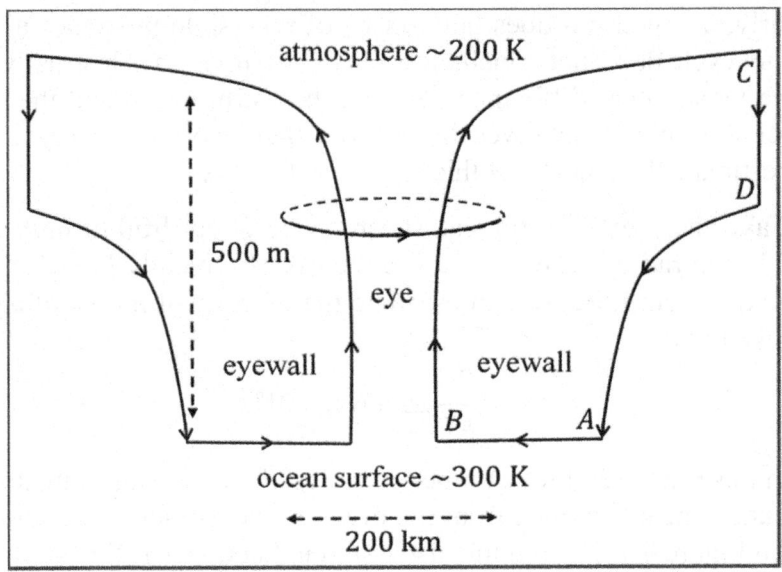

Figure 4.10 Schematic cross-section of a hurricane as a heat engine. The hot reservoir is the ocean surface, the cold reservoir is high up in the atmosphere.

4.8 Heat engines are thermometers: the absolute thermodynamic temperature scale

We have now established that any reversible engine operating between the same two heat reservoirs, regardless of its operating substance and regardless of its operating cycle, has the same efficiency

$$\eta_{rev} = \eta_{Carnot}.$$

We have also seen in Section 4.3 that the efficiency of the Carnot engine is related to the operating temperatures by

$$\eta_{Carnot} = 1 - \frac{T_c}{T_h}$$

giving a connection between the efficiency and the temperature, at least for the Carnot engine. In this expression, the temperatures are defined on the ideal gas temperature scale discussed in Chapter 2, and quantified through $\theta_0 \equiv \lim_{P \to 0} \left(\frac{PV}{R} \right)$ which goes to the same limit for any gas in thermal equilibrium with each other. That is, we depend upon the behaviour of gases in the limit of zero pressure to define the temperatures T_c and T_h in this equation. Wouldn't it be nice to be able to define a temperature scale that is independent of the use of any material substance? Can we use a heat engine as a thermometer for this?

We have, in general,

$$\eta = 1 - \frac{q_c}{q_h}$$

where the ratio of the heat absorbed and lost to the reservoirs determines the efficiency of an engine. Since we can directly measure the efficiency η of the engine, we can directly measure the ratio q_c/q_h. We now know that for any reversible engine, this ratio is the same, depending only upon the temperatures of the reservoirs which we denote generally by θ_c and θ_h. Therefore, we have

$$\frac{q_c}{q_h} = f(\theta_c, \theta_h)$$

where $f(\theta_c, \theta_h)$ is a function of its two arguments. Consider three temperatures $\theta_1 < \theta_2 < \theta_3$ and three reversible engines operating with reservoir temperatures $(\theta_1, \theta_2), (\theta_2, \theta_3)$ and (θ_1, θ_3). The ratios of q_c/q_h for these engines are, respectively,

$$\frac{q_1}{q_2} = f(\theta_1, \theta_2),$$

$$\frac{q_2}{q_3} = f(\theta_2, \theta_3),$$

$$\frac{q_1}{q_3} = f(\theta_1, \theta_3).$$

The ratio q_1/q_3 for the third engine can also be written as

$$\frac{q_1}{q_3} = \frac{q_1}{q_2}\frac{q_2}{q_3} = f(\theta_1, \theta_2)f(\theta_2, \theta_3),$$

so that

$$f(\theta_1, \theta_2)f(\theta_2, \theta_3) = f(\theta_1, \theta_3),$$

since the ratio q_1/q_3 cannot be dependent upon θ_2. Hence, the form of the function $f(\theta_1, \theta_2)$ must be

$$f(\theta_1, \theta_2) = \frac{g(\theta_1)}{g(\theta_2)}$$

in order for the factors of $g(\theta_2)$ to cancel out in the product $f(\theta_1, \theta_2)f(\theta_2, \theta_3)$. We thus have in general for a reversible engine that

$$\frac{q_c}{q_h} = f(\theta_c, \theta_h)$$

$$= \frac{g(\theta_c)}{g(\theta_h)}$$

$$= \frac{\theta_c}{\theta_h}$$

where in the third line, we have taken the function $g(\theta)$ to simply be equal to its argument. That is, why not *define* $g(\theta)$ as the temperature? If we do this, and we combine this with what we have found for the efficiency of the Carnot engine in terms of the ideal gas temperatures of its reservoirs, we have

$$\frac{\theta_c}{\theta_h} = \frac{T_c}{T_h}.$$

That is:

a) we can define the ratio of the reservoir temperatures through the efficiency of the reversible engine:

$$\frac{\theta_c}{\theta_h} = 1 - \eta.$$

This gives the absolute thermodynamic temperature scale. We have arrived at this scale using the properties of abstract reversible heat engines, which in turn are deduced solely from the fundamental dynamics of heat and work encapsulated in the Kelvin and Clausius statements. The absolute thermodynamic temperature is a basic property of Nature.

b) as it turns out, the ideal gas temperature scale is the same as this fundamental temperature scale.

If we further arbitrarily assign a numerical value to the temperature difference between two specific constant-temperature systems, we would then be able to establish a complete temperature scale. This seems slightly abstract, so we use a numerical example to illustrate.

Consider liquid water in equilibrium with water vapour at a pressure of one bar. This is a constant temperature system, at the temperature of boiling water, as long as we maintain the equilibrium. Similarly, we have another constant temperature system, at the temperature of melting ice, if we maintain equilibrium between ice and liquid water at a pressure of one bar. We run a reversible engine (whatever the operating cycle and whatever the operating substance) using these two constant temperature systems as the reservoirs at temperatures θ_h and θ_c, respectively. For these two reservoirs, the efficiency would be 0.267989, which can be measured directly through q_h and q_c. If we arbitrarily set the temperature difference $\Delta\theta = \theta_h - \theta_c = 100$, we would have completely set the numerical value of θ_h and θ_c. We determine these values from:

$$\frac{\theta_c}{\theta_h} = 0.267989$$

$$\theta_h - \theta_c = 100,$$

so that we have

$$\theta_h = 373.15$$

$$\theta_c = 273.15$$

which are immediately recognizable as the temperatures of the normal boiling and freezing points of water at one bar in terms of the ideal gas temperature scale. Therefore, selecting a temperature scale θ where there are 100 units between the boiling and freezing points of water, and using the efficiency of a reversible heat engine to measure temperature, we have arrived at the absolute temperature scale. Note that this is without reference to any operating substance for the engine. Given our choice of 100 units between the normal freezing and boiling points of water, our absolute temperature scale is also the same as the ideal gas temperature scale we introduced in Chapter 2. From this point on, when we write T, we are referring to this universal absolute temperature scale.

On the absolute temperature scale, we obtain a temperature of absolute zero when the temperature is $-273.15°C$. We point out here that from the equation for the efficiency of a reversible heat engine,

$$\eta = 1 - \frac{T_c}{T_h}$$

a 100% efficiency can be achieved if we have a low temperature reservoir at absolute zero temperature. We will discuss in Chapter 6 why this is not possible to achieve.

5

The Second Law

Even though we examined the efficiency of engines, you would have noticed that our discussion in the last chapter was most general, in terms only of the heat energy supplied to and lost by an engine, and the work extracted from it. No engineering detail of any machinery was examined. That was sufficient, and preferred, because we are trying to get to the fundamental law that governs heat and work interactions in general, rather than just the effectiveness of the specific construction of any engine. It was Carnot's genius to recognize that behind the problem of the efficiency of real steam engines there lies a fundamental principle that governs all of Nature. Although heat and work are both forms of energy, you will realise that heat is different from the other forms of energy. Work always needs to be done to pump heat from colder objects to hotter ones. In the reverse direction, a hot object can lose heat to a cold object without requiring any work to be done. No engine can ever completely convert an amount of heat to work. The other way around, you can always completely dissipate work into heat. There is an asymmetry in the direction of natural phenomena involving Heat. Our task in this chapter is to identify the one quantity that allows us to characterise the spontaneous direction of any process in Nature.

5.1 Why explore $dS = \frac{dQ}{T}$?

In Example 3.2 we examined an ideal gas undergoing the reversible cycle $ABCA$ illustrated again in Figure 5.1. Along each step of the cycle, we calculated the quantity

$$\Delta S = \int dS \equiv \int \frac{dQ}{T},$$

giving us

$$\Delta_{AB}S = (C_V + R)\ln\frac{V_B}{V_A}$$

$$\Delta_{BC}S = -C_V \ln\frac{V_B}{V_A}$$

$$\Delta_{CA}S = -R\ln\frac{V_B}{V_A}.$$

Hence, the sum of dS over the entire cycle is zero. We denote this by

$$\oint dS = \oint \frac{dQ}{T} = 0$$

where the circle through the integral symbol denotes integration over a cyclic path, from the start around the entire cycle, back to the initial state. $\oint dS$ for this reversible cycle is zero.

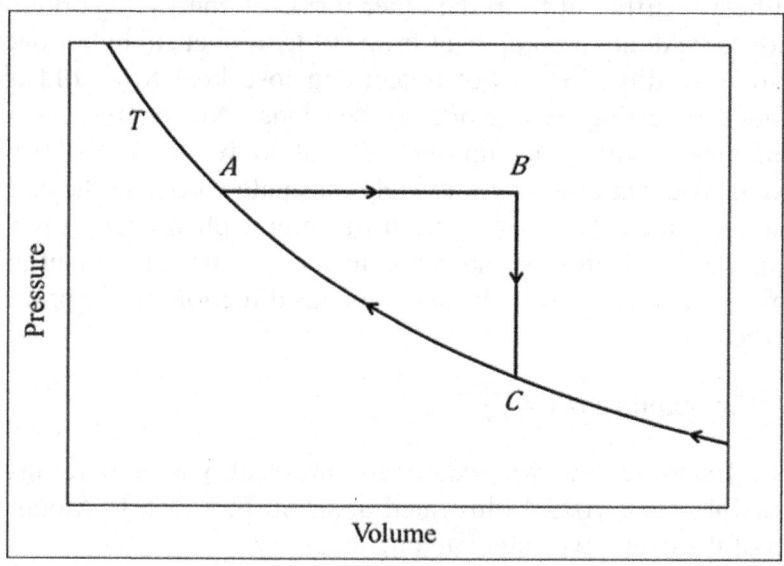

Figure 5.1 A reversible cycle ABCA, where AB is an isobaric (constant pressure) process, BC is an isochoric (constant volume) process, and CA is an isothermal process at constant temperature T.

Similarly, in Chapter 4 we have examined the heat and work interactions of the Carnot cycle using an ideal gas; see Figure 5.2. All the steps in the Carnot cycle are reversible. Let's look at $\oint dS$ for this.

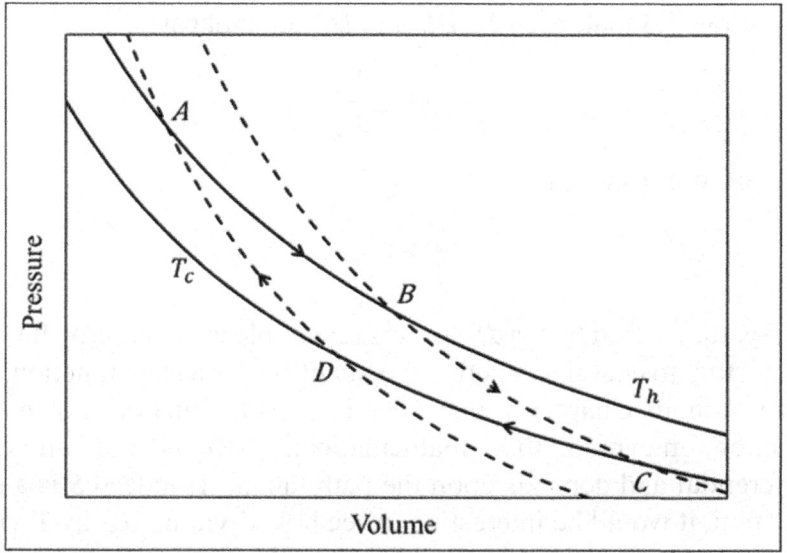

Figure 5.2 The ideal gas Carnot cycle on a PV diagram. Revisit Section 4.1.

Along AB, heat is supplied to the system from a reservoir at temperature T_h. Along CD, heat is lost from the system to a reservoir at temperature T_c. BC and DA are adiabatic, and thus no heat interaction occurs along these steps. Hence, for the Carnot cycle

$$\oint dS = \oint \frac{dQ}{T} = \int_A^B \frac{dQ}{T} + \int_C^D \frac{dQ}{T}$$

$$= \frac{1}{T_h} \int_A^B dQ + \frac{1}{T_c} \int_C^D dQ$$

$$= \frac{1}{T_h} \Delta_{AB} Q + \frac{1}{T_c} \Delta_{CD} Q$$

$$= R \ln \frac{V_B}{V_A} + R \ln \frac{V_D}{V_C}$$

using the $\Delta_{AB}Q$ and $\Delta_{CD}Q$ calculated in Section 3.1. We have seen in Section 3.3 that, because BC and DA are adiabats,

$$\frac{V_B}{V_A} = \frac{V_C}{V_D}.$$

Therefore, for the Carnot cycle,

$$\oint dS = 0.$$

Again, we find that $\oint dS$ for this reversible cycle is zero. It seems interesting to examine whether the quantity S is a state function. We have seen in Chapter 1 that heat is a path function, not a state function, meaning that mathematically dQ is not an exact differential and depends upon the path taken. If indeed S is a state function, it would be interesting to see how dividing dQ by T yields a state function in these reversible cycles. To get to this, we will use the conceptual device of heat engines and heat pumps introduced in the previous chapter to analyse cyclic thermodynamic transformations in general, both reversible and irreversible ones.

5.2 For all cyclic processes $\oint \left(\frac{dQ}{T} \right) \leq 0$

Consider a general cyclic process schematically illustrated in Figure 5.3. In each step i of the cycle, the system gains an amount of heat q_i from a reservoir R_i at temperature T_i. Note that the quantity q_i can be either positive or negative. We use the sign convention here that if heat is transferred to the system from the reservoir, then q_i is positive. Vice versa, if heat is transferred from the system to the reservoir, then q_i is negative. In general, some of the q_i's are positive and some are negative. Overall, the system performs an amount of work W_{sys} *done by* the system for each cycle of the process.

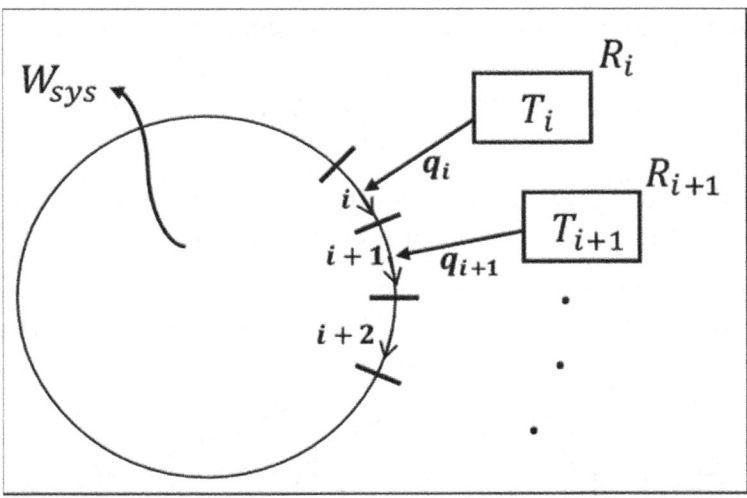

*Figure 5.3 A schematic illustration of a generalized cyclic process. In each step i, the systems gains heat from a reservoir R_i at temperature T_i. The total work **done by** the system is W_{sys} for each cycle. Note the direction of the arrows for the heat and work interactions.*

To analyse this system, we connect each reservoir R_i to a reversible heat engine/heat pump C_i. Each of these reversible heat engines/pumps exchanges an amount of heat q_i with its reservoir R_i, and exchanges an amount of heat q_{0i} with the common reservoir R_0 that is at temperature T_0. This forms the super-system illustrated in Figure 5.4. The heat flow q_i from R_i to the system is the same as the heat flow from C_i to R_i, so that after each cycle, ΔQ_i is equal to zero and each of the reservoirs R_1 to R_n is back to its initial state at the start of the cycle. At the same time, each reversible engine/pump C_i absorbs an amount of heat equal to q_{0i} from the common reservoir R_0 at temperature T_0. In each cycle of the super-system, the heat absorbed from R_0 is

$$Q_0 = \sum_{i=1}^{n} q_{0i},$$

and the total work done by the super-system is equal to

$$W_{tot} = W_{sys} + \sum_{i=1}^{n} w_i$$

where the second term on the right-hand side is the sum of the work done by the reversible heat engines/pumps C_1 to C_n.

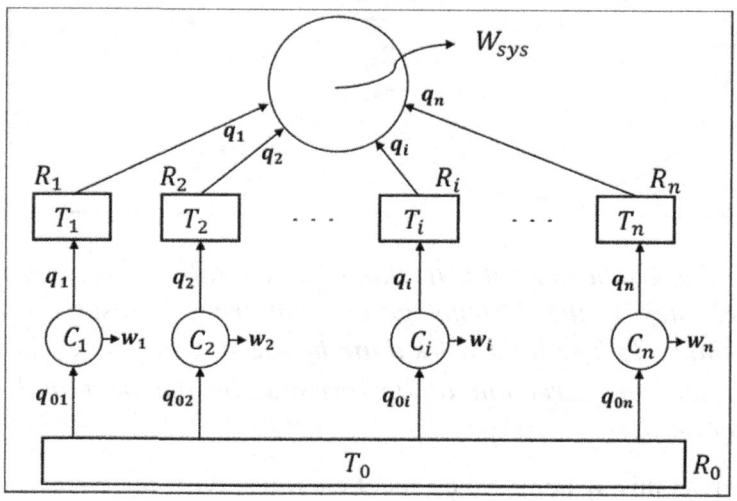

Figure 5.4 This is a super-system in which each step, from $i = 1$ to $i = n$, of the cyclic process illustrated in Figure 5.3 is now connected through a reversible engine C_i to a common reservoir R_0 at temperature T_0.

Examining Figure 5.4 and applying the First Law to the system, we immediately have

$$W_{sys} = \sum_{i=1}^{n} q_i.$$

Applying the First Law to the engines/pumps C_1 to C_n,

$$\sum_{i=1}^{n} w_i = \sum_{i=1}^{n} (q_{0i} - q_i) = \sum_{i=1}^{n} q_{0i} - W_{sys}$$

so that

$$W_{tot} = \sum_{i=1}^{n} q_{0i}$$

Hence, in each cycle, there is no change to the universe except for the common reservoir R_0 losing heat $\sum_i q_{0i}$ and the super-system performing work equal to $\sum_i q_{0i}$. The amount of heat lost by the reservoir R_0 is equal to the total amount of work done by the super-system. If W_{tot} is a positive quantity, then the super-system draws net heat from a reservoir R_0 and completely converts that energy into work. This, of course, violates the Kelvin statement, so W_{tot} cannot be positive. Hence, we conclude that

$$W_{tot} = \sum_{i=1}^{n} q_{0i} \leq 0.$$

But we have seen for each of the reversible heat engines that

$$\frac{q_i}{T_i} = \frac{q_{0i}}{T_0}.$$

Therefore, we also have

$$\sum_{i=1}^{n} q_{0i} = T_0 \sum_{i=1}^{n} \frac{q_i}{T_i}.$$

Combining this with the constraint from the Kelvin statement, we have the inequality

$$\sum_{i=1}^{n} \frac{q_i}{T_i} \leq 0$$

for each cycle. This applies for any generic cyclic process.

We have not specified whether the cyclic process in Figure 5.3 is reversible. We now consider what we get if the cyclic process is specifically a reversible one. Then, we can reverse all the heat interactions q_i, changing the sign of each of these quantities. Applying the same argument as above, we end up with

$$\sum_{i=1}^{n} \frac{(-q_i)}{T_i} \leq 0$$

$$\Rightarrow \sum_{i=1}^{n} \frac{q_i}{T_i} \geq 0.$$

This result holds only if the cyclic process is reversible. Thus, for reversible cyclic processes, $\sum_i \frac{q_i}{T}$ can only be either greater than or equal to zero. But we have seen above, for any cyclic processes, reversible or irreversible, $\sum_i \frac{q_i}{T}$ must be less than or equal to zero. In order for both of these results to be true, we must have:

a) for reversible cyclic processes

$$\sum_{i=1}^{n} \frac{q_i}{T_i} = 0$$

b) for irreversible cyclic processes

$$\sum_{i=1}^{n} \frac{q_i}{T_i} < 0.$$

These are the most important conclusions in this chapter; everything else really just follows from them.

If we consider dividing the cycle into infinitesimal steps, these sums become integrals over the cycle,

$$\sum_{i=1}^{n} \frac{q_i}{T_i} \rightarrow \oint \frac{dq}{T}$$

where the circle in the integral sign means that we integrate the quantity dq/T over the entire cyclic process. Hence, for reversible cycles, we have

$$\oint \left(\frac{dQ}{T}\right)_{rev} = 0$$

while for irreversible cycles, we have

$$\oint \left(\frac{dQ}{T}\right)_{irrev} < 0.$$

This is what we claimed at the start of this section. The result for reversible cycles helps us understand the two examples we discussed in Section 5.1, namely that $dS = (dQ/T)_{rev}$ is an exact differential making S a state function. It also is relevant to the inconsistency we encountered in Example 3.3 *if* we assume in that example that $dS = \frac{dQ}{T}$ for the *irreversible* adiabatic expansion. We will further explore this.

5.3 S is a state function

Consider any arbitrary cyclic process illustrated in Figure 5.5 in which a system transforms from a state 0 (at thermodynamic equilibrium) to another state 1 (also at thermodynamic equilibrium) along a reversible path A, and then back to state 0 along a different reversible path B. Since $0 \to 1 \to 0$ is a reversible process,

$$\oint \left(\frac{dQ}{T}\right)_{rev} = \int_0^1 \left(\frac{dQ}{T}\right)_A + \int_1^0 \left(\frac{dQ}{T}\right)_B$$

where the first and second integrals on the right-hand side account for the change of dQ/T along paths A and path B, respectively.

Since the cycle is reversible, the discussion in Section 5.2 gives us

$$\int_0^1 \left(\frac{dQ}{T}\right)_A + \int_1^0 \left(\frac{dQ}{T}\right)_B = 0.$$

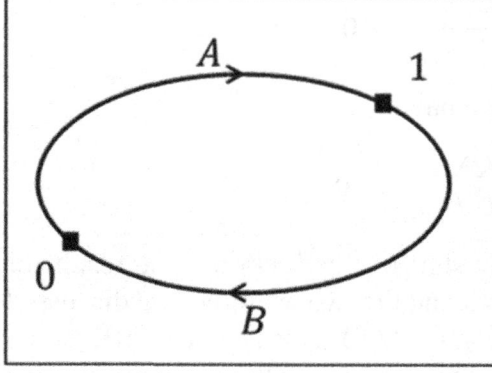

Figure 5.5 Equilibrium states 0 and 1 are denoted by the squares. A and B are reversible processes, thus making the cycle reversible.

Upon rearranging and switching the order of the limits for the integral along B, we obtain

$$\int_0^1 \left(\frac{dQ}{T}\right)_A = -\int_1^0 \left(\frac{dQ}{T}\right)_B = \int_0^1 \left(\frac{dQ}{T}\right)_B.$$

No matter which reversible path we choose to connect the two states, the integral of dQ/T along the path from 0 to 1 is the same.

If we now define the quantity dS along *any reversible* path as

$$dS \equiv \left(\frac{dQ}{T}\right)_{rev}$$

we have for the reversible paths A and B in Figure 5.5 the following equalities:

$$\Delta_{01}S = \int_0^1 dS_A = \int_0^1 \left(\frac{dQ}{T}\right)_A$$

$$= \int_0^1 dS_B = \int_0^1 \left(\frac{dQ}{T}\right)_B,$$

where $\Delta_{01}S$ is the change in S when we go from state 0 to state 1. Thus, the value of $\Delta_{01}S$ depends only upon the end-points of the connecting path, not on any specific connecting path. Because of this we can write:

$$\int_0^1 \left(\frac{dQ}{T}\right)_A = S_1 - S_0,$$

where A is *any* reversible path that is taken to go from state 0 to state 1. If we now choose state 0 as the standard state from which to measure the value of S for any state of the system, then

$$S_1 = \int_0^1 \left(\frac{dQ}{T}\right)_A + S_0.$$

The additive constant S_0 depends only upon the arbitrary choice of state 0. The integral is independent of the choice of path A used in its calculation, as long as the path is reversible, starts at 0 and ends at 1. Hence, the value of S_1 depends only upon state 1.

To generalise this result, we consider any other state, say 2, of the system. The value of S at state 2 is equal to

$$S_2 = \int_0^2 \left(\frac{dQ}{T}\right)_C + S_0$$

where C is an arbitrary reversible path that connects state 0 to state 2. Therefore, the difference in S for states 1 and 2 is equal to

$$\Delta_{12}S = S_2 - S_1$$

regardless of the path connecting the two states. This difference depends only upon the states 1 and 2. Thus, the quantity S is a state function and the infinitesimal dS is an exact differential. S is the Entropy of the system.

It is worth reflecting on what we have obtained here. Simply defining

$$dS \equiv \left(\frac{dQ}{T}\right)_{rev}$$

makes entropy S a state function, even though we have seen that the heat interaction of a system is a path-dependent quantity. The entropy of any state depends only upon the state, and not upon the process taken by the system to arrive at that state. In order to calculate the entropy change when a system transforms from state 1 to state 2, we calculate

$$\Delta_{12}S = S_2 - S_1 = \int_1^2 \left(\frac{dQ}{T}\right)_{rev}$$

along any reversible path from 1 to 2 for which the integral is convenient to calculate. Even though the change in state may have been effected through an irreversible process, the difference in ΔS going from 1 to 2 is always equal to this integral because entropy is a state function.

5.4 The Clausius Inequality

We reconsider the cyclic process in Figure 5.5, except now, letting the process A be irreversible. Since now the cycle is irreversible, the results of Section 5.2 tell us that

$$\oint \left(\frac{dQ}{T}\right)_{irrev} = \int_0^1 \left(\frac{dQ}{T}\right)_{A \atop irrev} + \int_1^0 \left(\frac{dQ}{T}\right)_{B \atop rev} < 0.$$

To remind ourselves, this equation is a succinct summary of empirical observations summarised by the Kelvin and Clausius statements, and not simply results logically deduced from some *a priori* postulate. Using the results in the previous section for the integral of dQ/T for the reversible path B, we can rearrange this equation to obtain

$$\int_0^1 \left(\frac{dQ}{T}\right)_{A \atop irrev} < \int_0^1 \left(\frac{dQ}{T}\right)_{B \atop rev} = S_1 - S_0.$$

Thus, when a system undergoes an irreversible process from state 0 to state 1, the change in the entropy is always larger than the integral of dQ/T along the irreversible process. For an infinitesimal change of state, we thus have

$$dS > \left(\frac{dQ}{T}\right)_{irrev}.$$

Combining this with the result of the previous section, we have the Clausius Inequality:

$$dS \geq \frac{dQ}{T},$$

where the equality holds for reversible processes while the inequality holds for irreversible processes. This is the differential form for the Clausius Inequality. The integral form is

$$\Delta S \geq \int_1^2 \frac{dQ}{T}$$

for a change in state from 1 to 2; the equality holding for a reversible process and the inequality holding for an irreversible process. With this, we can now completely understand the inconsistency we noticed in Example 3.3; taking $dS = \frac{dQ}{T}$ for an irreversible process was the error that led to inconsistency. You should revisit that example now.

The Clausius Inequality is an important result that will lead to general conditions for phase and chemical equilibrium which we will discuss in later chapters. For now, we can immediately apply it to characterise processes in Nature. You might also be concerned about the definition for S if the absolute temperature goes to zero for which we would have mathematical difficulty with $\frac{dQ}{T}$. We will discuss what happens close to absolute zero temperature in Chapter 6 and Section 7.13. There we will find that it is physically not possible to get to absolute zero temperature.

5.5 Reversible, spontaneous and impossible processes

Consider an isolated system, that is, a system where neither energy nor material is exchanged with its surroundings. An example is perhaps our observable Universe. We immediately have $dQ = 0$. Thus, any reversible process taking place in an isolated system is characterized by the equality above. For any process in the Universe,

$$\Delta S_{Universe} = 0.$$

Examples of such processes that we have already encountered are:

- reversible isothermal volume change in a gas
- reversible adiabatic volume change in a gas
- the reversible cycle discussed in Example 3.1
- the Carnot cycle

These are examples of idealised processes you can use to analyse real processes using Thermodynamics. Each of these, and any other reversible process, is conceived to be at thermodynamic equilibrium at every single point along it. Hence, none of these processes are real, even though we can quite well imagine them in our analysis. When a system is at equilibrium, all macroscopic forces are balanced so that the system cannot change its thermodynamic state. To analyse the change in state along a reversible process, we think of an infinitesimal change in some thermodynamic quantity, a temperature change dT for example, at each point of the process. Because of the infinitesimal change dT, the state of the system responds by changing its volume, for example, by an infinitesimal amount dV. A key point in thinking about this is that at any point along a reversible process, the response of the system is reversed when the infinitesimal driving change is reversed; any change is reversible only in the infinitesimal limit. Because its steps are infinitesimal, any reversible process must be infinitely slow, maintaining the system and its surroundings at equilibrium throughout the process. Reversible processes are, hence, also quasistatic. Any reversible process would take an infinitely long time to occur. Thus, the Clausius Inequality tells us that any (idealised) reversible processes we use to analyse the thermodynamics of an isolated system is accompanied by zero change in the entropy of the system.

How about processes that we actually observe in Nature? Such as the cooling down of a cup of hot tea in an air-conditioned classroom? Or a ball that gradually decreases its bouncing off the floor and finally becomes stationary? Or the Joule experiment in which the weight falls down and the water warms up? Nobody has ever observed the reverse of any of these processes. For that cup of hot tea sitting in an isolated universe, we have $dQ = 0$ for the entire Universe, treating the Universe as an isolated system. But we have $dS > \left(\frac{dQ}{T}\right)_{irrev}$. Hence, observations of the innumerable cups of hot tea cooling to the temperature of the environment are all encapsulated in

$$\Delta S_{Universe} > 0.$$

Any irreversible change in Nature leads to an increase in the entropy of the universe. Indeed, any change that we actually observe in Nature is irreversible. We can think of *all* natural processes as occurring spontaneously because they are driven by the net increase in the entropy of the Universe. This includes processes such as the freezing of water to form ice in your refrigerator. This seems to decrease the entropy of the Universe because heat is taken out from the water. But at the same time, more heat is dumped into the surroundings by the refrigerator compressor and by the generation of the electricity to power the compressor. The decrease in the entropy of a part of the Universe, the water, is more than compensated for by the increase in entropy in other parts of the Universe.

Wouldn't it be nice if in Nature the Joule experiment can spontaneously go in reverse? The water in the apparatus cools down slightly and the weight is raised to a height. If this were to occur, we can tap on such processes to continuously provide the energy to drive all human activities. Besides, the reverse of Joule's experiments, other examples of such processes are

- a cup of water at room temperature spontaneously absorbs heat from the room to get hot enough to brew tea with while the room cools slightly, with no other change in the rest of the universe
- the protein molecules in a scrambled egg spontaneously rearrange themselves to form raw egg white, with no other change in the rest of the universe (or laboratory)

These are all processes you might expect can occur because they are physically (or chemically) possible according to the fundamental time-reversible laws of mechanics. If you analyse the molecular motion of the parts of the system in each of these processes, you will find that this motion obeys the laws of (classical or) quantum mechanics. All these processes obey the law of conservation of energy. They are all consistent with the First Law of Thermodynamics. These are the fantastical processes which you would laugh at if you watch a video of a natural process, such as a clown falling off his bicycle into a muddy pond, played backward in time. The characteristic of all these processes is

$$\Delta S_{universe} < 0.$$

These are impossible processes. Or at least, the probability that any of these processes occurs within the lifetime of the Universe is vanishing small. We invoke probability in this statement. As you will learn in your future study of Statistical Mechanics, the Second Law is a probabilistic law.

In summary, the Clausius Inequality organises all physical processes into three types:

$\Delta S_{universe} = 0$: reversible processes; the idealised processes used to analyse the thermodynamics of real systems

$\Delta S_{universe} > 0$: natural, spontaneous, irreversible processes; all processes that occur in Nature

$\Delta S_{universe} < 0$: impossible processes.

The Second Law of Thermodynamics was therefore stated concisely by Clausius as:

"The entropy of the universe tends towards a maximum."

As Richard Feynman put it, the Second Law provides the distinction between past and future[1]. Time passes and entropy increases when natural processes occur. If we play in reverse a video of a natural process such a Joule experiment, we will soon notice that it is going in the direction opposite to your experience. However, if you zoom in to follow the trajectory of a molecule of the gas colliding with the paddle wheel, you would see that it is responding to its intermolecular interactions with the other molecules exactly as prescribed by the laws of mechanics. You cannot tell whether the video is played forward or backward in time by looking at the trajectory of the molecule: the fundamental laws of mechanics are the same whether time is proceeding in the forward or reverse direction. Amongst the fundamental laws of Nature, the Second Law is the only one consistent with our sense of the passage of time, giving us the so-called "Arrow of Time". In this sense, all that happens in Nature are processes that increase the entropy of the universe.

5.6 Using the Second Law

5.6.1 ΔS in a reversible isothermal expansion

§   *Example 5.1 Reversible isothermal expansion*

Calculate the entropy change for the gas ΔS_{gas} and the surroundings ΔS_{surr} for an ideal gas undergoing isothermal expansion. Show that entropy change of the universe $\Delta S_{universe}$ is equal to zero for this process.

---

[1] R.P. Feynman, *Character of Physical Law* (MIT Press, 1965).

We have examined this previously, applying the First Law to relate ΔU, ΔW and ΔQ in the first process of Example 3.1. We also took a sneak peek by calculating ΔS although we did not know the significance of the entropy at that point. Review that example if you did not completely understand it.

An ideal gas is in thermal contact with a heat reservoir at temperature T. It is expanded reversibly from initial volume V_i to final volume V_f by decreasing the external pressure infinitesimally slowly so that $P = P_{ext}$ at all times. The gas is in equilibrium at all times; its temperature is maintained equal to T at all times. Since the process is isothermal and the gas is an ideal gas $dU = 0$. Thus, applying the First Law,

$$dQ_{gas} = -dW = \frac{RT}{V} dV,$$

$$dS_{gas} = \frac{dQ_{gas}}{T} = \frac{R}{V} dV,$$

where we have added a subscript gas to emphasize that dQ_{gas} is the heat going into the gas. Hence, the increase in the entropy of the gas is

$$\Delta S_{gas} = \int_{V_i}^{V_f} \frac{R}{V} dV = R \ln \frac{V_f}{V_i}$$

which is a positive quantity. As the gas expands and increases in its entropy, heat is transferred from the isothermal surroundings into the gas:

$$dQ_{gas} = -dQ_{surr}.$$

If dQ_{gas} is positive, then dQ_{surr} is negative. Therefore, the entropy change of the surroundings at each infinitesimally step of the reversible process is

$$dS_{surr} = \frac{dQ_{surr}}{T} = -\frac{dQ_{gas}}{T}$$

$$= -\frac{R}{V} dV.$$

This is integrated from the initial state to the final state of the process to give

$$\Delta S_{surr} = -\int_{V_i}^{V_f} \frac{R}{V} dV = -R \ln \frac{V_f}{V_i}.$$

This is exactly the negative of ΔS_{gas}, making the total entropy change of the universe

$$\Delta S_{universe} = \Delta S_{gas} + \Delta S_{surr} = 0,$$

as we expect for a reversible process. It is an idealised quasistatic process that helps us analyse the thermodynamic change when a gas is slowly and isothermally expanded. Thinking in terms of the energy, as the gas expands it does work on the external pressure. However, its temperature stays the same, and thus it has to absorb heat from the surroundings in order to keep its internal energy constant. Therefore, the gas gains heat from the surroundings. In terms of entropy, the gas increases in entropy as it expands because it absorbs heat from the surroundings. At the same time the entropy of the surroundings decreases because the surroundings lose heat. These entropy changes are exactly the same because the temperatures of the gas and the surroundings are the same. §

Additionally, some amount of physical insight is gained from considering the form of the expression

$$\Delta S_{gas} = R \ln \frac{V_f}{V_i}.$$

This tells us that the entropy of an ideal gas increases logarithmically with its volume at constant temperature. Qualitatively, this comes from the larger number of possible positions that a gas molecule can occupy when the volume of a gas is increased. We will discuss this further in a later chapter when we have firmed up our conceptual grasp of the physical meaning of entropy.

5.6.2 ΔS in an irreversible adiabatic expansion

§   *Example 5.2 Ideal gas irreversible adiabatic expansion*

You have an ideal gas in a box with thermally insulated walls. If the gas undergoes an irreversible expansion in a vacuum, calculate the entropy change for the gas, the surroundings and the universe.

This process was previously discussed in Example 3.3 as an application of the First Law. A gas is contained in a compartment of volume V_i in an insulating box. The partition is removed allowing the gas to expand to fill up the entire volume V_f of the box with impermeable rigid, insulated walls. Initially the empty compartment is a vacuum. We illustrate this below:

Since the gas is in an insulated box, the process is adiabatic. The expansion of the gas is against zero pressure because it expands into vacuum. Furthermore, the volume of the box does not change because the walls are rigid. Therefore,

$$\Delta Q = 0$$

$$\Delta W = 0.$$

Hence, applying the First Law,

$$\Delta U = \Delta Q + \Delta W = 0,$$

which for an ideal gas means that the temperature change is zero because the internal energy of an ideal gas is dependent only upon its temperature. Therefore, the state change of the gas is

$$(V_i, T) \rightarrow (V_f, T).$$

These are the same initial and final states of a reversible isothermal expansion discussed in the previous example.

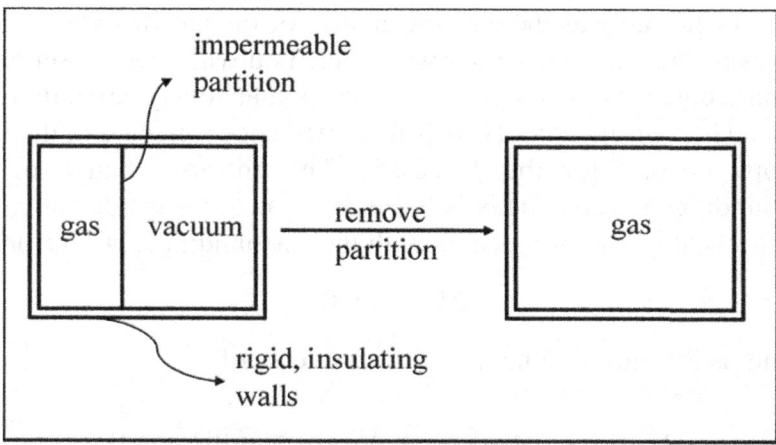

Figure 5.6 An ideal gas is initially in the left compartment. Upon removing the partition, it fills the entire box which has impermeable, rigid and insulating wall.

Because entropy is a state function, we know that the entropy change ΔS for the gas depends only upon its initial and final states. Hence, the entropy change for this process is obtained in the same way as in the previous example even though the two processes are not the same. Why do we not directly calculate the entropy change in this irreversible process? We are not able to do that because during any irreversible process, the system is not in equilibrium, and therefore the thermodynamic state of the system during the process is not defined; indeed, the process can take any one of an infinite number of possible irreversible paths between any two states.

Hence, in order to calculate the entropy change (or the change in any other state function), we look for a reversible path to go from the initial state to the final state. In this case, we can use the reversible path described in Example 5.1 above, of an isothermal expansion from V_i to V_f. Using that path, the entropy change of the gas in this irreversible adiabatic expansion is equal to

$$\Delta S_{gas} = R \ln \frac{V_f}{V_i},$$

which is the same as the entropy change of the gas in Example 5.1. However, this process is not reversible. You will never observe the gas molecules in any macroscopic box spontaneously collecting into just one half of the box. How is this difference reflected in the total entropy change for the process? The entropy change of the surroundings is zero. This is because the process is adiabatic, and thus no heat interaction occurs with the surroundings. Therefore,

$$\Delta S_{surr} = 0,$$

giving us the entropy change of the universe of

$$\Delta S_{universe} = \Delta S_{gas} + \Delta S_{surr} = R \ln \frac{V_f}{V_i} > 0.$$

Thus, an ideal gas will spontaneously expand adiabatically. The reverse process, with $\Delta S_{universe} < 0$ is impossible.

Two points to reiterate here. First, in both this and the previous example, the entropy of the system increases during the process. However, in the previous example the entropy of the surroundings decreases, exactly balancing the increase in the entropy of the gas. In this example the increase in the entropy of the gas is not compensated for by a decrease in the entropy of the surroundings, so that this process is irreversible. In deciding whether a process is reversible or otherwise, we have to pay attention to both the entropy of the system and the entropy of the surroundings so that we can apply the Clausius Inequality to the Universe:

$$\Delta S \geq \int_{1}^{2} \frac{dQ}{T}.$$

Second, to calculate the entropy change of a system undergoing an irreversible process, we identify a reversible process that has the same initial and final states as the irreversible process. Then, because entropy is a state function,

$$\Delta S_{irrev} = \Delta S_{rev} = \int\limits_{initial}^{final} \left(\frac{dQ}{T}\right)_{rev}$$

where the integrand on the right-hand side is for that reversible process. The thermodynamic state for a system is ill-defined during any irreversible process, but the net change in a state function between the final and initial states can be calculated using an appropriate reversible process. §

5.6.3 ΔS in a reversible adiabatic expansion

§   *Example 5.3 Ideal gas reversible expansion*

An ideal gas expands adiabatically and reversibly. Calculate the entropy change for the gas, the surroundings and the universe.

In an adiabatic expansion dQ_{gas} is zero. The entropy change of the gas for each infinitesimal step in the reversible expansion is thus

$$dS_{gas} = \frac{dQ_{gas}}{T} = 0$$

so that the entropy change for the system is $\Delta S_{gas} = 0$. At the same time, no heat interaction occurs with the surroundings for an adiabatic process. Therefore, the entropy change of the surroundings ΔS_{surr} is also equal to zero. Thus, the entropy of the universe stays constant

$$\Delta S_{universe} = \Delta S_{gas} + \Delta S_{surr} = 0.$$

If we examine this argument carefully, we see that the two conditions for the process, namely being adiabatic and being reversible, require that the total entropy change of the universe is zero. This is regardless of the substance composing the system, i.e., the system does not need to be an ideal gas or any specific material. We conclude that, in general all reversible adiabatic processes are isentropic. §

Let us compare the entropy change here to that in Example 5.1. The calculation in Example 5.1 gave $\Delta S_{gas} = R \ln\frac{V_f}{V_i}$. This result was obtained by setting dQ equal to PdV because of the First Law, where physically the work done by the gas is compensated for by absorbing heat from the surroundings in order to keep the gas at a constant temperature. Here, the gas also expands from V_i to V_f, so it may seem puzzling that we are able conclude that ΔS_{gas} is zero here simply by (correctly!) arguing that dQ is zero. In particular, why isn't ΔS_{gas} also equal to $R \ln \frac{V_f}{V_i}$ here, and hence not equal to zero?

As it turns out, in the adiabatic expansion here there are two contributions to the entropy change. One is from the increase in the volume, which is exactly $R \ln \frac{V_f}{V_i}$, while the other is due to the temperature change which occurs here but not in the isothermal process in Example 5.1 In Example 7.8, we will learn to calculate ΔS when volume and temperature both change in a process. For now, we note that what happens physically in this reversible adiabatic expansion is as follows. The entropy of the gas increases due to the larger number of possible positions that a gas molecule can occupy in the larger volume. However, at the same time the entropy of the gas decreases because its temperature drops in a reversible adiabatic expansion. These changes exactly cancel out.

5.6.4 ΔS in irreversible heat transfer

§   *Example 5.4 Irreversible heat transfer*

Consider two objects A and B, for example two volumes of liquid water, of the same mass at different temperatures T_A and T_B. When they are simply placed in thermal contact, heat spontaneously flows from the hotter to the colder one; this is an irreversible process which you have frequently observed. Calculate the entropy change of A, B and the universe.

We consider an infinitesimal amount of heat dQ gained by A. Then the heat *gained* by B will $-dQ$ because of the First Law. Therefore, the entropy changes are

$$dS_A = \frac{dQ}{T_A}$$
$$dS_B = -\frac{dQ}{T_B},$$

with no change to the rest of the universe. We ignore the volume change of A and B, taking the work interactions to be zero. This gives a total entropy change of the universe equal to

$$dS_{universe} = dS_A + dS_B$$

$$= \left(\frac{1}{T_A} - \frac{1}{T_B}\right) dQ$$

$$= \left(\frac{T_B - T_A}{T_A T_B}\right) dQ.$$

The Second Law requires that $dS_{universe}$ is greater than zero. If this is the case, then arbitrarily taking dQ to be a positive quantity, we get T_B greater than T_A. Thus, the Second Law "ensures" that when heat flows spontaneously, it is from the hotter object to the colder object, in agreement with empirical observations. This means that when two objects at different temperatures are placed in thermal contact, the entropy of the universe keeps increasing until the temperatures of the two objects are the same. The entropy of the universe is then at a maximum.

Let the initial temperatures be T_{Ai} and T_{Bi}. If A and B have equal heat capacities C, equilibrium is reached when the temperatures T_A and T_B are both equal to a final temperature $T_f = (T_{Ai} + T_{Bi})/2$, the average of the initial temperatures. How do we calculate the change in the entropy from the initial state to the final state? This is an irreversible process, so to calculate the entropy change, we imagine using a reversible process to increase the temperature of the cooler object A from T_{Ai} to T_f and to reduce the temperature of the hotter object B from T_{Bi} to T_f. To heat up A reversibly, we imagine placing A into thermal contact with a series of heat reservoirs at temperatures $T_{Ai} + \Delta T$, $T_{Ai} + 2\Delta T$, $T_{Ai} + 2\Delta T$, ... , T_f. In the limit when ΔT goes to zero, we have an infinite number of reservoirs, with the temperature of each reservoir an infinitesimal amount higher than the preceding one. This allows us to reversibly increase the temperature of A. Then the entropy change for A in this reversible process is given by

$$\Delta S_A = \int_{T_{Ai}}^{T_f} \left(\frac{dQ}{T_A}\right)_{rev}$$

$$= \int_{T_{Ai}}^{T_f} \frac{C dT_A}{T_A}$$

paying attention to the subscript rev for (dQ/T_A). Similarly, the entropy change for B is

$$\Delta S_A = \int_{T_{Bi}}^{T_f} \left(\frac{dQ}{T_B}\right)_{rev}$$

$$= \int_{T_{Bi}}^{T_f} \frac{C dT_B}{T_B}.$$

Then the net change in the entropy for the two objects in this reversible process is

$$\Delta S_A + \Delta S_B = C \ln \frac{(T_{Ai} + T_{Bi})^2}{4 T_{Ai} T_{Bi}}.$$

Note that the entropies of the reservoirs used in the reversible processes are also changed. However, we only use these reversible processes as a calculation aid to obtain the entropy change of A and B for the original irreversible process. When the two objects are simply placed into thermal contact to equilibrate irreversibly, $\Delta S_A + \Delta S_B$ is the change in the entropy of the universe. Hence, for the original irreversible process, we have

$$\Delta S_{universe} = C \ln \frac{(T_{Ai} + T_{Bi})^2}{4T_{Ai}T_{Bi}}.$$

Since $\frac{(T_{Ai}+T_{Bi})^2}{4T_{Ai}T_{Bi}}$ is always larger than unity, $\Delta S_{universe}$ is always positive.

We illustrate this with numerical values to get a sense of the magnitudes. Let the hotter and colder body initially be at 373.15 K and 273.15 K, respectively, so that the equilibrium temperature is equal to the mean of the initial temperatures, 323.15 K. Let the heat capacities of A and B be 4.184 kJ K^{-1}, which is the heat capacity of one kg of liquid water. Then the heat transferred from B to A is equal to 209.2 kJ. And the increase in the entropy of the universe due to this irreversible heat process is 101.39 JK^{-1}. $\S$

5.7 Harnessing irreversible processes to do useful work

When two objects at different temperatures are placed into thermal contact to equilibrate irreversibly, the only change is an increase in the entropy of the universe. Nothing "useful" gets achieved. As Carnot realised two hundred years ago, any two objects at different temperatures has the potential to do useful work. In order to tap on this potential to do work, we only need to connect these objects up to drive a heat engine, extracting heat from the hotter object and losing some heat to the colder object. What is the maximum amount of work we can extract?

We know from our discussion of heat engines, that we obtain the maximum work by connecting these objects up to any reversible engine, for example a Carnot engine. This is illustrated in Figure 5.7. This system is much the same as the heat engine configurations that we have encountered previously, except now A and B not infinitely large heat reservoirs at constant temperatures. Because they are objects with finite heat capacities, the temperatures of A and B change as heat is transferred between them and the heat engine. We first figure out how the temperatures of A and B are related when the Carnot engine operates between them.

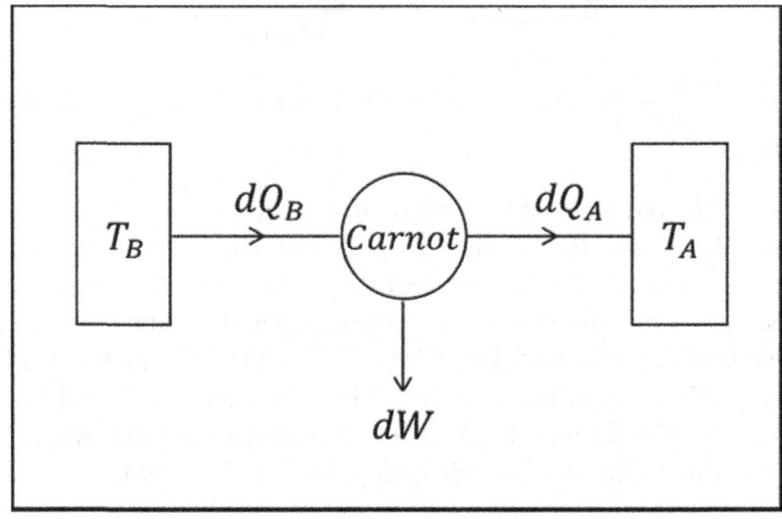

Figure 5.7 A Carnot engine is run between the two objects A and B. For an infinitesimal step during which the instantaneous temperatures are T_A and T_B, the engine absorbs heat dQ_B from the hotter reservoir B and loses heat dQ_A to the colder reservoir A, producing work dW.

Let the heat capacities of both A and B be equal to C. Then, the heat lost to A by the engine is related to the change in temperature of A by

$$dQ_A = CdT_A.$$

The heat extracted from B by the engine is related to the change in temperature of B by

$$dQ_B = -CdT_B,$$

noting the minus sign; as the engine extracts heat from B, T_B decreases. For a Carnot engine, we have

$$\frac{dQ_A}{T_A} = \frac{dQ_B}{T_B},$$

so that, the temperature changes dT_A and dT_B are related by

$$\frac{CdT_A}{T_A} = -\frac{CdT_B}{T_B}.$$

Integrating this, the temperatures T_A and T_B at any instant are related by

$$\ln\frac{T_A}{T_{Ai}} = -\ln\frac{T_B}{T_{Bi}}$$
$$\frac{T_A}{T_{Ai}} = \frac{T_{Bi}}{T_B},$$

where T_{Ai} and T_{Bi} are the initial temperatures of A and B. When A and B reach the same temperature, the heat engine can no longer operate because there is no temperature differential in the system. This equilibrium temperature T_{f2} is reached when

$$T_{f2} = T_A = T_B = \sqrt{T_{Ai}T_{Bi}}.$$

Therefore, when we operate a Carnot engine between A and B, the temperature of A is raised from its initial value of T_{Ai} to the final value of $\sqrt{T_{Ai}T_{Bi}}$. At the same time, the temperature of B is reduced from its initial value of T_{Bi} to this same final temperature. You can show algebraically that the final temperature of $\sqrt{T_{Ai}T_{Bi}}$ is lower than the final temperature $(T_{Ai} + T_{Bi})/2$ you would obtain if the two objects A and B were simply placed into thermal contact to equilibrate, as in Example 5.4 above. Physically, this makes sense because we are able obtain some work from the system. Hence, the amount of heat lost by B to the engine has to be larger than the amount of heat gained by A from the engine. When A and B are simply placed into thermal contact, the amount of heat lost by B is equal to the amount of heat gained by A.

What about the change in the entropy of the universe? Because we use a reversible engine, we expect the change in the entropy of the universe to be zero. The engine operates cyclically, so that its ΔS_{engine} is zero simply because entropy is a state function. The system has no heat interactions with any other parts of the universe, so we only need to examine the entropy change of the objects A and B. The change in the entropy of A is equal to

$$\Delta S_A = \int_{T_{Ai}}^{\sqrt{T_{Ai}T_{Bi}}} \frac{CdT_A}{T_A}$$

$$= C \ln \sqrt{\frac{T_{Bi}}{T_{Ai}}}$$

and the change in the entropy of B is

$$\Delta S_B = \int_{T_{Bi}}^{\sqrt{T_{Ai}T_{Bi}}} \frac{CdT_B}{T_B}$$

$$= C \ln \sqrt{\frac{T_{Ai}}{T_{Bi}}}$$

$$= -\Delta S_A.$$

Hence, $\Delta S_A + \Delta S_B = 0$, as we expected. We can now figure out how much work we can obtain from equilibrating A and B through the workings of a Carnot engine. Applying the First Law, and using the relationships above for temperatures and the heat flows, we obtain

$$dW = dQ_B - dQ_A$$

$$= dQ_B \left(1 - \frac{T_A}{T_B}\right)$$

$$= -CdT_B\left(1 - \frac{T_{Ai}T_{Bi}}{T_B^2}\right)$$

for the infinitesimal work obtained when the temperature of B is equal to T_B. To find the total amount of work done, we integrate over T_B from the initial temperature of T_{Bi} to the final temperature of $\sqrt{T_{Ai}T_{Bi}}$. This gives

$$\Delta W = C\left(\sqrt{T_{Bi}} - \sqrt{T_{Ai}}\right)^2.$$

Since no engine is more efficient than the Carnot engine, this is the maximum amount of work that is theoretically possible to extract given two objects with the same heat capacity C at different initial temperatures T_{Ai} and T_{Bi}. At the end of the process, the two objects are at the same temperature $\sqrt{T_{Ai}T_{Bi}}$, and we are able to store useful energy of the amount ΔW, perhaps in the form of a raised weight or in a battery.

§   *Example 5.5 Maximum useful work from heat flow*

We again illustrate with numerical values. As in Section 5.6.4 we have a system consisting of one kg of liquid water A at 373.15 K and a second kg of liquid water B at 273.15 K. The heat capacity of liquid water is 4.184 kJ $K^{-1}kg^{-1}$. What is the maximum amount of work you can extract from this system?

To extract this maximum amount of work, we use the most efficient engine, namely, any reversible engine. The final equilibrium temperature reached is now the geometric mean of T_A and T_B, which is 319.26 K. As shown above, this is lower than the equilibrium temperature of 323.15 K when the two objects are simply placed into thermal contact to equilibrate. Then the heat extracted from B is equal to 225.48 kJ while the heat lost to A is 192.92 kJ. The difference is the maximum amount of work that can be extracted from the initial temperature difference between the two masses of water. This is equal to 32.56 kJ. If this amount of energy is converted completely into heat (through friction, for example), then that heat would be sufficient to raise the temperatures of the two bodies to the final temperature of 323.15 K found in Example 5.4. §

We highlight an important point by comparing Examples 5.4 and 5.5. In Example 5.4, the potential of the two objects to perform useful work is lost when the objects are simply placed into thermal contact to equilibrate. At the end of the process, the two objects are in equilibrium at the same temperature. If the universe consists of only these two objects, then its entropy has increased to the maximum possible. No further spontaneous thermodynamic change is possible in that universe. If the two objects A and B form the entirety of a universe, that universe would have come to its endpoint.

In Example 5.5, however, we have made use of the initial temperature difference between A and B to extract mechanical work, perhaps in raising a weight which can subsequently be used in a Joule's experiment or some other useful endeavour. This work, equal to 32.56 kJ, is the potential that was lost when the irreversible process in Example 5.4 generated entropy, with all the heat lost from B going into heat in A. The entropy change in Example 5.5 is zero because we used a reversible engine, which gives us the maximum possible amount of work that can be extracted given the initial temperature difference. Any real engine will be able to extract some work, although less than the Carnot engine in Example 5.5. Any real engine that is capable of extracting some work will suffice to produce less entropy than in Example 5.4. In other words, the entropy increase is a measure of the loss of potential to extract useful work from the initial temperature difference. We will return to this theme again when we discuss how to extract useful work from spontaneous chemical reactions.

§ Example 5.6 How many air-conditioners can you operate using the thermal energy of a hot spring?

The Sembawang hot spring in Singapore produces $150 \, kg \, min^{-1}$ of 70°C water. How much power can we extract from a Carnot engine operating between the hot spring and an infinitely large cold reservoir of 30°C?

The maximum work obtainable from the hot spring is

$$dW = dQ_{spr} - dQ_{surr} = dQ_{spr} \left(1 - \frac{T_{surr}}{T_{spr}} \right)$$

where Q_{spr} and Q_{surr} are the heat lost by hot spring and gained by surroundings, at steadily decreasing temperature T_{spr} and constant temperature T_{surr}, respectively. Integrating this for T_{spr} decreasing from 343 K to 303 K, and with the heat capacity of water C equal to be 4.184 kJ K^{-1} kg^{-1}, we obtain

$$\Delta W = \int_{303}^{343} C dT - T_{surr} \int_{303}^{343} \frac{C dT}{T_{spr}} = 10.16 \text{ kJ kg}^{-1}.$$

Given the flowrate, this is a power of 25.4 kW; an air-conditioner requires a power of about 1 kW.

5.8 A possible fate of the universe

In Example 5.4, when the universe that consists of the two objects A and B reaches its state of maximum entropy, it will be in equilibrium. Because it has reached equilibrium, no spontaneous change is possible in that universe. Nothing further will happen in that universe. This is the heat death scenario that was suggested as a possible fate of the Universe by Kelvin. The Second Law tells us that the entropy of the universe always increases as it heads toward such a final equilibrium state. If different parts of the universe are at different temperatures, for example a star is hot compared to the cold cosmic microwave background radiation, then the universe is not yet at thermal equilibrium. Natural processes can still occur spontaneously as Time passes. Therefore, if we can consider the Universe to be an isolated thermodynamic system, it will keep changing spontaneously to equalize the temperatures of all its parts, continuously increasing its entropy in the process, until it gets to equilibrium. At heat death, the Universe will have reached thermodynamic equilibrium throughout, with no further macroscopic change possible. The Second Law tells us that an isolated Universe evolves along a one-way street.

6

Absolute Zero

By relating entropy to heat and temperature in reversible processes, the Second Law of Thermodynamics provides a means of calculating the entropy change of a system transforming from one state to another. From this, we have:

$$\Delta S = S_A - S_0 = \int_{state\ 0}^{state\ A} \frac{dQ}{T}$$

where S_A and S_0 are the entropies of state A and a suitable reference state 0. This gives ΔS, the entropy of any substance in any state *relative* to an appropriate standard reference state for the substance. In order to calculate the absolute value of entropy, we turn to the Third Law of Thermodynamics.

There are a few different but closely related statements of the Third Law; amongst the laws of Thermodynamics, it is the only one where a precise formulation or articulation is still being actively examined. The First and Second Laws of Thermodynamics are succinct statements with well-established universal applications to the properties of all matter and radiation. In a similar spirit, the Third Law of Thermodynamics is aimed at summarizing the thermodynamic behavior of all matter at temperatures approaching absolute zero, which is an active research field.

A bit of a digression to set the context. First, how cold is absolute zero? By now you know that the temperature of any sample of matter – gas, liquid, solid – is a measure of how fast its atoms move. At room temperature molecules in the air move at about 500 m s^{-1}, as you have seen in Example 1.2. If you cool some gases down to sufficiently low temperatures, the atoms slow down to almost a standstill to form a state of matter known as a Bose-Einstein condensate (BEC). For example, for a gas of rubidium atoms this happens upon cooling to about 0.17 of a millionth of a Kelvin above absolute zero. This was first done only in 1995, which is rather recent in comparison with fundamental experiments examining the other laws of Thermodynamics. The BEC is an interesting state of matter because the atoms in the BEC lose their individual nature and the collection of atoms behave as a single quantum mechanical object. Today, the coldest temperatures achieved is some trillionths of a degree (10^{-12} K) above zero, which is so cold that even light travelling through a BEC slows down to a few meters per second.

Second, is it possible to reach absolute zero temperature? As it turns out, Nature being quantum mechanical implies that this is not possible using a finite number of thermodynamic transformations. We will discuss this later in this chapter.

6.1 Planck, Boltzmann, absolute entropy

To discuss the Third Law we start with the Planck statement of the law:

"In the limit as the temperature $T \to 0$, the entropy $S \to 0$ for any system in internal equilibrium."

That is, Planck assigned the value of zero to the entropy of any system in the limit as the temperature approaches zero. Entropies measured or referenced to $S \to 0$ for $T \to 0$ are referred to as absolute entropy. Setting $S \to 0$ in the limit $T \to 0$ is motivated by and consistent with statistical and quantum mechanics. To understand how this comes about we discuss the concepts of microstate and macrostate.

With the Clausius inequality, we have an equation, a recipe, to relate the entropy of a system to its heat interactions and temperature. But what physical quantity is entropy, really? To answer this, we connect the macroscopic picture provided by Thermodynamics with the underlying microscopic understanding of the fundamental interactions in a piece of matter. In chemical and physical systems typically encountered – single molecules, gases, liquids, solids, plasmas, radiation – the energy in the system is distributed over a large number of quantum states. For example, in a gas whose thermodynamic state is defined by a fixed temperature T and pressure P, the molecules can be distributed in many translational, rotational, vibrational, electronic and nuclear spin states. The macrostate of the gas is specified by the values of T and P. The state of each molecule in a gas is characterized by the set of quantum numbers, one from each of these modes of excitation that it is in. Additionally, the spatial position and orientation of the molecules also add to the manifold of states that the molecule can be in. The set of all quantum numbers, molecular positions and orientations for all the molecules in the gas specifies its microstate.

As the molecules in the gas collide and interact with each other while maintaining constant T and P, the microstate of the gas changes because the quantum numbers, position and orientation for each molecule change. Each of these microstates is consistent with the macrostate determined by the values of T and P. In other words, the number of microstates is the number of possible molecular arrangements of translational velocities, rotations, vibrations and spatial position for the molecules in a gas. In statistical mechanics, it was shown by Boltzmann that the entropy of a system is defined as the natural logarithm of the number of microscopic states which the system can be in given that it is in a particular thermodynamic (macroscopic) state:

$$S = k_B \ln W$$

where k_B is the Boltzmann constant and W is the number of microstates. The fundamental physical picture of entropy is that it is proportional to the natural logarithm of the number of molecular arrangements of a system. The constant of proportionality, which is the Boltzmann constant, relates this logarithm to the units we use for energy and temperature. From quantum mechanics, it is known that W goes to 1 as the temperature goes to zero because all systems go into the quantum mechanical ground state. Even for systems with degenerate (multiple) ground states, the time taken for a system to transition from one ground state to another degenerate ground state becomes infinitely long when $T \to 0$. Thus, this combination of fundamental results from statistical mechanics and quantum mechanics tells us that we may set $S \to 0$ as $T \to 0$.

It may seem mysterious that a quantity defined thermodynamically as $\left(\frac{dQ}{T}\right)$ is equal to the logarithm of the number of microstates available to the system in a specific macrostate. Suffice it here to note that the number of microstates, which are the microscopic arrangements of energy and matter accessible to a system, is determined by the work and heat interactions of the macrostate. There is thus an intuitive link between the two definitions of entropy. A basic result of statistical mechanics is that the Boltzmann definition of entropy is rigorously connected to the thermodynamics definition. Although the laws of Thermodynamics can be formulated completely independently of the nature of matter at its microscopic level, it is good to know that a macroscopic quantity as fundamental as entropy is indeed deeply connected to the microscopic nature of matter.

With the Planck statement of the Third Law assigning the value of zero to the entropy of any equilibrium system at absolute zero, the absolute value of entropy for any system can be calculated by integrating $dS = \frac{dQ}{T}$ starting from absolute zero. We illustrate this by showing how we calculate the absolute entropy of a gas at some temperature T and pressure P (holding the pressure constant for convenience). When we heat up a substance changing its temperature from T_i to T_f, its entropy change is

$$\Delta S = \int_{T_i}^{T_f} \frac{dQ}{T}.$$

For a constant pressure process, we have $dQ = C_P dT$ by the definition of the heat capacity at constant pressure C_P, which can be measured through calorimetry. Hence, we have the following entropy change at constant pressure:

$$\Delta S = \int_{T_i}^{T_f} \frac{C_P}{T} dT.$$

Substances can also absorb/release heat during a phase change even though the temperature remains constant, for example, when water boils. For such transformations

$$\Delta S = \frac{\Delta Q}{T} = \frac{\Delta H}{T}$$

where ΔH is the enthalpy change of the phase transformation.

Therefore, the absolute entropy of a gas at temperature T (and constant pressure P) is:

$$S(T, P) = S(T \to 0, P) + \int_0^{T_m} \frac{C_P^{solid}}{T} dT + \frac{\Delta_{fus} H}{T_m}$$
$$+ \int_{T_m}^{T_b} \frac{C_P^{liquid}}{T} dT + \frac{\Delta_{vap} H}{T_b} + \int_{T_b}^{T} \frac{C_P^{gas}}{T} dT$$

where C_P^{solid}, C_P^{liquid} and C_P^{gas} are the constant pressure heat capacities of the solid, liquid and gas phases; $\Delta_{fus} H$ and $\Delta_{vap} H$ are the enthalpies of fusion and vaporization; T_m and T_b are the melting and boiling temperatures at pressure P. This calculation is illustrated graphically in Figure 6.1. Discontinuities (dotted lines) in the absolute entropy occur at the melting and boiling points when the substance undergoes phase changes. These result in jumps in entropy equal to $\frac{\Delta_{fus} H}{T_m}$ and $\frac{\Delta_{vap} H}{T_b}$, respectively.

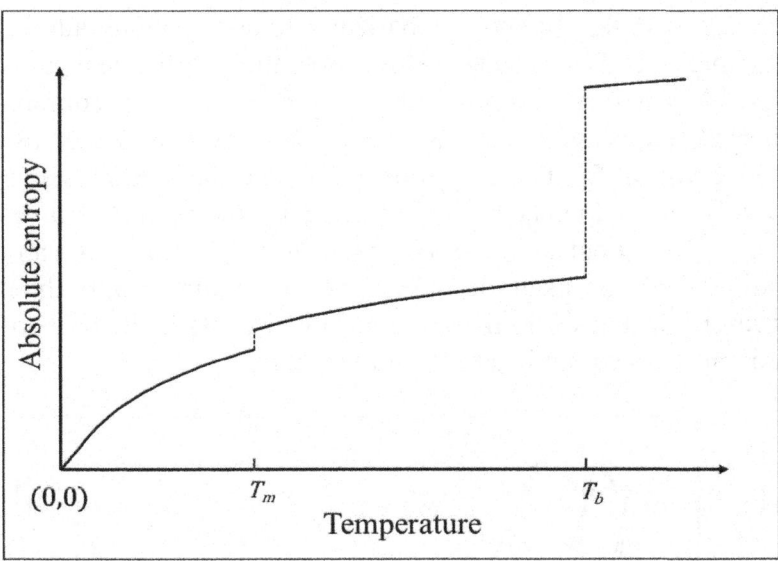

Figure 6.1: *The absolute entropy of any substance increases monotonically from $S = 0$ at $T = 0$, with jumps when phase changes occur. T_m and T_b are the melting and boiling temperatures.*

6.2 Nernst: Can you reach absolute zero?

In practice, there is somewhat of a problem in carrying out this calculation: it is not possible to get to absolute zero temperature! To understand why we think so, let us start by considering how you might cool down an ideal gas. Consider the *P-V* diagram in Figure 6.2 with ideal gas isotherms (solid lines) at temperatures T_1, T_2, T_3 and adiabats (dashed lines) A_1, A_2, A_3. States a, c, e, g are at pressure P_1, and states b, d, f are at higher pressure P_2. Starting from state a at pressure P_1, an ideal gas is compressed isothermally to state b at pressure P_2 while in thermal contact with an initial cold reservoir at T_1.

Work is done on the gas, but the internal energy of the ideal gas stays constant because the temperature stays constant; thus, heat has to be lost to the reservoir. At state b, the system is insulated and then expanded slowly and adiabatically to state c along adiabat A_1 back to pressure P_1. At state c, the temperature of the ideal gas cools to T_2. A small fraction of the gas can now be compressed isothermally to state d at pressure P_2, using the rest of the gas as a heat reservoir at T_2. Thus, by repeated isothermal compression and slow adiabatic expansion, as illustrated by the path $a \rightarrow b \rightarrow c \rightarrow d \rightarrow e \rightarrow f \rightarrow g$ between any two pressures P_1 and P_2, we can cool a small fraction of a sample of gas to a lower temperature than any cold reservoir that we initially have. This process is the basis of the Claude process which is used to liquefy air.

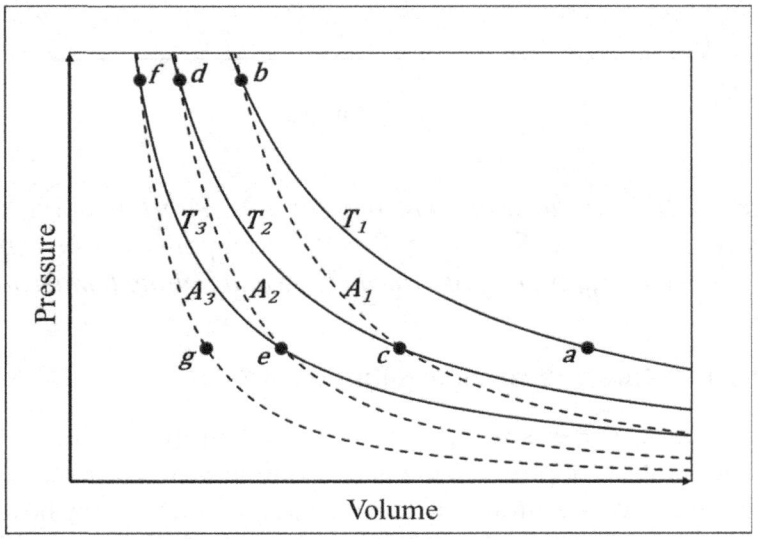

Figure 6.2: *Cooling of an ideal gas from T_1 to below T_3 by repeated isothermal compressions and adiabatic expansions following the path $a \rightarrow b \rightarrow c \rightarrow d \rightarrow e \rightarrow f \rightarrow g$. Solid lines are isotherms, dashed lines are adiabats. States a, c, e, g are at pressure P_1, and states b, d, f are at higher pressure P_2. Compare with the same process sketched in an S-T diagram in Figure 6.3.*

It is instructive to examine what this process looks like on a *S-T* diagram. This is shown in Figure 6.3 where the same path $a \rightarrow b \rightarrow c \rightarrow d \rightarrow e \rightarrow f \rightarrow g$ is shown with solid lines for the isothermal sections and dashed lines for the adiabatic sections. The points a, c, e, g are along the isobar P_1 while b, d, f are along isobar P_2. This clearly illustrates the back and forth transformations between two sets of ideal gas states each characterized by a specific value of pressure. We next examine a similar series of transformations between two set of states each characterized by the magnetic spin in a solid-state paramagnet. This is the process of adiabatic demagnetization, which we will use to discuss how to cool systems down to temperatures close to absolute zero. We are not able to discuss this using the ideal gas system in Figures 6.2 and 6.3 because the ideal gas model, not being quantum mechanical, is not a good physical model near absolute zero.

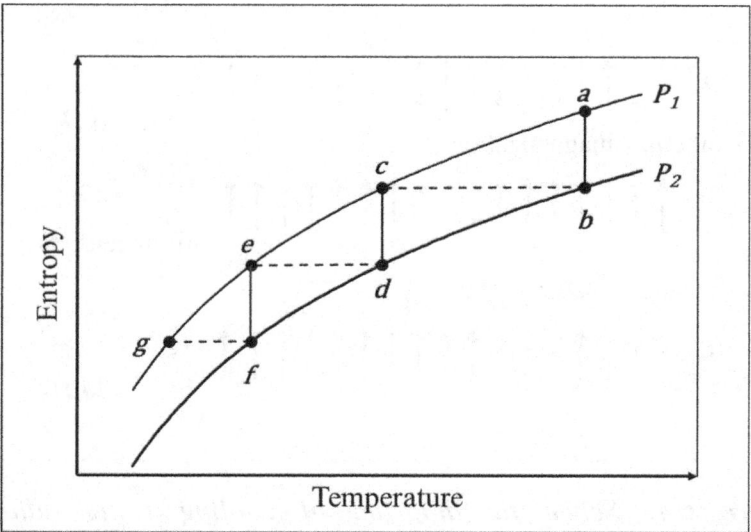

Figure 6.3: *The same process illustrated in Figure 6.2, sketched here in an S-T diagram. Solid lines are isotherms, dashed lines are adiabats. The curve aceg is an isobar at pressure P_1, and the curve bdf is an isobar at pressure P_2.*

Cooling using adiabatic demagnetization is illustrated in Figure 6.4. For simplicity, we consider a spin-$\frac{1}{2}$ system where the spin is either spin-up or spin-down. We start a solid-state spin paramagnet in state a at temperature T_1 in a low magnetic field H_1 with randomly distributed up- and down-spins; at low H_1, the net magnetization M_1 is low. In the first step the solid-state paramagnet is isothermally magnetized to state b by increasing the applied field to H_2 while maintaining contact with a heat reservoir at the coldest convenient starting temperature T_1, for example, the temperature of liquid helium. State b with higher magnetization M_2 is represented by the more ordered row of spins aligned to the applied field. The spin energy has decreased because the spins are more aligned to the applied field, and thus the paramagnet loses heat to the reservoir, just as the first isothermal compression step $a \rightarrow b$ in the ideal gas system in Figure 6.3.

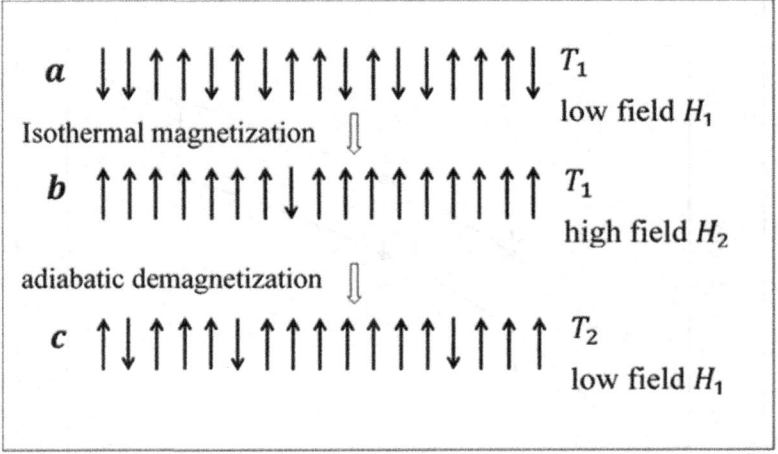

Figure 6.4: Schematic diagram of cooling using adiabatic demagnetization. The process is analogous to the process of cooling an ideal gas illustrated in Figures 6.2 and 6.3.

The change in the magnetic field from H_1 to H_2 is analogous to the transformation from pressure P_1 at state a to a higher pressure P_2 at state b. The system is now insulated and then the applied magnetic field slowly decreased from H_2 back to H_1. This step of slow adiabatic demagnetization is analogous to the adiabatic expansion from b to c in the ideal gas system in Figure 6.3. As the spins in the solid become less aligned to the smaller applied field and thus more disordered, the spin entropy increases. Since the process is adiabatic, as the spin entropy increases, the lattice entropy due to the vibrations of the atoms must decrease because a slow adiabatic process is isentropic; as we have seen in Chapter 5 a reversible adiabatic process has zero entropy change. Thus, the temperature decreases. At this point, the process of isothermal magnetization is repeated on a small fraction of the system bringing it to state d, with the rest of the system serving as the heat bath at temperature T_2, which you should note is already colder than the cold reservoir temperature T_1 that we started out with.

By repeating the process illustrated in Figure 6.4 we can reach progressively colder temperatures, albeit with progressively smaller portions of the solid-state paramagnet that we started out with. This method of alternating isothermal and adiabatic transformations between two sets of thermodynamic states characterized here by the applied magnetic field can be generalized to states characterized by any thermodynamic variable. The key point is that the total entropy of the system approaches zero as $T \to 0$ for all values of the applied field. This total entropy of the solid-state paramagnet consists of the spin entropy and the lattice entropy, the latter being the entropy due to the vibrations of the atoms in the solid-state.

In Figure 6.5, we sketch this total entropy as a function of the temperature for the process illustrated in Figure 6.4. The upper curve is a set of states at applied magnetic field H_1 and the lower curve a set of states at a higher applied magnetic field H_2. These curves meet at zero entropy and zero absolute temperature. The step $b \to c$ is the slow adiabatic demagnetization step taking the system from H_2 to H_1. In the analogous ideal gas adiabatic expansion in Figure 6.3, the gas cools. In Section 9.9 we will show that adiabatic demagnetization is accompanied by a cooling of the solid-state paramagnet.

We have emphasized that the First and Second Laws were formulated as summaries of experimental observations. This is similarly the case for the Third Law which Nernst first articulated as a statement to summarize his observations that the entropy change for many chemical reactions tends to zero as temperature is decreased.

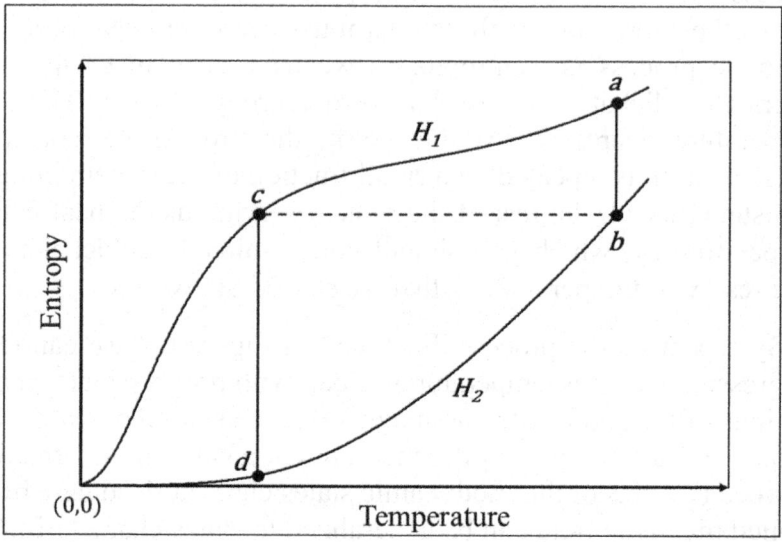

Figure 6.5: *Entropy-temperature diagram for the process illustrated in Figure 6.4 to cool a paramagnetic using adiabatic demagnetization.*

Figure 6.5 is accurate in the sense that it is consistent with experiments; it has not been possible to cool systems down to absolute zero because the entropy change of the isothermal magnetization steps in the figure goes to zero as temperature approaches absolute zero. Additionally, when the spins in the solid-state paramagnet tend toward perfect alignment as $T \rightarrow 0$, the (Boltzmann) entropy tends to zero since there is then only one microstate, which is consistent with assigning a value of $S = 0$. Thus, the two curves in Figure 6.5 are sketched as intersecting at $T = 0$ at a value of $S = 0$. This gives the following limit:

$$\lim_{T \to 0} \Delta S = \lim_{T \to 0} S(H_1, T) - S(H_2, T) = 0.$$

This equation is Nernst's conception of the Third Law, which is stated as follows:

"As T approaches absolute zero, the change in the entropy ΔS of a system undergoing any isothermal thermodynamic transformation goes to zero."

It is a manifestation of this statement of the Third Law that repeated isothermal magnetization and adiabatic demagnetization steps can never get to the intersection point at absolute zero in a finite number of steps. We illustrate this situation schematically here:

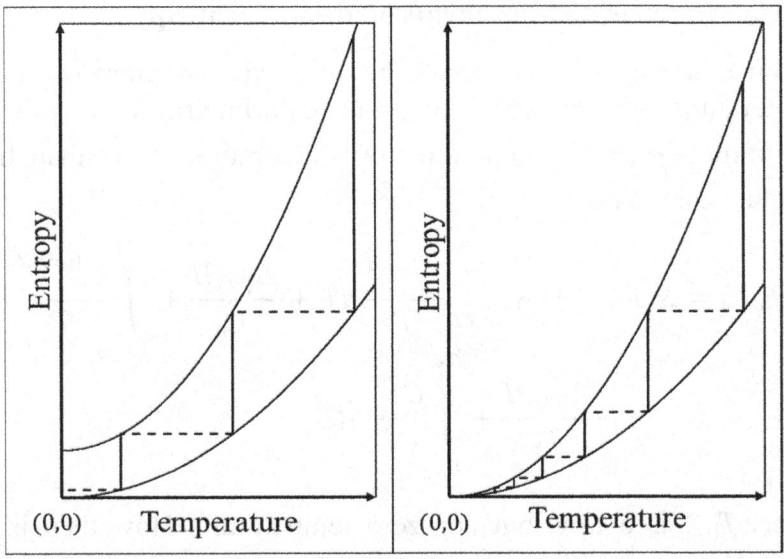

Figure 6.6: *Illustrating Nernst's unattainability principle. Upper curves at low magnetic field, lower curves at high magnetic field.*

In the left panel, which is not consistent with calorimetric measurements, ΔS does not goes to zero as $T \to 0$ when an isothermal transformation occurs from the upper curve at H_1 to the lower curve at H_2. In the three iterations of isothermal magnetization followed by adiabatic demagnetization shown in the diagram, the temperature of the system can be brought to absolute zero. In the right panel, which is consistent with calorimetric measurements (and quantum and statistical mechanics), ΔS approaches zero as $T \to 0$ for isothermal transformations from the upper curve to the lower curve. An infinite series of steps needs to be taken to reach absolute zero. In this case, you can get as close as you wish to absolute zero but you can never reach it. Thus, the Nernst statement of the Third Law implies the unattainability principle: it is not possible to reach absolute zero of temperature.

6.3 Calorimetric measurements of absolute entropy

Since absolute zero cannot be reached, the entropy at any temperature can be obtained from calorimetric data only by integrating $\frac{dQ}{T}$ from a non-zero low temperature rather than from absolute zero. That is:

$$S(T,P) = S(T_{low}, P) + \int_{T_{low}}^{T_m} \frac{C_P^{solid}}{T} dT + \frac{\Delta_{fus} H}{T_m} + \int_{T_m}^{T_b} \frac{C_P^{liquid}}{T} dT$$

$$+ \frac{\Delta_{vap} H}{T_b} + \int_{T_b}^{T} \frac{C_P^{gas}}{T} dT,$$

where T_{low} is a low, but non-zero temperature. How then is the absolute entropy $S(T_{low}, P)$ determined?

At temperatures close to absolute zero the thermal energy of a solid is due to each atom in a solid vibrating about its average position in a perfect crystal lattice and interacting with its neighbours through interatomic potentials. Thus, the thermal energy of a crystal lattice at low temperatures is due to lattice vibrational modes. Treating these lattice vibrations quantum mechanically, the Debye model of lattice vibrations accurately reproduces the experimental T^3 dependence of the heat capacity of non-metallic crystals. In the case of metals, the thermal excitation of the conduction (valence) electrons results in an additional specific heat contribution that is linearly proportional to the temperature T. Using the Debye model of heat capacity for lattice vibrations with the additional heat capacity due to the thermal excitation of the conduction electrons, the entropy of a solid at temperatures close to absolute zero is given by

$$S(T_{low}) = S(0) + \int_0^{T_{low}} \frac{(AT^3 + BT)}{T} dT = S(0) + \frac{A}{3} T_{low}^3 + BT_{low}$$

where the material parameters A and B in the specific heat $C(T) = AT^3 + BT$ can be determined from calorimetric measurements. With the Third Law setting $S(0)$ equal to zero gives

$$S(T_{low}) = \frac{A}{3} T_{low}^3 + BT_{low}$$

In other words, taking $S(0)$ equal to zero is consistent with experimental and quantum mechanical results that the heat capacity $C(T)$ goes to zero at absolute zero. A practical result of this is that measurements of the heat capacity at temperatures close to zero directly gives the entropy of a solid. Thus, calorimetric data, and some bit of understanding of material properties close to absolute zero, can be used to calculate the absolute entropy even though a temperature of absolute zero cannot be reached experimentally. This is illustrated in Figure 6.7 where heat capacity data for potassium is accurately fitted with $C(T) = AT^3 + BT$ for a T^3 dependence for lattice vibrations and a T dependence for the contribution of the conduction electrons.

The ideal gas heat capacity is a constant, and hence does not go to zero at absolute zero. From this, you realize that the ideal gas model is not consistent with the Third Law. This is not unexpected because the ideal gas model is based upon classical physics. The kinetic energy of the random translational and rotational motion of its molecules is proportional to the temperature, and the rate of momentum change upon collision of its molecules with the container wall being proportional to the pressure. Thus, as $T \rightarrow 0$, the classical thermal energy goes to zero, and hence the classical momentum of each gas molecule goes to zero. This is not consistent with the uncertainty principle in quantum mechanics, meaning that the ideal gas model is not a good physical model for any substance as temperature approaches absolute zero. It should be noted that in the Debye model, the energy in lattice vibrations approach the quantum mechanical zero-point vibrational energy as temperature goes to zero; the Debye model is a quantum mechanically accurate description of solids. Absolute entropies can be determined from calorimetry if we properly account for the quantum mechanical behavior of matter as the temperature approaches absolute zero.

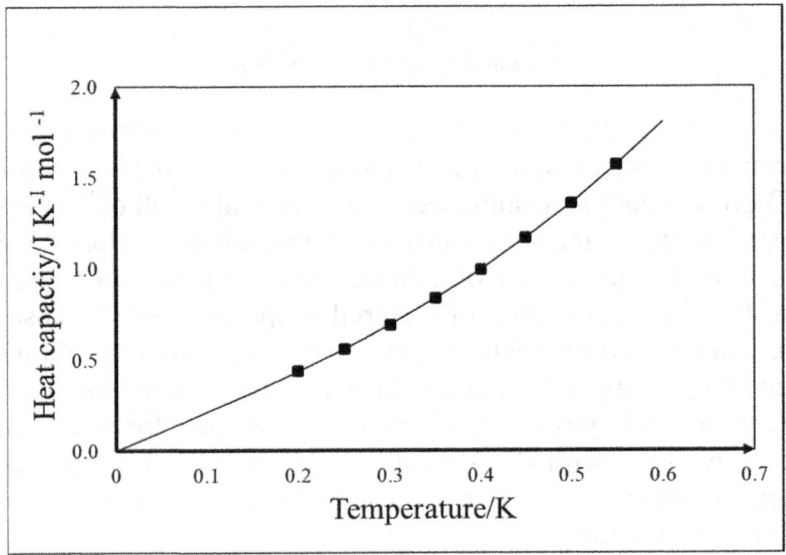

Figure 6.7: *Heat capacity of potassium at low temperatures. Black squares are data points and the curve is a fit to $C(T) = AT^3 + BT$. Integrating C/T from absolute zero to a low temperature T_{low} within the range of the fit then gives the absolute entropy at T_{low}.*

6.4 Standard absolute entropy

Entropies determined relative to $S = 0$ at absolute zero are referred to as absolute entropies or Third Law entropies. Standard Third Law entropies are frequently tabulated for use to determine the thermodynamic equilibrium state. The CRC Handbook of Chemistry and Physics, reports absolute entropies $S^\emptyset$ for a number of elements and compounds at the standard temperature of 298.15 K and standard pressure of 10^5 Nm^{-2}. Some of these values are reproduced in the Table below, along with the corresponding standard enthalpy of formation $\Delta_f H^\emptyset$ (see Section 3.3).

substance	physical state	$\Delta_f H^\emptyset$ kJ mol^{-1}	$S^\emptyset$ J mol^{-1} K^{-1}
C	graphite	0	5.74
C	diamond	1.895	2.377
S	rhombic	0	31.80
S	monoclinic	0.33	32.6
H_2O	Gas	-241.826	188.835
H_2O	liquid	-285.830	69.95
H_2	gas	0	130.680
O_2	gas	0	205.152

The absolute entropy of graphite is larger than that of diamond by 3.363 J mol^{-1} K^{-1}, so you might hasten to correctly conclude that graphite is more stable than diamond. That would not be the correct argument because the Second Law says that spontaneous processes are characterized by

$$\Delta S_{universe} > 0,$$

so that in addition to the entropy of the system (the carbon) you need to account for the entropy of the surroundings. Consider the allotropic transformation

$$C(diamond) \rightarrow C(graphite)$$

in which your diamond converts ("degenerates") into graphite at 298.15 K and 10^5 N m^{-2} = 1 bar. For this process, the change in the entropy of one mole (larger than the Hope diamond) of carbon is

$$\Delta S_{sys} = (5.74 - 2.377) = 3.363 \, \text{J K}^{-1}$$

Since the enthalpy of formation of diamond is larger than that of graphite, the enthalpy change of the system for the transformation from diamond to graphite is negative. For a process at constant pressure, the enthalpy change is equal to the heat exchanged. Thus, heat is *lost* by the system to the surroundings as diamond converts into graphite. From the standard enthalpy in the Table, the entropy of the surroundings increases by

$$\Delta S_{surr} = \frac{\Delta Q_{surr}}{T} = -\frac{\Delta H_{sys}}{T} = \left(\frac{1895}{298.15}\right) = 6.356 \, J \, K^{-1}.$$

This release of heat to the surroundings is due to the stronger, more stable chemical bonds in graphite than in diamond. Thus, the overall change in the entropy of the Universe is

$$\Delta S_{universe} = 9.719 \, J \, K^{-1}$$

per mole of carbon transformed. Diamond is indeed less stable than graphite because the "degeneration" of diamonds to graphite increases both the entropy of the system and the entropy of the surroundings; there is no thermodynamic hope for diamonds. Of course, this is considering only the thermodynamics. The extremely slow kinetics of the transformation due to a large activation energy is what saves your diamonds from transforming into pencil lead.

§   *Example 6.1 Rhombic sulphur to monoclinic sulphur*

Consider the standard enthalpy and entropy data in the Table above. Of the two sulphur allotropes, rhombic and monoclinic sulphur, which is more stable at standard temperature and pressure?

Consider one mole of rhombic sulphur converting to monoclinic sulphur:

$$S(rhombic) \rightarrow S(monoclinic)$$

The rhombic crystal has an entropy that is lower than the monoclinic crystal so that the entropy change of the system is

$$\Delta S_{sys} = (32.6 - 31.80) = 0.8 \, J \, K^{-1}$$

However, the enthalpy of formation of the monoclinic crystal is larger so that the entropy change of the surroundings due to heat *absorbed* by the system is

$$\Delta S_{surr} = \frac{\Delta Q_{surr}}{T} = -\frac{\Delta H_{sys}}{T} = -\left(\frac{330}{298.15}\right) = -1.107 \text{ J K}^{-1}.$$

There is a *decrease* in the entropy of the surroundings due to the heat lost from the surroundings to the system when rhombic sulphur transforms to monoclinic sulphur. Hence, the entropy change of the universe is

$$\Delta S_{universe} = 0.8 - 1.107 = -0.307 \text{ J K}^{-1}.$$

The conversion of rhombic sulphur to monoclinic sulphur is not spontaneous at standard temperature and pressure, even though the former has a lower entropy. This is due to the difference in the chemical bond strengths, with monoclinic sulphur having stronger bonds and thus lower energy. The relative chemical bond strengths is the dominant thermodynamic factor here. §

§   *Example 6.2 Combustion of hydrogen*

The reaction of hydrogen gas and oxygen gas to form water vapour is

$$2H_2(g) + O_2(g) \rightarrow 2H_2O(g).$$

Two moles of $H_2(g)$ and one mole of $O_2(g)$ has a higher entropy than two moles of $H_2O(g)$. Nonetheless, this reaction is spontaneous at standard temperature and pressure. Deduce this from data in the Table above.

The combustion of hydrogen gas to form water vapour is an explosive process. At standard temperature and pressure, the equilibrium lies far to the product side of the chemical reaction:

$$2H_2(g) + O_2(g) \rightarrow 2H_2O(g).$$

From the data above, the change in the entropy of the system is actually negative, that is, the product is lower in entropy than the reactants by

$$\Delta S_{sys} = 2(188.835) - 2(130.680) - 205.152 = -88.842 \text{ J K}^{-1}$$

per mole of the reaction. A decrease in entropy is expected from the elementary consideration that three moles of gas has been converted into two moles of gas at the same temperature and pressure. For the same temperature and pressure, the number of microstates is considerably higher for three moles of gas than for two moles of gas.

At the same time, the change in the entropy of the surroundings per mole of the reaction is

$$\Delta S_{surr} = \frac{\Delta Q_{surr}}{T} = -\frac{\Delta H_{sys}}{T} = \left(\frac{2 \times 241826}{298.15}\right) = 1622.18 \text{ J K}^{-1}.$$

This increase in the entropy of the surroundings is rather large. Thus overall, the entropy of the universe increases by

$$\Delta S_{universe} = 1622.176 - 88.842 = 1533.33 \text{ J K}^{-1}$$

per mole of the reaction: hydrogen and oxygen react spontaneously at standard temperature and pressure.   §

This large increase in entropy ΔS of the surroundings is due to the heat lost by the system to the surroundings. Because enthalpy is $H = U + PV$, there are two contributions to this:

$$\Delta H_{sys} = \Delta U_{sys} + P\Delta V_{sys}.$$

Treating the reactants and products as ideal gases,

$$\Delta V_{sys} = \frac{RT}{P}\Delta n$$

where Δn is the change in the number of moles per mole of reaction; here it is -1. Thus, one contribution to the change in the enthalpy of the system is

$$P\Delta V_{sys} = RT\Delta n.$$

This gives a contribution to the increase in the entropy of the surroundings of

$$-\frac{P\Delta V_{sys}}{T} = -R\Delta n = 8.314 \text{ J K}^{-1}.$$

Most of the decrease in the enthalpy of the system comes from the ΔU term. From average values of the bond energies, we have

$H - H$ bond energy in hydrogen: 432.0 kJ mol^{-1}

$O - O$ bond energy in oxygen: 498.6 kJ mol^{-1}

$O - H$ bond energy: 458 kJ mol^{-1}.

Counting the number of bonds formed and broken per mole of this reaction gives

$$\Delta U_{sys} = -(4 \times 458) + (2 \times 432.0) + 498.6 = -469.4 \text{ kJ}.$$

Remember that the bond energy is the energy required to *break* the bond. It is the energy we need to put into the molecule to bring it to the dissociated state. This gives a contribution to the increase in the entropy of the surroundings of

$$-\frac{\Delta U_{sys}}{T} = 1574.38 \text{ J K}^{-1}.$$

Thus, using approximate bond energies and treating the gases as ideal gives an estimate of

$$\Delta S_{surr} \cong 1574.38 + 8.314 = 1582.69 \text{ J K}^{-1},$$

which is reasonably close to the value of 1622.18 J K^{-1} obtained from tables of standard state thermodynamic data. Our calculations suggest that the reaction is spontaneous principally because two molecules of water are energetically more stable, has lower energy, than two molecules of hydrogen plus a molecule of oxygen. The decrease in the energy of the system is manifested as heat lost to the surroundings, leading to an increase in S_{surr}. Again, there is a kinetic barrier, an activation energy, that the reaction needs to overcome in order to proceed, but the instant a mixture of hydrogen and oxygen gases is sparked, it reacts rather explosively in getting toward its thermodynamic equilibrium state.

Fundamental Equations

7.1 The four Laws of Thermodynamics

Up to this point we have focused on the conceptual foundations of Thermodynamics. The Zeroth Law puts the idea of temperature and thermal equilibrium on firm ground. The First Law gives us the conservation of energy, and tells us that the internal energy of a system is a state function. The Second Law defines entropy, which is a state function, and leads us to the Clausius inequality. This tells us that the total entropy of the Universe is always increasing,

$$dS_{universe} > 0,$$

giving an asymmetric direction to Time for all macroscopic processes in Nature. The Third Law summarizes experimental observations that the entropy change for all transformations between equilibrium states goes to zero as temperature goes to absolute zero. Consistent with Boltzmann's definition of entropy from the microscopic point of view, it provides the reference value of $S = 0$ at $T = 0$ to calculate the absolute value of entropy.

If nothing else, you should take away from this course that these four laws allow you to understand thermal interactions – heat, work, energy, entropy, etc – of all types of matter and radiation, from the chemical reactions in your test-tube to the nucleosynthesis reactions going on in the core of stars, to the cosmic microwave background radiation that is cooling down due to the expansion of the Universe.

While the Zeroth Law and the Third Law define and give meaning to thermal equilibrium and absolute entropy, it is the First Law and the Second Law which provide the conceptual basis for the framework that will enable you to conveniently apply Thermodynamics. We set this up in this chapter.

The Second Law tells us that there is a direction that natural processes all take. A cup of hot tea in a cool room spontaneously cools down to room temperature. Solutions of hydrochloric acid and sodium hydroxide in the laboratory spontaneously neutralize to form a sodium chloride solution when mixed. The reverse processes do not occur. We have seen that the direction taken by all processes in Nature increases the entropy of the universe. When we examine the thermodynamics of a system, we would like a framework that focuses on the appropriate thermodynamic variables for only the system rather than have to do an entropy accounting for the entire universe. This is where we start.

7.2 Criteria for spontaneous processes

Explicitly indicating the system and surroundings with subscripts, we have from the Second Law:

$$dS_{universe} = dS_{sys} + dS_{surr} \geq 0$$

for all processes, reversible and irreversible. The heat gained by the system is equal to the negative of the heat gained by the surroundings, that is

$$dQ_{sys} = -dQ_{surr}.$$

Note that this is true regardless of whether the process is reversible or irreversible. Remember, however, that in an irreversible process we cannot directly calculate dQ_{sys} using the change in the thermodynamic properties of the system as the process occurs. This is because thermodynamic properties are not well-defined during irreversible processes.

With the surroundings at a temperature T_{surr}, the entropy change of the surroundings is given by

$$dS_{surr} = \frac{dQ_{surr}}{T_{surr}}.$$

Combining this with the total entropy change of the universe,

$$dS_{sys} + dS_{surr} \geq 0$$

$$dS_{sys} \geq -\frac{dQ_{surr}}{T_{surr}},$$

and hence,

$$dS_{sys} \geq \frac{dQ_{sys}}{T_{surr}} \quad \Rightarrow \quad T_{surr}dS_{sys} \geq dQ_{sys}$$

where the equality holds for reversible processes while the inequality holds for irreversible processes. This is really just the Clausius inequality $dS \geq dQ/T$, but we have taken some care tracking the subscripts "*sys*" and "*surr*". Similarly, the differential work done on the system (assuming only PV-work) is:

$$dW = -P_{surr}dV_{sys}$$

where P_{surr} is the pressure exerted by the surroundings on the system as it changes volume by dV_{sys}. We substitute our expressions for dQ_{sys} and dW above into the First Law $dU - dW - dQ_{sys} = 0$ to get:

$$dU + P_{surr}dV_{sys} - T_{surr}dS_{sys} \leq 0,$$

and dropping the subscripts for V_{sys} and S_{sys}, we have the following inequality governing the heat and work interactions of a system with its surroundings at temperature T_{surr} and pressure P_{surr}:

$$dU + P_{surr}dV - T_{surr}dS \leq 0.$$

This is the starting point for us to develop the general framework for Thermodynamics, combining the First and Second Laws. When the interaction is reversible,

$$P_{surr} = P_{sys}$$

$$T_{surr} = T_{sys}$$

at all times since the system is always at equilibrium in a reversible process. Hence, we end up with the following two important conclusions:

$$dU + P_{surr}dV - T_{surr}dS < 0 \qquad \textbf{(I)}$$

for irreversible processes, and

$$dU + PdV - TdS = 0 \qquad \textbf{(II)}$$

for reversible processes where we have equated P_{surr} to P_{sys} and dropped the "*sys*" subscript. The same for the temperature T. To summarize: we obtained **I** and **II** by making using of the First Law $dU = dW + dQ$, and the Second Law $dS_{universe} \geq 0$, noting that for reversible processes the system is always in equilibrium so that its temperature and pressure are equal to those of the surroundings.

If a process satisfies inequality **I**, it is a spontaneous process driving a system towards equilibrium. For example, **I** is satisfied in a chemical reaction system that is still proceeding towards equilibrium.

The equality **II** is called the fundamental equation of Thermodynamics. It relates all thermodynamic variables to each other for any system in equilibrium or undergoing any reversible process.

For processes that satisfy neither **I** nor **II**, we have $dU + P_{surr}dV - T_{surr}dS > 0$. A macroscopic system cannot undergo such a process. These are the impossible processes mentioned in Section 5.5. If any such process were possible, it would be the basis for building a perpetual motion machine but, as you might guess, there is no such machine.

We use inequality **I** to figure out the conditions for spontaneous processes for systems operating under different physical constraints:

1) Isolated system: no heat or work interaction is possible, so that the internal energy and the volume of the system do not change: $dU = 0$ and $dW = 0$. Hence, **I** is reduced to

$$-T_{surr} dS < 0 \quad \Rightarrow \quad dS > 0.$$

We have already found this result in Chapter 3: any spontaneous process in Nature increases the entropy of the Universe. In the laboratory a bomb calorimeter is frequently used to measure the heats of reaction. The system consisting of the reacting substances and the water inside the calorimeter is enclosed in a rigid and thermally insulating jacket. It is thus an isolated system, at least to the extent that the calorimeter is effectively thermally insulated and does not change in volume when reaction occurs. Equilibrium is reached when its entropy is maximized.

2) Constant volume and entropy: we have $dV = 0$ and $dS = 0$. Thus, **I** gives:

$$dU < 0,$$

which means that the internal energy is minimized at equilibrium. Consider a ball bouncing on a table. If we take the system to include the ball and the gravitational field of the Earth, then the internal energy of the system consists of the kinetic energy of the ball and its gravitational potential energy. As the ball bounces on the table it heats up the table through the vibrations of the atoms of the table. If the ball is made of an thermally insulating and hard substance with its atoms held tightly in fixed positions, it does not also heat up as a result of the collisions with the table. If, as a consequence, the distribution of random velocities and random positions of the ball does not change at all, then the number of microstates remains the same, and Boltzmann tells us that its entropy does not change. Thus, while the kinetic and gravitational potential energy of the ball is lost as heat to the table, dS is zero for the ball. Equilibrium is reached when the kinetic and gravitational potential energy of the ball is completely lost as heat and the ball is motionless. The internal energy of the system is minimized, dV is zero for the ball so that it does not do any PV-work and dS is zero for the ball because it does not gain any heat. This example illustrates that the difference between mechanics and thermodynamics lies in the central role that heat plays in thermodynamics. In a purely mechanical frictionless system with a perfectly elastic table, the ball will keep bouncing forever because its energy will oscillate between kinetic energy and gravitational potential energy, and none of it will be lost as heat to the table.

3) Constant pressure and entropy: since S and P_{surr} are constant, we have

$$T_{surr} dS = 0$$

and

$$P_{surr} dV = d(P_{surr} V).$$

Therefore, **I** can be rewritten as

$$dU + d(P_{surr} V) < 0 \quad \Rightarrow \quad d(U + P_{surr} V) < 0.$$

Thus, equilibrium is reached when the quantity $U + P_{surr} V$ is minimized. In order for mechanical equilibrium between the system and the surroundings P_{surr} must be equal to the pressure of the system. Thus, at equilibrium $H = U + PV$ is minimized. Hence, for systems at constant entropy and pressure, equilibrium is reached when the enthalpy is minimized. That is:

$$dH < 0$$

is the condition for spontaneous processes at constant S and P.

It is not convenient to hold the entropy of a system constant in laboratory experiments. It is much easier to directly hold variables such as P, V, and T constant.

4) Constant volume and temperature: for systems at constant V and constant $T_{surr} = T$, we have

$$P_{surr}dV = 0$$

and

$$-T_{surr}dS = -d(T_{surr}S) = -d(TS)$$

so that **I** becomes

$$dU - d(TS) < 0.$$

If we define $A \equiv U - TS$, this gives us the condition for a spontaneous process at constant volume and temperature to be:

$$dA < 0.$$

Equilibrium for systems at constant V and T is reached when the quantity A is minimized. A is called the Helmholtz free energy. Since U, T and S are state functions, the Helmholtz free energy is a state function. Spontaneous processes at constant V and T are always accompanied by a decrease in the Helmholtz free energy of the system.

5) Constant pressure and temperature: for systems at constant $P_{surr} = P$ and constant $T_{surr} = T$, we have

$$P_{surr}dV = d(P_{surr}V) = d(PV)$$

and

$$-T_{surr}dS = -d(T_{surr}S) = -d(TS)$$

so that **I** becomes

$$dU + d(PV) - d(TS) < 0.$$

If we define $G \equiv U + PV - TS = H - TS$, this gives us the condition for a spontaneous process at constant pressure and temperature to be:

$$dG < 0.$$

Since H, T and S are state functions, the Gibbs free energy is a state function. Equilibrium for systems at constant P and T is reached when the quantity G is minimized. G is called the Gibbs free energy. Spontaneous processes at constant P and T are always accompanied by a decrease in the Gibbs free energy of the system. Since many chemical reactions, in the laboratory or in industry or in living organisms, occur at controlled/regulated constant temperature and pressure, the Gibbs free energy is a rather useful quantity for dealing with the thermodynamics in much of Chemistry and Biology.

7.3 What is the maximum work that can be extracted from any system?

It was essentially this question that set Sadi Carnot on the path to the discovery of the Second Law. This led to

$$\eta = 1 - \frac{T_c}{T_h}$$

as the maximum possible efficiency of an engine operating cyclically between a reservoir at T_c and a second hotter reservoir at T_h. In Section 5.7, we have already encountered this question, analyzing the maximum work we can extract from two objects at different temperatures using only the concept of entropy. Now that we have the definitions of the Helmholtz and Gibbs free energies A and G, we can calculate more conveniently the maximum amount of work that a system can perform when it transforms from one thermodynamic state to another, not necessarily in a cyclic process.

In any purely mechanical system, the work done on the system is equal to its change in energy:

$$dW = dU.$$

When a mass m is lifted to a height h, the increase in the gravitational potential energy is equal to the mechanical work done lifting the mass:

$$dW = dU = mgh.$$

The change in the energy of a system is solely equal to the work done on it. This is not the case for a thermodynamic system because heat exchange dQ is generally involved. In thermodynamic systems, we have

$$dW = dU - dQ.$$

For example, think of steam engines, Carnot engines, the internal combustion engine in a car where the working substance – steam, an ideal gas, a mixture of air and petrol – undergo substantial heat exchange with the surroundings. And we also know that $dQ \leq TdS$ from the Second Law, giving us

$$dW = dU - dQ \geq dU - TdS.$$

If we restrict all heat exchange in a state transformation to take place at a fixed temperature T, then we have

$$dU - TdS = d(U - TS)$$

so that

$$dW \geq d(U - TS).$$

The work done by the system is equal to $-dW$. Therefore, we have the

$$-dW \leq -d(U - TS) \quad \Rightarrow \quad -dW \leq -dA.$$

Therefore, for any system where all the steps involving heat exchange occur at a fixed temperature, the maximum amount of work that can be extracted from the system is equal to the decrease in the Helmholtz free energy of the system. This is why A is referred to as a *free* energy: the decrease in A is the maximum amount of energy that is available to be extracted as work.

Notice that in the above we have not specified the type of mechanical work dW that is extracted from the system. Let us say that we have an expanding/contracting volume of gas which we have connected up to some mechanical device to extract work by making it lift a weight. Let us distinguish volume expansion work dW_{PV} from other forms of mechanical work, i.e.,

$$dW = dW_{PV} + dW_{other}.$$

From the First Law,

$$dW = dW_{PV} + dW_{other} = dU - dQ,$$

so that

$$dW_{other} = dU - dW_{PV} - dQ = dU + PdV - dQ,$$

replacing dW_{PV} by $-PdV$ for expansion or compression work. Hence, we have

$$dW_{other} \geq dU + PdV - TdS$$

$$dW_{other} \geq d(U + PV - TS),$$

for constant P and T. Therefore, the maximum amount of non-PV work that can be obtained from a system undergoing a state transformation where all heat exchanges take place at a fixed temperature and all work exchanges take place at a fixed pressure is:

$$-dW_{other} \leq -d(U + PV - TS)$$
$$-dW_{other} \leq -dG.$$

The decrease in the Gibbs free energy of a system is the maximum amount of non-volume expansion/contraction work that can be extracted from a system undergoing any thermodynamic transformation at constant T and P. What insight into real reactions does this idea of maximum work provide us? We look at examples, but first we introduce the terminology of standard Gibbs free energy of formation to help us organize thermodynamics data.

7.4 The Gibbs free energy of formation

Consider the transformation, by whatever chemical process needed, of graphite and gaseous oxygen to form gaseous carbon dioxide:

$$C(graphite) + O_2(g) \rightarrow CO_2(g).$$

The change in the Gibbs free energy for this transformation is called the Gibbs free energy of formation $\Delta_f G$ of carbon dioxide, with the subscript f denoting "formation". Specifically, if the initial and final states of this process are at a temperature of 298.15 K and a pressure of 1 bar, we refer to the change in Gibbs free energy as the standard Gibbs free energy of formation, $\Delta_f G^{\emptyset}$, with the superscript $\emptyset$ denoting the standard state of 298.15 K and 1 bar. The temperature of 298.15 K and pressure of 1 bar are selected for convenience rather than for fundamental considerations, and we have encountered these reference conditions when we discussed standard enthalpies of formation (Section 3.3) and standard absolute entropies (Section 6.4). Note that to define this standard Gibbs free energy of formation, the reference state of carbon is chosen as graphite which is the most stable state of the element at 298.15 K and 1 bar. Similarly, gaseous O_2 is chosen as the reference state for oxygen because it is the most stable thermodynamic state of the element oxygen at 298.15 K and 1 bar. Tabulations of standard Gibbs free energies of formation are rather useful for calculating the Gibbs free energy change for any reaction under standard state conditions. We illustrated this below.

§   *Example 7.1 Standard Gibbs free energy of formation*

Crystalline silver and chlorine gas react to form crystalline silver chloride as follows:

$$Ag(s) + \frac{1}{2}Cl_2(gas) \rightarrow AgCl(s)$$

Given the standard enthalpy of formation of $AgCl(s)$ and the standard absolute entropies of the reactants and product, calculate the standard Gibbs free energy of formation of crystalline silver chloride.

From calorimetric measurements, the standard entropies of the reactants and products are

$$S^{\emptyset}(Ag(s)) = 42.55 \text{ J K}^{-1}\text{mol}^{-1}$$

$$S^{\varnothing}(Cl_2(gas)) = 223.07 \text{ J K}^{-1}\text{mol}^{-1}$$

$$S^{\varnothing}(AgCl(s)) = 96.2 \text{ J K}^{-1}\text{mol}^{-1}.$$

Bomb calorimetry measurement of the reaction gives the standard enthalpy of formation of crystalline silver chloride

$$\Delta_f H^{\varnothing}(AgCl(s)) = -127.07 \text{ kJ mol}^{-1}.$$

Since crystalline silver and gaseous chlorine are the most stable forms of the elements at standard temperature and pressure, their standard enthalpies of formation are zero. Therefore, the standard Gibbs free energy of formation is

$$\Delta_f G^{\varnothing}(AgCl(s)) = \Delta_f H^{\varnothing}(AgCl(s)) - T\Delta_f S^{\varnothing}(AgCl(s))$$

$$= \Delta_f H^{\varnothing}(AgCl(s)) - T\left\{ S^{\varnothing}(AgCl(s)) \right.$$

$$\left. - S^{\varnothing}(AgCl(s)) - \frac{1}{2}S^{\varnothing}(AgCl(s)) \right\}$$

$$= -109.82 \text{ kJ mol}^{-1}. \qquad\qquad §$$

We note that although we generally think of energy as the quantity powering processes, it is really the Gibbs free energy which measures the driving force for thermodynamic processes at constant temperature and pressure. When we discussed criteria for spontaneous processes, we have pointed out that the difference between mechanics and thermodynamics is the important role of heat interactions leading to entropy changes of the system and surroundings. We accounted for ΔS_{sys} and ΔS_{surr} separately. At constant temperature and pressure, the Gibbs free energy automatically accounts for both. As we have seen in Section 7.3, it is the decrease in the Gibbs free energy that is equal to the maximum amount of non-PV work that can be extracted from a system at constant T and P. With this, let us estimate the efficiency of biological systems.

§ *Example 7.2 Efficiency of aerobic respiration in organisms*

Glucose is the primary metabolic fuel in all organisms. In the biochemical pathways of respiration, the energy from glucose oxidation is used to synthesize adenosine triphosphate (ATP), which in turn is used as the direct source of energy to power metabolism in organisms. Estimate the efficiency of organisms in using glucose oxidation to synthesize ATP.

The oxidation of glucose in aqueous solution by gaseous oxygen to form carbon dioxide gas and liquid water is:

$$C_6H_{12}O_6(aq) + 6O_2(g) \rightarrow 6CO_2(g) + 6H_2O(l).$$

Let us consider how we calculate the change in the Gibbs free energy when this process takes place at standard temperature and pressure. The standard Gibbs free energies of formation of the reactants and products from their respective most stable states of the elements are:

$$\Delta_f G^\varnothing\left(C_6H_{12}O_6(aq)\right) = -917.2 \text{ kJ mol}^{-1}$$
$$\Delta_f G^\varnothing\left(O_2(g)\right) = 0 \text{ kJ mol}^{-1}$$
$$\Delta_f G^\varnothing\left(CO_2(g)\right) = -386.2 \text{kJ mol}^{-1}$$
$$\Delta_f G^\varnothing\left(H_2O(l)\right) = -237.2 \text{ kJ mol}^{-1}.$$

We have seen above that the standard state is defined as 298.15 K and 1 bar. For oxygen gas at this temperature and pressure, $\Delta_f G^\varnothing$ is zero, because it is the most stable form of the element at the reference state. For an aqueous solution it is necessary to add to the temperature and pressure specification of the reference state, a reference concentration of 1 mol dm^{-3}. Thus, $\Delta_f G^\varnothing\left(C_6H_{12}O_6(aq)\right)$ is the sum of the Gibbs free energy of the formation of solid glucose from the elements carbon, hydrogen, and oxygen in their respective reference states, plus the Gibbs free energy of dissolving the solid glucose in water to form a 1 mol dm^{-3} solution, with these processes all carried out at 298.15 K and 1 bar.

Therefore, the standard Gibbs free energy change per mole of this reaction going forward is

$$\Delta_{rxn}G^{\emptyset} = 6\Delta_f G^{\emptyset}[CO_2(g)] + 6\Delta_f G^{\emptyset}[H_2O(l)]$$
$$-\Delta_f G^{\emptyset}[C_6H_{12}O_6(aq)] - 6\Delta_f G^{\emptyset}[O_2(g)]$$
$$= 6(-386.2) + 6(-237.2) - (-917.2)$$
$$= -2823.2 \text{ kJ mol}^{-1}.$$

From this, the maximum non-PV work that can be obtained from the aerobic oxidation of glucose is 2823.2 kJ mol^{-1}. In organisms, this "work" is used to phosphorylate adenosine diphosphate (ADP) to form adenosine triphosphate (ATP), schematically:

$$ADP + P \rightarrow ATP.$$

For each mole of this reaction going in the forward direction, the change in the Gibbs free energy is about 28 kJ mol^{-1} at standard temperature and pressure. The decrease in Gibbs free energy from the oxidation of one mole of glucose is sufficient to drive the formation of about 100 moles of ATP. In organisms these two reactions are coupled through the biochemical processes of respiration which forms about 30 moles of ATP per mole of glucose metabolized. Hence, your respiration is approximately 30% efficient at powering the formation of ATP from the oxidation of glucose. §

§   *Example 7.3 How high can Spiderman leap?*

Many superheroes can apparently leap quite high. Does the Second Law put a limit on their, presumably ATP-fueled, jumping abilities? The average superhero weighs 80 kg and is 40% muscle, and there is about 8×10^{-3} mol kg^{-1} of ATP available for use in muscles. Assume that superheroes have the efficiency of reversible engines.

The amount of ATP in the muscles of a superhero is thus $80 \times 0.4 \times (8 \times 10^{-3}) = 0.256$ moles. When this is metabolized to ADP, the decrease in the Gibbs free energy is

$$\Delta_{leap}G = -28 \times 0.256 = -7.168 \text{ kJ} = -7168 \text{ J}.$$

Thus, assuming that all the muscles in the body can be brought to bear, the superhero can leap up to a maximum height h_{max} where

$$mgh_{max} = -\Delta_{leap}G.$$

This gives h_{max} equal to 9.13 m, which is impressive but far short of Hollywood's estimates. Humans are clearly not reversible engines. Pole vault, which leverages almost all the muscles in the body, has a record of 6.23 m.   §

§   *Example 7.4 Maximum work from fuel cells*

In a hydrogen fuel cell the spontaneous chemical reaction

$$H_2(gas) + \frac{1}{2}O_2(gas) \rightarrow H_2O(liq)$$

is used to produce electricity. Hydrogen and oxygen are the input, and only water is produced. Thus, these fuel cells are important because of environmental and sustainability considerations.

What is the maximum work that can be extracted from a hydrogen fuel cell? Take the operating temperature and pressure to be 298.15 K and one bar.

The standard reaction enthalpy is

$$\Delta_{rxn}H^\emptyset = \Delta_f H^\emptyset[H_2O(l)] = -285.83 \text{ kJ mol}^{-1}$$

and standard reaction entropy is

$$\Delta_{rxn}S^\emptyset = S^\emptyset[H_2O(l)] - S^\emptyset[H_2(g)] - \frac{1}{2}S^\emptyset[O_2(g)]$$

$$= 69.91 - 130.684 - \left(\frac{1}{2} \times 205.138\right)$$

$$= -163.34 \text{ J K}^{-1} \text{ mol}^{-1},$$

using values from standard thermodynamic data tables of enthalpy of formation and absolute entropy. Thus, the standard Gibbs free energy of reaction is

$$\Delta_{rxn}G^{\varnothing} = \Delta_{rxn}H^{\varnothing} - T\Delta_{rxn}S^{\varnothing}$$

$$= -285.830 + \left(\frac{298.15 \times 163.34}{1000}\right)$$

$$= -237.13 \text{ kJ mol}^{-1}.$$

Operating under standard state conditions, a fuel cell can yield a maximum of $237.13 \text{ kJ mol}^{-1}$ of electrical energy per mole of hydrogen gas consumed. §

If you spark a balloon filled with $H_2(g)$ and $O_2(g)$, an explosion occurs. The chemical reaction involved is the same one that powers the hydrogen fuel cell. What is the difference between these two systems? Given the same reaction, the system entropy decreases by $163.34 \text{ J K}^{-1} \text{ mol}^{-1}$. Now, in order for a reaction to proceed spontaneously,

$$\Delta S_{univ} = \Delta S_{surr} + \Delta S_{sys} = \frac{\Delta Q_{surr}}{T} + \Delta S_{sys} \geq 0.$$

Hence, the minimum amount of heat ΔQ_{surr}^{min} that the system needs to lose to the surroundings is $-T\Delta S_{sys} = 48.70 \text{ kJ mol}^{-1}$ in order for ΔS_{surr} to compensate for the system entropy decrease. However, the reaction enthalpy is $\Delta_{rxn}H^{\varnothing}$ so that the actual amount of heat lost to the surroundings in an explosion which is *not* harnessed to perform any work is equal to $285.83 \text{ kJ mol}^{-1}$. In such an unharnessed explosion, $285.83 - 48.70 = 237.13 \text{ kJ mol}^{-1}$ of energy which could have been extracted to do useful work is, instead, simply lost as heat to the surroundings. Notice that this amount of energy is

$$-\Delta_{rxn}H^{\varnothing} - \Delta Q_{surr}^{min} = -\Delta_{rxn}H^{\varnothing} + T\Delta_{rxn}S^{\varnothing}$$

which is just $-\Delta_{rxn}G^{\varnothing}$ as we have seen in Section 7.3. Efficient fuel cells extract as large a fraction of this energy as possible to do electrical work. Thermodynamics shows that there is a fundamental limit to the amount of useful work that can be extracted. No matter how clever your fuel cell design is, Thermodynamics imposes a maximum equal to $-\Delta_{rxn}G^\varnothing$ for the reaction used. Beware of startup-technologies that promise better performance than Nature.

§   *Example 7.5 Maximum work from gasoline*

Gasoline is a mixture of a large number of different hydrocarbons, but let us consider it to be just benzene. For a given maximum amount of energy you can extract, compare the mass of the hydrogen in a fuel cell to the mass of gasoline needed in an internal combustion engine.

The complete combustion of benzene is:

$$C_6H_6(g) + \frac{15}{2}O_2(g) \rightarrow 6CO_2(g) + 3H_2O(g)$$

for which

$$\Delta_{rxn}H^\varnothing = -3169.45 \text{ kJ mol}^{-1}$$

$$\Delta_{rxn}S^\varnothing = 41.085 \text{ J K}^{-1} \text{ mol}^{-1}$$

$$\Delta_{rxn}G^\varnothing = -3181.70 \text{ kJ mol}^{-1}.$$

At standard T and P, the maximum work that can be extracted from an internal combustion engine is 3181.70 kJ mol^{-1} of benzene. That molar mass of benzene is 78 g mol^{-1}. Thus, the maximum amount of energy extractable from gasoline is

$$\frac{3181.70}{78} = 40.79 \text{ kJ g}^{-1}.$$

For the hydrogen fuel cell, this is

$$\frac{237.13}{2} = 118.57 \text{ kJ g}^{-1}.$$

Operating at 100% of thermodynamically allowed efficiency, 1 kg of hydrogen fuel produces the same amount of energy as approximately 3 kg of gasoline. These numbers do not consider the relative efficiencies of the hydrogen fuel-cell engine and the internal combustion engine. The gasoline car has to carry a larger load of fuel, but the hydrogen fuel cell car has to store the hydrogen gas at high pressures considering the high molar volume of hydrogen gas relative to liquid gasoline.

The change in the Gibbs free energy gives us a quantitative means of deciding which direction natural processes take at constant temperature and pressure. Analyzing the work done in terms of the change in the Gibbs free energy of a system enables us to quantify the maximum amount of work that can theoretically be extracted from any process in nature. This was what Sadi Carnot did with steam engines.

7.5 Thermodynamic potentials H, A, G are extensive variables

We collect the definitions for H, A and G here because these are central quantities in Thermodynamics:

$$H \equiv U + PV$$
$$A \equiv U - TS$$
$$G \equiv U + PV - TS = H - TS$$

Since these quantities are minimized when systems tend toward equilibrium, they are referred to as thermodynamic potentials. Of course, each is the appropriate thermodynamic potential for a particular set of constraints. For example, G is minimized at equilibrium for systems held at constant T and P. We saw in Section 7.2 that for systems at constant V and S, the internal energy U is minimized at equilibrium.   The internal energy U is the thermodynamic potential for a system at constant V and S.

We have seen in Section 2.1 that U and V are extensive variables while P is intensive. Hence, the enthalpy H is extensive.   The heat required for a specific state change is an extensive variable, so that the entropy is an extensive variable. Therefore, the Helmholtz free energy A is extensive. Finally, the Gibbs free energy is also extensive. If the system size is scaled up/down by a factor of λ, the value of each of H, A, G, U and S is also changed by the same factor. We will revisit this point of scaling with system size when we consider open systems in the next chapter.

7.6 The fundamental equations

From Section 7.2 we have equation **II** applicable for reversible processes:

$$dU + PdV - TdS = 0.$$

Note that all the quantities $- U, P, V, T, S -$ in this equation are state functions. Thus, we can use it to calculate, for example, ΔU, in any irreversible transformation from any initial equilibrium state to any final equilibrium state. All we need to do is to come up with a convenient reversible path connecting the initial and the final states, and then integrate $dU = TdS - PdV$ along this reversible path. If we had limited our discussion above to just reversible processes, we can rather quickly arrive at the fundamental equation. Using the First Law:

$$dU = dQ + dW.$$

For systems doing only PV-work, $dW = -PdV$. Then, using the Second Law through the Clausius inequality for reversible processes $dS = dQ/T$, we get to:

$$dU = TdS - PdV$$

as above. In this expression the change in internal energy dU is written in terms of the change in entropy dS and the change in pressure dV; the natural variables for internal energy are entropy and volume.

Given **II**, and the definitions of H, A and G, which we have found are useful in relation to the criteria for spontaneous processes, we get the following fundamental equations for systems that are at equilibrium:

$H \equiv U + PV$:
$$dH = dU + PdV + VdP$$
$$= (TdS - PdV) + PdV + VdP$$
$$= TdS + VdP.$$

The natural variables for enthalpy are entropy and pressure.

$A \equiv U - TS$:
$$dA = dU - TdS - SdT$$
$$= (TdS - PdV) - TdS - SdT$$
$$= -SdT - PdV$$

The natural variables for the Helmholtz free energy are temperature and volume.

$G \equiv U + PV - TS = H - TS$:
$$dG = dH - TdS - SdT$$
$$= (TdS + VdP) - TdS - SdT$$
$$= -SdT + VdP$$

The natural variables for the Gibbs free energy are temperature and pressure. Including the fundamental equation **II** for internal energy, these results are collected in the table below.

State function	Fundamental equation	Natural variables
$U(S,V)$	$dU = TdS - PdV$	S, V
$H(S,P)$	$dH = TdS + VdP$	S, P
$A(T,V)$	$dA = -SdT - PdV$	T, V
$G(T,P)$	$dG = -SdT + VdP$	T, P

Each of these four equations is a fundamental equation of thermodynamics. Knowing any one of the functions $U(S,V)$ or $H(S,P)$ or $A(T,V)$ or $G(T,P)$ as a function of its respective natural variables enables us to calculate <u>all</u> thermodynamic properties of a system.

7.7 Using the fundamental equations

7.7.1 Maxwell relations

We examine the structure of these fundamental equations further. Consider U as a function of S and V. We can thus write, purely mathematically:

$$dU = \left(\frac{\partial U}{\partial S}\right)_V dS + \left(\frac{\partial U}{\partial V}\right)_S dV.$$

The subscript for each of the differentials indicates the variable which is held constant. Comparing the structure for the differential of a function U that depends upon S and V to what we have from the First and Second Laws, namely:

$$dU = TdS - PdV,$$

we immediately have

$$\left(\frac{\partial U}{\partial S}\right)_V = T$$

and

$$\left(\frac{\partial U}{\partial V}\right)_S = -P.$$

Doing the same for the other fundamental equations, we have:

$$dU \Rightarrow \left(\frac{\partial U}{\partial S}\right)_V = T; \quad \left(\frac{\partial U}{\partial V}\right)_S = -P$$

$$dH \Rightarrow \left(\frac{\partial H}{\partial S}\right)_P = T; \quad \left(\frac{\partial H}{\partial P}\right)_S = V$$

$$dA \Rightarrow \left(\frac{\partial A}{\partial T}\right)_V = -S; \quad \left(\frac{\partial A}{\partial V}\right)_T = -P$$

$$dG \Rightarrow \left(\frac{\partial G}{\partial T}\right)_P = -S; \quad \left(\frac{\partial G}{\partial P}\right)_T = V.$$

Consider further the differential of the internal energy in terms of its natural variables:

$$dU = \left(\frac{\partial U}{\partial S}\right)_V dS + \left(\frac{\partial U}{\partial V}\right)_S dV.$$

Taking the derivative with respect to V of the coefficient of dS, we get

$$\left[\frac{\partial}{\partial V}\left(\frac{\partial U}{\partial S}\right)_V\right]_S = \frac{\partial^2 U}{\partial V \partial S}.$$

But from mathematics, this mixed second derivative is equal to the derivative with respect to S of the coefficient of dV, which is

$$\left[\frac{\partial}{\partial S}\left(\frac{\partial U}{\partial V}\right)_S\right]_V = \frac{\partial^2 U}{\partial S \partial V}.$$

Since $\dfrac{\partial^2 U}{\partial V \partial S} = \dfrac{\partial^2 U}{\partial S \partial V}$ mathematically, we thus have a general relationship between the derivatives of the coefficients of dS and dV in the fundamental equation $dU = TdS - PdV$.

From the fundamental equation for dU, the coefficient of dS is T:

$$\left(\frac{\partial U}{\partial S}\right)_V = T.$$

Thus, the derivative of this with respect to V keeping S constant gives:

$$\left[\frac{\partial}{\partial V}\left(\frac{\partial U}{\partial S}\right)_V\right]_S = \left(\frac{\partial T}{\partial V}\right)_S.$$

The coefficient of dV in the fundamental equation for dU is P:

$$\left(\frac{\partial U}{\partial V}\right)_S = -P.$$

Thus, the derivative of this with respect to S keeping V constant gives:

$$\left[\frac{\partial}{\partial S}\left(\frac{\partial U}{\partial V}\right)_S\right]_V = -\left(\frac{\partial P}{\partial S}\right)_V.$$

Equating these mixed second derivatives, the coefficients T and P in the fundamental equation for U are related by:

$$\left(\frac{\partial T}{\partial V}\right)_S = -\left(\frac{\partial P}{\partial S}\right)_V.$$

This is known as a Maxwell relation. There is one Maxwell relation from each of dU, dH, dA and dG. These are as follows:

$$dU \Rightarrow \left(\frac{\partial T}{\partial V}\right)_S = -\left(\frac{\partial P}{\partial S}\right)_V$$

$$dH \Rightarrow \left(\frac{\partial T}{\partial P}\right)_S = \left(\frac{\partial V}{\partial S}\right)_P$$

$$dA \Rightarrow \left(\frac{\partial S}{\partial V}\right)_T = \left(\frac{\partial P}{\partial T}\right)_V$$

$$dG \Rightarrow \left(\frac{\partial S}{\partial P}\right)_T = -\left(\frac{\partial V}{\partial T}\right)_P.$$

When each of dU, dH, dA and dG is written in terms of the differentials of its natural variables, the coefficients of these variables are just the thermodynamic variables T, S, P and V. The Maxwell relations connect derivatives of these variables to each other. Consider, for example, the Maxwell relations from the fundamental equations for dA and dG. Derivatives of the entropy, which can be determined from calorimetric data, are related to derivatives from PVT equation of state data.

7.7.2 Applying the Maxwell relations

In this section we consider three examples of how to use the Maxwell relations. From $dU = TdS - PdV$, we have

$$\left(\frac{\partial U}{\partial V}\right)_T = T\left(\frac{\partial S}{\partial V}\right)_T - P$$

$$= T\left(\frac{\partial P}{\partial T}\right)_V - P,$$

where we have used the Maxwell relation from dA in the second line. In Section 2.8 we have encountered the derivative $\pi_T = \left(\frac{\partial U}{\partial V}\right)_T$. It is called the internal pressure and is related to the intermolecular interactions in the system. How do we get an intuitive sense of this quantity? It is the derivative of the internal energy with respect to volume, keeping temperature constant. Now, the internal energy is the sum of the intermolecular potential energy and the kinetic energy arising from random thermal motion of the molecules, the latter of which is constant at constant temperature. Therefore, the internal pressure is determined by how the intermolecular potential energy changes with volume when temperature is kept constant. But this is related through the Maxwell relation to $\left(\frac{\partial P}{\partial T}\right)_V$. Since we can calculate $T\left(\frac{\partial P}{\partial T}\right)_V$ from the equation of state, we can completely determine the internal pressure from the equation of state. For an ideal gas, $PV = nRT$ so that

$$\left(\frac{\partial P}{\partial T}\right)_V = \frac{nR}{V}.$$

Hence, for an ideal gas, the internal pressure is given by

$$\left(\frac{\partial U}{\partial V}\right)_T = T\left(\frac{\partial P}{\partial T}\right)_V - P = \frac{nRT}{V} - P = 0,$$

that is, the internal energy of an ideal gas is not dependent upon its volume. We have made use of this property of ideal gases previously in Chapters 1 and 2, but our basis at that point was the equipartition theorem applied to a microscopic model of ideal gases for which the particles only have kinetic energy and zero intermolecular potential energy. Here, solely from $PV = nRT$, which is a *macroscopic* description of an ideal gas, we can also draw the same conclusion.

Given the equation of state of a real gas, we can calculate its internal pressure and thus obtain information about its intermolecular interactions. The Maxwell relation is what allows us to make this connection.

Similarly, from $dH = TdS + VdP$, we have

$$\left(\frac{\partial H}{\partial P}\right)_T = T\left(\frac{\partial S}{\partial P}\right)_T + V$$

$$= -T\left(\frac{\partial V}{\partial T}\right)_P + V$$

where we have used the Maxwell relation from dG in the second line. We have previously encountered $\left(\frac{\partial H}{\partial P}\right)_T$ in Sections 3.5.3 and 3.6 in our discussion of the Joule-Thomson coefficient of gases: $\mu_{JT} \equiv \left(\frac{\partial T}{\partial P}\right)_H = -\left(\frac{\partial H}{\partial P}\right)_T / C_P$. Whether a gas cools down or warms up when it expands isenthalpically is dependent upon the sign of its Joule-Thomson coefficient, and thus upon the sign $\left(\frac{\partial H}{\partial P}\right)_T$. As we have described qualitatively in Section 3.6, μ_{JT} is determined by the nature of the intermolecular interactions of the gas. As in the case of the internal pressure $\left(\frac{\partial U}{\partial V}\right)_T$ above, for ideal gases, which have no intermolecular interactions, we expect $\left(\frac{\partial H}{\partial P}\right)_T$ to be zero. Again, we can derive this result solely from the equation of state of the ideal gas:

$$\left(\frac{\partial H}{\partial P}\right)_T = -T\left(\frac{\partial V}{\partial T}\right)_P + V = -\frac{nRT}{P} + V = 0.$$

We make a third quick application of a Maxwell relation to obtain the general equation for the difference between the heat capacities C_P and C_V. Treating the entropy as a function of T and V, we have:

$$dS = \left(\frac{\partial S}{\partial T}\right)_V dT + \left(\frac{\partial S}{\partial V}\right)_T dV.$$

Using this expression to take the derivative of S with respect to temperature while keeping the pressure constant, and then multiplying by T gives us:

$$T\left(\frac{\partial S}{\partial T}\right)_P = T\left(\frac{\partial S}{\partial T}\right)_V + T\left(\frac{\partial S}{\partial V}\right)_T \left(\frac{\partial V}{\partial T}\right)_P.$$

First, the $T\left(\frac{\partial S}{\partial T}\right)_P$ term. We have the fundamental equation for enthalpy: $dH = TdS + VdP$. This gives immediately,

$$\left(\frac{\partial H}{\partial T}\right)_P = T\left(\frac{\partial S}{\partial T}\right)_P$$

where the second term VdP drops out because we are taking the derivative at constant pressure. We see from our discussion of enthalpy in Section 3.2 that the left-hand side is just the heat capacity at constant pressure C_P. Therefore, we have

$$T\left(\frac{\partial S}{\partial T}\right)_P = C_P.$$

Similarly, using the fundamental equation for internal energy and $\left(\frac{\partial U}{\partial T}\right)_V = C_V$, we have

$$T\left(\frac{\partial S}{\partial T}\right)_V = C_V.$$

Using heat capacities in place of the derivatives of entropy with respect to temperature, we thus obtain:

$$C_P = C_V + T\left(\frac{\partial S}{\partial V}\right)_T \left(\frac{\partial V}{\partial T}\right)_P$$
$$= C_V + T\left(\frac{\partial P}{\partial T}\right)_V \left(\frac{\partial V}{\partial T}\right)_P$$

where in the second line we have used the Maxwell relation from dA. This gives us a general expression for the difference in the heat capacities in terms of derivatives that can be obtained from the PVT equation of state:

$$C_P - C_V = T\left(\frac{\partial P}{\partial T}\right)_V \left(\frac{\partial V}{\partial T}\right)_P,$$

where on the left-hand side are quantities from calorimetric data, while the right-hand side are quantities from *PVT* data. This illustrates the point made after Example 1.3 that Thermodynamics is a general framework for interrelating different kinds of macroscopic properties for any system. It is good to keep this "big picture" in mind as you learn Thermodynamics.

§   ***Example 7.6 Ideal gas $C_P - C_V$***

Calculate the difference $C_P - C_V$ for an ideal gas using its equation of state.

From the ideal gas equation of state $PV = nRT$, we have

$$\left(\frac{\partial P}{\partial T}\right)_V = \frac{nR}{V}$$

and

$$\left(\frac{\partial V}{\partial T}\right)_P = \frac{nR}{P}$$

from which we get

$$C_P - C_V = T\left(\frac{\partial P}{\partial T}\right)_V \left(\frac{\partial V}{\partial T}\right)_P = nR. \qquad §$$

§   ***Example 7.7 van der Waals gas $C_P - C_V$***

Calculate the difference $C_P - C_V$ for a van der Waals gas using its equation of state.

For the van der Waals equation of state, the derivatives $\left(\frac{\partial P}{\partial T}\right)_V$ and $\left(\frac{\partial V}{\partial T}\right)_P$ can be readily obtained by first differentiating the equation of state term-by-term to get:

$$(V - nb)dP - \frac{2an^2}{V^3}(V - nb)dV + \left(P + \frac{an^2}{V^2}\right)dV = nRdT.$$

Then substituting for $\left(P + \frac{an^2}{V^2}\right)$ in terms of V and T, we get

$$(V - nb)dP + \left[-\frac{2an^2}{V^3}(V - nb) + \frac{nRT}{(V - nb)}\right]dV = nRdT.$$

From this we can write, solely in terms of V and T:

$$\left(\frac{\partial P}{\partial T}\right)_V = \frac{nR}{(V - nb)}$$

$$\left(\frac{\partial V}{\partial T}\right)_P = \frac{(V - nb)}{T}\left[1 - \frac{2an(V - nb)^2}{V^3RT}\right]^{-1}.$$

The difference between the heat capacities C_P and C_V is more complicated than for the ideal gas:

$$C_P - C_V = nR\left[1 - \frac{2an(V - nb)^2}{V^3RT}\right]^{-1}.$$

When the volume is large or when the temperature is high, the van der Waals gas approaches ideal gas behavior and $C_P - C_V \to nR$. §

7.8 Natural variables

We introduced the natural variables for each thermodynamic potential when we discussed the fundamental equations. For the internal energy U, the natural variables are S and V, for the enthalpy the natural variables are S and P, and so on. We also mentioned in Section 7.6 the point that if we know any of the thermodynamic potential as a function of its natural variables, then we know all the thermodynamic properties of a system – complete thermodynamic knowledge. Here we discuss this a little more. What is the difference knowing U as a function of S and V compared to knowing U as a function of, say T and V? The fundamental equation for internal energy $dU = TdS - PdV$ is simple in that the coefficients of the differentials are just T and P. If U is expressed as a function of some other set of variables, the corresponding coefficients are not this simple. Similarly, for the other fundamental equations. However, it is not just a matter of simplicity.

We have $dU = TdS - PdV$, so that if we know $U = U(S, V)$, we immediately have

$$T = \left(\frac{\partial U}{\partial S}\right)_V$$

$$-P = \left(\frac{\partial U}{\partial V}\right)_S$$

by taking derivatives of the function $U(S, V)$. The first gives us temperature while the second gives pressure, both as functions of S and V. We can use these equations to eliminate the variable S, leaving a function, which may be quite complicated, relating only P, T and V. Hence, we have complete knowledge of the equation of state relating P, V and T.

§   ***Example 7.8 Ideal gas internal energy***

Derive an expression for the internal energy of an ideal gas in terms of its natural variables entropy and volume.

For one mole of ideal gas, $U = C_V T$ and $PV = RT$, so that the fundamental equation for entropy is:

$$dS = \frac{1}{T}dU + \frac{R}{V}dV$$

$$= \frac{C_V}{U}dU + \frac{R}{V}dV.$$

Aside from the constants, the right-hand side is explicitly expressed in terms of the natural variables of the entropy. Thus, we can integrate to obtain:

$$S(U, V) - S(U_0, V_0) = C_V \ln\left(\frac{U}{U_0}\right) + R \ln\left(\frac{V}{V_0}\right),$$

with (U_0, V_0) specifying a reference state denoted by the subscript 0. Or, writing this in terms of the internal energy,

$$U(S,V) = U_0 \left(\frac{V_0}{V}\right)^{\frac{R}{C_V}} exp\left(\frac{S - S_0}{C_V}\right)$$

where S_0 is $S(U_0, V_0)$. §

Either of the two expressions, $S(U,V)$ or $U(S,V)$, gives the complete thermodynamics for an ideal gas. We can get an expression for the entropy in terms of the more familiar set of variables (T,V), by using $S(U,V)$ and substituting for U in terms of T. With this the entropy per mole of the ideal gas is

$$S(T,V) - S(T_0, V_0) = C_V \ln\left(\frac{T}{T_0}\right) + R \ln\left(\frac{V}{V_0}\right).$$

In this form, you see from the first term on the right-hand side that the ideal gas model is problematic at absolute zero temperature. This is due essentially to the ideal gas having a constant heat capacity C_V at all temperatures. This is not consistent with quantum mechanics which require that the heat capacity C_V of any real substance goes to zero as temperature approaches absolute zero. However, at sufficiently high temperatures this expression is not a bad description of real gases.

Let's now consider that we know U as a function of T and V instead of S and V. Then we have

$$dU = \left(\frac{\partial U}{\partial T}\right)_V dT + \left(\frac{\partial U}{\partial V}\right)_T dV.$$

We have seen that $\left(\frac{\partial U}{\partial T}\right)_V$ is equal to the heat capacity at constant volume for all systems, as long as we are considering systems for which there is only volume expansion/compression work. Thus,

$$dU = C_V dT + \left(\frac{\partial U}{\partial V}\right)_T dV.$$

Therefore, we see that we can obtain complete information on the constant volume heat capacity C_V if we know $U = U(T, V)$. What about the equation of state? Consider the coefficient of dV, i.e., $\left(\frac{\partial U}{\partial V}\right)_T$. If we know $U(T, V)$ we can calculate $\left(\frac{\partial U}{\partial V}\right)_T$ by simply taking the derivative of $U(T, V)$ with respect to V, keeping T constant. But we have encountered this quantity in the previous section:

$$\pi_T = \left(\frac{\partial U}{\partial V}\right)_T = T\left(\frac{\partial P}{\partial T}\right)_V - P.$$

The derivative $\left(\frac{\partial U}{\partial V}\right)_T$ is completely determined given $U(T, V)$, which leaves us with an equation that relates P, V and T, so it seems rather possible for us to get the equation of state.

However, notice that this is a differential equation because of the term $T\left(\frac{\partial P}{\partial T}\right)_V$. If you have a differential equation, you will need to integrate it in order to obtain an algebraic function relating P, V and T. In the process of integrating it, you will need to bring in an integration constant, which requires knowledge of the value of pressure P at some specific value of temperature T. Hence, in order to calculate the PVT equation of state, you will need an additional piece of information to determine this integration constant. Thus, unlike $U(S, V)$, the function $U(T, V)$ does not contain the complete thermodynamic information to determine the equation of state.

We have shown that $U(T, V)$ does not give complete thermodynamic information, but we have not really demonstrated that $U(S, V)$ allows you to calculate all thermodynamic properties completely. We will plug this gap to some extent later in this chapter by working with the fundamental equation for G, which is especially important and convenient for systems at constant temperature and pressure.

7.9 $\left(\frac{\partial U}{\partial V}\right)_T$, $\left(\frac{\partial H}{\partial P}\right)_T$ and intermolecular interactions

In the above sections, we have illustrated the use of Maxwell relations to calculate the derivatives $\left(\frac{\partial U}{\partial V}\right)_T$ and $\left(\frac{\partial H}{\partial P}\right)_T$ from the equation of state. In Chapters 2 and 3, when we first encountered the derivative $\left(\frac{\partial U}{\partial V}\right)_T$, we emphasized that having the equations:

$$dU = dQ - PdV$$
$$dU = C_V dT + \left(\frac{\partial U}{\partial V}\right)_T dV$$

does NOT mean that you can equate term by term to obtain:

$$dQ = C_V dT$$

$$-PdV = \left(\frac{\partial U}{\partial V}\right)_T dV.$$

This is simply because dQ generally depends upon both the change in volume dV and the change in temperature dT, so that there is no one-to-one correspondence of the terms in the two equations above for dU.

Now that we have a general framework for thermodynamics, we can re-examine these derivatives. Why might these derivatives be interesting in understanding the properties of materials? We have seen that the internal energy of an ideal gas consists solely of the kinetic energy of random thermal motion of its molecules, and thus depends only upon its temperature, that is $U = U(T)$. This is because ideal gas molecules do not interact with each other. For the same reason ideal gases neither warm up nor cool down in Joule-Thomson expansions. These two properties can, respectively, be traced to $\left(\frac{\partial U}{\partial V}\right)_T$ and $\left(\frac{\partial H}{\partial P}\right)_T$ being zero for ideal gases. For real gases, however, these derivatives are not zero; real gases have intermolecular interactions and these derivatives give insight into these interactions.

With the internal pressure $\left(\frac{\partial U}{\partial V}\right)_T$ equal to $T\left(\frac{\partial P}{\partial T}\right)_V - P$, we have

$$dU = dQ + dW = TdS - PdV$$

$$= \left(\frac{\partial U}{\partial T}\right)_V dT + \left(\frac{\partial U}{\partial V}\right)_T dV$$

$$= C_V dT + \left[T\left(\frac{\partial P}{\partial T}\right)_V - P\right]dV$$

$$= \left[C_V dT + T\left(\frac{\partial P}{\partial T}\right)_V dV\right] - PdV.$$

In the third line we have used $\left(\frac{\partial U}{\partial T}\right)_V = C_V$ and the square brackets is equal to the internal pressure π_T. In the fourth line we have regrouped the terms so that the PV-work term is separated out as in the first line. Comparing the first and the second lines, one might be tempted to hastily conclude $\left(\frac{\partial U}{\partial V}\right)_T = -P$, and then by incorrectly eliminating the second terms in these two lines, proceed to write $TdS = \left(\frac{\partial U}{\partial T}\right)_V dT = C_V dT$. This is not correct because, in general, S depends on both T and V; the TdS term in the first line of the equation has contributions from both dT and dV terms in the second line. Comparing the first and fourth lines and eliminating $dW = -PdV$, we can see that in a reversible process,

$$dQ = TdS = C_V dT + T\left(\frac{\partial P}{\partial T}\right)_V dV,$$

underscoring the point that the change in entropy generally has contributions from changes in both temperature and volume.

It is instructive to examine the grouping of terms in the third line of the equation above for a process in which a temperature change dT and a volume change dV occurs:

$$dU = C_V dT + \left[T \left(\frac{\partial P}{\partial T} \right)_V - P \right] dV.$$

The first term on the right-hand side gives the portion of the internal energy change that depends only upon the temperature change of the system. Hence, $C_V dT$ is the amount of energy that goes into the kinetic energy of random molecular motion (translation, rotation, vibration) because it is the random molecular motion that depends upon the temperature.]

The second term gives the change in the potential energy of intermolecular interactions when the volume changes by dV. The $-PdV$ portion of this second term is the work done on the system by the external pressure when the volume changes. What about the term $T \left(\frac{\partial P}{\partial T} \right)_V dV$? In an ideal gas $T \left(\frac{\partial P}{\partial T} \right)_V dV$ cancels the $-PdV$ term, so only the first term on the right-hand side remains. For ideal gas molecules, the only form of internal energy is kinetic energy since ideal gas molecules do not interact. Thus, $dU = C_V dT$; the internal energy for an ideal gas depends only upon dT.

In general, for a real gas (or any liquid or solid) in which intermolecular interaction is not zero, $T \left(\frac{\partial P}{\partial T} \right)_V - P$ is not equal to zero. In real gases with non-zero intermolecular interactions, an isothermal volume change dV, which means a change in the average intermolecular distance, results in a change in the average potential energy of the intermolecular interaction. Thus, the second term in the equation above,

$$\pi_T dV = \left(\frac{\partial U}{\partial V} \right)_T dV = \left[T \left(\frac{\partial P}{\partial T} \right)_V - P \right] dV,$$

gives the change in the average potential energy of intermolecular interactions for an isothermal volume change. This gives the meaning of the internal pressure π_T. You can think of π_T as a "spring" in the system arising from the intermolecular interactions; expanding or compressing this spring changes the internal potential energy of the system. How large is this internal pressure? We consider a numerical example.

§   ***Example 7.9 van der Waals gas internal pressure***
Approximating water vapour as a van der Waals gas, calculate its internal pressure at a temperature of 298.15 K and a pressure of 1 bar.

Consider   the   van   der   Waals   equation   of   state $\left(P + \frac{an^2}{V^2}\right)(V - nb) = nRT$ where a accounts for the attractive interatomic potential and b for the repulsive interaction arising from the finite size of the molecules. Using this equation of state, the internal pressure is given by

$$\pi_T = \left(\frac{\partial U}{\partial V}\right)_T = T\left(\frac{\partial P}{\partial T}\right)_V - P = \frac{an^2}{V^2}.$$

Since $\frac{an^2}{V^2}$ is positive, the internal energy increases with volume when temperature is kept constant. This is because with expansion, the average intermolecular distance increases, so that the potential energy of intermolecular interaction goes up given the attractive interatomic potential.

We calculate some numerical values. For water vapour, the van der Waals coefficients a and b are, respectively, equal to 5.536×10^{-1} N m^4 mol^{-2} and 3.05×10^{-5} m^3 mol^{-1}.   Hence,   for   a pressure equal to 1 bar $= 10^5$ N m^{-2} and temperature equal to 298.15 K, the van der Waals equation of state for one mole of gas gives

$$\left(10^5 + \frac{a}{V^2}\right)(V - b) = 8.314 \times 298.15.$$

Solve this cubic equation with the given values of a and b to get

$$V = 24.594 \text{ dm}^3$$

and an internal pressure

$$\pi_T = \left(\frac{\partial U}{\partial V}\right)_T = T\left(\frac{\partial P}{\partial T}\right)_V - P = \frac{an^2}{V^2} = 915.246 \text{ N m}^{-2}.$$

For this van der Waals gas at a pressure of 10^5 N m^{-2}, we have

$$T\left(\frac{\partial P}{\partial T}\right)_V = 1.009152 \times 10^5 \text{ N m}^{-2}.$$

The internal pressure due to the intermolecular interactions is only approximately 1% of the pressure for water vapour at standard temperature and pressure. §

In the same way as above for $\left(\frac{\partial U}{\partial V}\right)_T$, we obtain $\left(\frac{\partial H}{\partial P}\right)_T$ as follows:

$$dH = TdS + VdP,$$

$$\left(\frac{\partial H}{\partial P}\right)_T = T\left(\frac{\partial S}{\partial P}\right)_T + V$$

$$= -T\left(\frac{\partial V}{\partial T}\right)_P + V$$

where in the third line we have used the Maxwell relation from the fundamental equation for Gibbs free energy. The quantity

$$\left(\frac{\partial H}{\partial P}\right)_T = -T\left(\frac{\partial V}{\partial T}\right)_P + V$$

shows that the isothermal change in enthalpy resulting from pressure variation is due to the effects of intermolecular interactions in the system; it is not zero in real gases, as we will calculate below. We have previously encountered this derivative in examining the Joule-Thomson coefficient in Chapter 2.

Here, we reiterate that writing the equations:

$$dH = dQ + VdP$$

$$dH = C_PdT + \left(\frac{\partial H}{\partial P}\right)_T dP$$

does NOT mean that you can equate term-by-term to obtain:

$$dQ = C_P dT$$

$$V dP = \left(\frac{\partial H}{\partial P}\right)_T dP.$$

We find, instead, that describing a reversible process in terms of dT and dP, we have

$$dQ = T dS = C_P dT - T \left(\frac{\partial V}{\partial T}\right)_P dP.$$

With the means of calculating $\left(\frac{\partial H}{\partial P}\right)_T$ from the equation of state, we are now able to obtain an explicit expression for the Joule-Thomson coefficient if we have the equation of state.

§   *Example 7.10 van der Waals gas Joule-Thomson coefficient*

The Joule-Thomson coefficient of a gas is determined by its intermolecular interactions. For the van der Waals equation of state, obtain the dependence of μ_{JT} upon the parameters for intermolecular attraction and repulsion.

For a van der Waals gas, we found in Example 7.6 that

$$\left(\frac{\partial V}{\partial T}\right)_P = \frac{(V - nb)}{T}\left[1 - \frac{2an(V - nb)^2}{V^3 RT}\right]^{-1}.$$

For any gas at sufficiently low pressures the volume of the molecules can be neglected compared to the volume of the gas. For a van der Waals gas expanding to low pressure, we thus have

$$T\left(\frac{\partial V}{\partial T}\right)_P \cong (V - nb)\left(1 + \frac{2na}{RTV}\right)$$

by approximating $(V - nb)$ with V. Hence, we get

$$\left(\frac{\partial H}{\partial P}\right)_T = -T\left(\frac{\partial V}{\partial T}\right)_P + V \cong nb - \frac{2na}{RT}.$$

Therefore, the Joule-Thomson coefficient for a van der Waals gas is

$$\mu_{JT} \equiv \left(\frac{\partial T}{\partial P}\right)_H = -\left(\frac{\partial H}{\partial P}\right)_T / C_P \cong \frac{1}{C_P}\left(\frac{2na}{RT} - nb\right),$$

for sufficiently low pressures. From this equation, it should be rather clear that the intermolecular interactions, quantified by the parameters a and b, play a central role in determining the Joule-Thomson coefficient. Indeed, to gain insight into the internal interactions of the gases as they expand/contract was the aim of the original Joule-Thomson experiment. This, led, of course to applications such as refrigeration and the liquefaction of gases. §

Both the van der Waals coefficients a and b are positive. Thus, μ_{JT} changes sign when the temperature is equal to the inversion temperature

$$T_{inv} = \frac{2a}{Rb}$$

If the temperature is above T_{inv}, then μ_{JT} is negative so that the gas cools upon Joule-Thomson expansion. On the other hand, below their inversion temperatures, gases warm up when expanded from higher to lower pressures. The sign of μ_{JT} results from the balance between the attractive and repulsive intermolecular interactions.

§   *Example 7.11 Inversion temperature of a van der Waals gas*

Approximating carbon dioxide as a van der Waals gas, calculate its inversion temperature.

CO_2 fire-extinguishers are loaded up to about 55 atm pressure. Estimate the temperature drop when you discharge one of these extinguishers into room pressure.

Approximating the PVT-behavior of carbon dioxide using the van der Waals equation of state, we have $a = 3.658 \times 10^{-1}$ N m^4 mol^{-2}

and $b = 4.29 \times 10^{-5}$ m^3 mol^{-1}.   Therefore,   the   inversion temperature is given by

$$T_{inv} = \frac{2a}{Rb} = \frac{2(3.658 \times 10^{-1})}{8.314(4.29 \times 10^{-5})} = 2051 \text{ K.}$$

Hence, the inversion temperature of carbon dioxide is approximately 2051 K. It is actually measured to be 968 K at one atmosphere pressure. The van der Waals estimate is not that great, but the calculation shows that at room temperature, throttling of carbon dioxide is accompanied by cooling of the gas; see Section 3.6 With a heat capacity $C_P = 37.11$ J K^{-1} mol^{-1}, the van der Waals approximation gives a room temperature Joule-Thomson coefficient of

$$\mu_{JT} \cong \frac{1}{C_P}\left(\frac{2na}{RT} - nb\right).$$

$$= \left(\frac{7.316 \times 10^{-1}}{8.314 \times 298} - 4.29 \times 10^{-5}\right)/37.11$$

$$= 6.801 \times 10^{-6} \text{ K N}^{-1} \text{ m}^2$$

$$= 0.69 \text{ K atm}^{-1}.$$

A more accurate value for μ_{JT} at 298 K is 1.11 K atm^{-1}. Carbon dioxide fire-extinguishes have pressures of about 55 atm. If you discharge of one these into room pressure, the temperature drop is approximately

$$\Delta T \cong \mu_{JT}\Delta P = 0.69 \times (-54) \cong -37 \text{ K.}$$

Starting off from room temperature, the temperature of the nozzle of the fire-extinguisher will drop below the freezing point of water, and ice will form from the water vapour in the air. This cooling is the basis for the Linde process of liquefying gases. §

§   ***Example 7.12 Joule-Thomson coefficient for hydrogen***
The van der Waals coefficients for hydrogen gas H_2 are

$$a = 2.453 \times 10^{-2} \ N \ m^4 \ mol^{-2}$$
$$b = 2.651 \times 10^{-5} \ m^3 \ mol^{-1}.$$

Estimate the Joule-Thomson coefficient hydrogen.   How does temperature change when compressed hydrogen is released?

In connection with hydrogen-fuel cell vehicles, the highest pressure compressed hydrogen storage tanks is about $10^8 \ N \ m^{-2}$ pressure. If hydrogen gas is released into atmosphere, it can ignite and burn if the gas temperature is greater than 585 °C. Does a hydrogen leak into atmosphere from a tank at $10^8 \ N \ m^{-2}$ pose an autoignition problem?

$$\mu_{JT} \cong \frac{1}{C_P}\left(\frac{2na}{RT} - nb\right)$$

$$= \left(\frac{4.906 \times 10^{-2}}{8.314 \times 298} - 2.651 \times 10^{-5}\right)/28.62$$

$$= -2.34 \times 10^{-7} \ K \ N^{-1} \ m^2.$$

This is a negative Joule-Thomson coefficient. Thus, a decrease in pressure from $10^8 \ N \ m^{-2}$ (about 1000 atmospheres!) to atmospheric pressure causes an increase in temperature of about 23.4 °C.   This does not get the hydrogen gas to anywhere close to 585 °C; hence, the Joule-Thomson effect, on its own, does not results in a risk of autoignition. Any spark in this system would, though.   §

7.10 How the Gibbs free energy depends on *T* and *P*

The fundamental equation for the Gibbs free energy tells us that:

$$\left(\frac{\partial G}{\partial T}\right)_P = -S$$

$$\left(\frac{\partial G}{\partial P}\right)_T = V.$$

We use these equations to examine the variation of the Gibbs free energy with temperature at constant pressure, and with pressure at constant temperature. Consider the gas and the liquid phases of any substance. The Third law, and Boltzmann's conception of entropy, tells us that entropy is always positive so that G always decreases with T at constant P. From the fundamental equation for enthalpy

$$\left(\frac{\partial H}{\partial T}\right)_P = T\left(\frac{\partial S}{\partial T}\right)_P = C_P \Rightarrow \left(\frac{\partial S}{\partial T}\right)_P = \frac{C_P}{T}$$

In order for any homogeneous, equilibrium state of a substance to be stable to temperature variations, its heat capacity has to be positive. Otherwise, any heat transferred from the surroundings to a system due to the latter's lower temperature will further decrease its temperature, leading to even more heat flowing into the system from the surroundings; then the system could not have been in equilibrium. With heat capacity always positive, we get

$$\frac{C_P}{T} = \left(\frac{\partial S}{\partial T}\right)_P \geq 0.$$

Therefore, with $\left(\frac{\partial G}{\partial T}\right)_P = -S$, the rate of change of G with temperature becomes more negative as T increases; the curve of the Gibbs free energy versus temperature is thus concave downward. The negative curvature is equal to

$$\left(\frac{\partial^2 G}{\partial T^2}\right)_P = -\left(\frac{\partial S}{\partial T}\right)_P = -\frac{C_P}{T}.$$

Hence, we have the dependence of G upon temperature as illustrated in Figure 7.1, at any fixed pressure P. The solid line is $G_{liquid}(T)$ and the dashed line is $G_{gas}(T)$. The dotted line indicates the boiling point temperature at pressure P. At temperatures below the boiling point, the Gibbs free energy of the liquid phase is lower than the Gibbs free energy of the gas phase. Thus, at equilibrium below this temperature, the Gibbs free energy of the system is minimized when it is all liquid. Conversely, above the boiling point, the Gibbs free energy of the gas is lower than that for the liquid, so at temperatures above the boiling point, the system minimizes its Gibbs free energy if it is all in the gas phase. At the boiling point, we have:

$$G_{liq}(T) = G_{gas}(T)$$

so that liquid and gas phases co-exist at equilibrium at this temperature and pressure P. We will discuss phase equilibrium in greater detail in the next chapter.

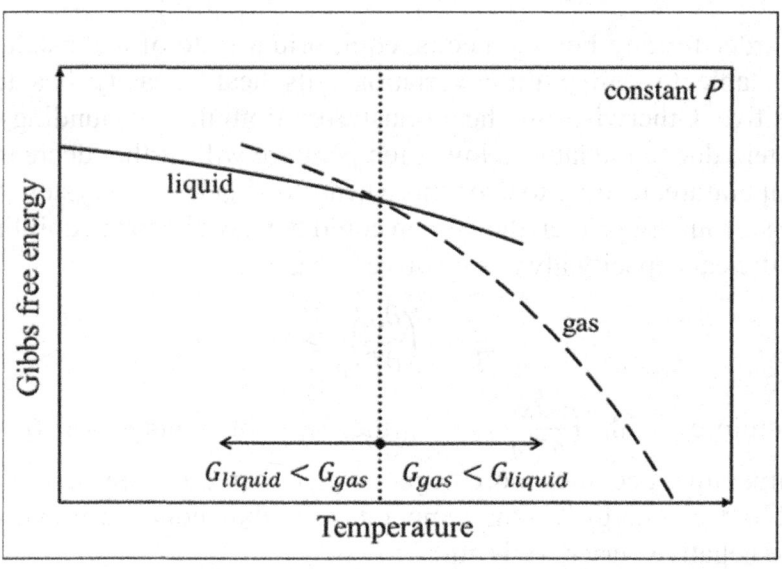

Figure 7.1: The temperature dependence of the Gibbs free energy G_{liq} for the liquid phase and G_{gas} for the gas phase of a substance. The solid line is for liquid, dashed line for gas.

The volume of a substance is always positive. With $\left(\frac{\partial G}{\partial P}\right)_T$ equal to the volume, the slope of the Gibbs free energy as a function of pressure is always positive. Additionally, the compressibility of any substance $\kappa_T = -\frac{1}{V}\left(\frac{\partial V}{\partial P}\right)_T$ is always positive so that any homogeneous substance at equilibrium is stable to variations in pressure: if the volume of a system decreases in response to an increase in the external pressure, the pressure of the system increases. If this were not the case, any decrease in volume in response to increased external pressure will result in an unstable state with further continuing decrease in volume. Similar to the dependence upon temperature, the curve for the Gibbs free energy versus pressure is also concave downward for this stability reason. The negative curvature is equal to

$$\left(\frac{\partial^2 G}{\partial P^2}\right)_T = \left(\frac{\partial V}{\partial P}\right)_T = -\kappa_T V$$

where κ_T is the isothermal compressibility. At fixed temperature, the dependence of the Gibbs free energy upon pressure is illustrated in Figure 7.2 for the liquid and gas phases of a substance.

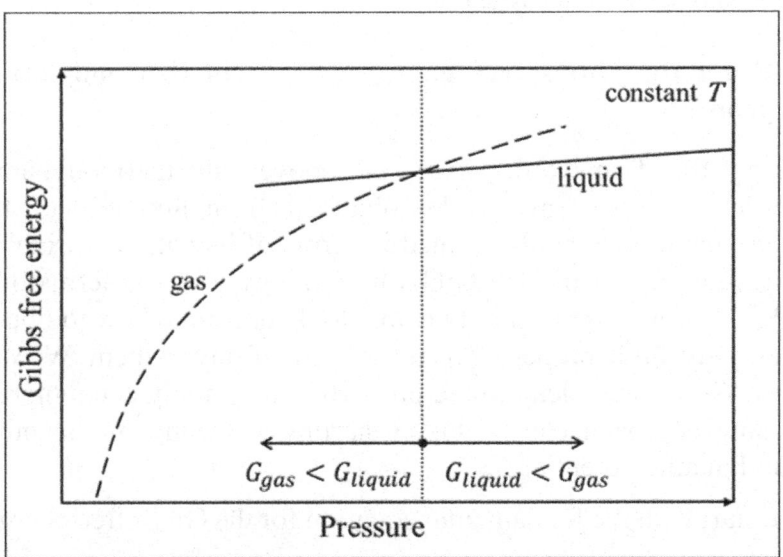

Figure 7.2: The pressure dependence of the Gibbs free energy G_{liq} for the liquid phase and G_{gas} for the gas phase of a substance. Solid line is for liquid, dashed line for gas.

The solid line is $G_{liq}(P)$ and the dashed line is $G_{gas}(P)$. The dotted line indicates the boiling point pressure at the fixed temperature T.

At pressures below the boiling point pressure, the Gibbs free energy of the gas phase is lower than the Gibbs free energy of the liquid phase. Thus, below this pressure, the Gibbs free energy of the system is minimized when it is all in the gas phase. Conversely, above the boiling point pressure, the Gibbs free energy of the liquid phase is lower than that for the gas phase, so at pressures above the boiling point pressure the system is all liquid. At the boiling point pressure, we have:

$$G_{liq}(P) = G_{gas}(P)$$

so that liquid and gas phases co-exist at equilibrium at the boiling point at (T, P) We will discuss phase equilibrium in greater detail in the next chapter, including the relationship between the temperature and pressure at the boiling point.

7.11 From the Gibbs free energy $G(T,P)$ to thermodynamic properties

Using the framework discussed above, all thermodynamic properties of a system can be obtained from any one of the thermodynamic potentials defined in terms of its natural variables. We illustrate this using the Gibbs free energy given in terms of T and P. Of course, we will not be able to demonstrate how to obtain *all* thermodynamic properties as there is an infinity of them. We will see how $G(T,P)$ can lead to the properties commonly encountered. The same approach can be taken starting with any of the other thermodynamic potentials.

We start with the fundamental equation for the Gibbs free energy:

$$dG = -SdT + VdP$$

and with the Gibbs free energy as a function of temperature and pressure: $G = G(T,P)$. Immediately from the fundamental equation, we have:

$$-S = \left(\frac{\partial G}{\partial T}\right)_P,$$

$$V = \left(\frac{\partial G}{\partial P}\right)_T.$$

Thus, taking derivatives of $G(T, P)$, we immediately have the entropy and the volume, $S(T, P)$ and $V(T, P)$. The latter, you recognize to be the equation of state, relating the quantities P, V and T. From the equation of state, the isothermal compressibility and the isobaric expansion coefficient can be obtained through

$$\kappa_T = -\frac{1}{V}\left(\frac{\partial V}{\partial P}\right)_T$$

$$\beta = \frac{1}{V}\left(\frac{\partial V}{\partial T}\right)_P.$$

§ **Example 7.13 Rise in sea-level**

Water has an isobaric thermal expansion coefficient of approximately $2 \times 10^{-4}\ K^{-1}$. Estimate the rise in sea-level if the average temperature of the oceans increases by 1 K. The volume of the oceans V and their surface area A are about $1.4 \times 10^9\ km^3$ and $3.6 \times 10^8\ km^2$.

The average depth of the ocean is approximately

$$l = \frac{V}{A} = 3.9 \text{ km}$$

Then, approximating the surface area of the ocean to be constant,

$$\beta = -\frac{1}{V}\left(\frac{\partial V}{\partial T}\right)_P$$

$$= -\frac{1}{Al}\left(\frac{\partial(Al)}{\partial T}\right)_P$$

$$= -\frac{1}{l}\left(\frac{\partial l}{\partial T}\right)_P.$$

Thus, an estimate for the rise in sea-level for a one Kelvin increase of the average ocean temperature is

$$\frac{\Delta l}{l} = \beta \Delta T.$$

Hence,

$$\Delta l \cong 0.78 \ m.$$

It is difficult to estimate the mean temperature of the oceans. Additionally, several other factors such as land glaciers breaking off and melting into the ocean play a role. But this is a worrying estimate. The measured sea-level rise from 1880 to today is about 0.16 m even though the temperature of the ocean has not yet equilibrated to the Earth's surface temperature. §

Similarly, the difference between the heat capacities can be determined using the equation we obtained in the section above when we discussed Maxwell relations:

$$C_P - C_V = T \left(\frac{\partial P}{\partial T}\right)_V \left(\frac{\partial V}{\partial T}\right)_P$$

both derivatives on the right-hand side can be directly obtained from the equation of state.

Then using the definition $G = H - TS$, we obtain the enthalpy $H(T, P)$, using the entropy. In turn, we can determine the internal energy $U(T, P)$ from $U = H - PV$ since we would then have both the enthalpy and the volume defined in terms of T and P. Then, using the definition $A = U - TS$, we obtain the Helmholtz free energy $A(T, P)$. Using any one of the following equations, we can determine the constant pressure heat capacity:

$$C_P = \left(\frac{\partial H}{\partial T}\right)_P = T \left(\frac{\partial S}{\partial T}\right)_P = -T \left(\frac{\partial^2 G}{\partial T^2}\right)_P$$

since $H(T,P)$, $S(T,P)$ and $G(T,P)$ are all known. From C_P and the difference $C_P - C_V$ determined above from the equation of state, we can then obtain the constant volume heat capacity.

How might you determine other derivatives, such as $\left(\frac{\partial U}{\partial T}\right)_V$ and $\left(\frac{\partial S}{\partial T}\right)_V$? Since we at working with T and P as the independent variables, we can write, for example,

$$dU = \left(\frac{\partial U}{\partial T}\right)_P dT + \left(\frac{\partial U}{\partial P}\right)_T dP.$$

This then gives:

$$\left(\frac{\partial U}{\partial T}\right)_V = \left(\frac{\partial U}{\partial T}\right)_P + \left(\frac{\partial U}{\partial P}\right)_T \left(\frac{\partial P}{\partial T}\right)_V$$

for the temperature derivative at constant volume for the internal energy, and similarly for any other quantity. Thus, all the derivatives on the right-hand side can be obtained starting from $G(T,P)$, the last derivative obtained from the equation of state. In this way, the functions $\left(\frac{\partial U}{\partial T}\right)_V$, $\left(\frac{\partial S}{\partial T}\right)_V$ and $\left(\frac{\partial^2 G}{\partial T^2}\right)_V$ can be determined, with the last taking the constant volume temperature derivative twice. With these functions, you can, for example, determine the constant volume heat capacity using:

$$C_V = \left(\frac{\partial U}{\partial T}\right)_V = T \left(\frac{\partial S}{\partial T}\right)_V = -T \left(\frac{\partial^2 G}{\partial T^2}\right)_V.$$

Using the approach described here, you can thus obtain any thermodynamic property of a system from any one of the thermodynamic potentials defined in terms of its natural variables.

This includes the entropy defined as $S = S(U,V)$ with the fundamental equation:

$$dS = \frac{1}{T}dU + PdV.$$

However, it is typically a bit of the reverse approach that is used in practice; heat capacities and equation of state data are much easier to measure than any of S, U, H, A, G. We describe below how calorimetric and PVT-data can be used to get a handle on the thermodynamics of a system.

As above, we also start with

$$S = -\left(\frac{\partial G}{\partial T}\right)_P,$$

$$V = \left(\frac{\partial G}{\partial P}\right)_T.$$

If we measure the equation of state data $V(T,P)$, we integrate the second equation with respect to pressure starting from any reference pressure P_0, holding temperature constant, to obtain:

$$G(T,P) = G(T,P_0) + \int_{P_0}^{P} V(T,P')\, dP'$$

where P' is the dummy variable for the integral over the pressure, which can be fitted/determined using the PVT data. Similarly, with data from calorimetry, we can establish $S(T,P)$. Then, by integrating this with respect to temperature starting from any reference temperature T_0, keeping pressure constant, we have

$$G(T,P) = G(T_0,P) - \int_{T_0}^{T} S(T',P)\, dT',$$

with T' being the dummy variable for the integral over the temperature. In this way the Gibbs free energy can be determined from calorimetry and PVT measurements.

7.12 The Gibbs-Helmholtz equation

The Gibbs free energy is defined in terms of enthalpy and entropy $G \equiv H - TS$, and the constant pressure derivative of G with respect to temperature is $\left(\frac{\partial G}{\partial T}\right)_P = -S$. This leads to the Gibbs-Helmholtz equation, which will be extensively used in our later discussions of phase and chemical equilibria. We will see that this equation gives the temperature dependence of $\frac{\Delta G}{T}$ where ΔG is the change in the Gibbs free energy of a system undergoing some process. First, why is this ratio of quantities relevant? From the definition of the Gibbs free energy, we have

$$\Delta G_{sys} = \Delta H_{sys} - T\Delta S_{sys}$$

for a process with heat and work interactions at constant P and T, with the subscript sys to remind ourselves that these are system properties. Thus,

$$\frac{\Delta G_{sys}}{T} = \frac{\Delta H_{sys}}{T} - \Delta S_{sys},$$

where ΔS_{sys} is the change in the entropy of the system. Additionally, at constant pressure $\Delta H_{sys} = \Delta Q_{sys}$, the heat gained by the system, so that the term $\frac{\Delta H_{sys}}{T}$ is equal to the negative of the change of the entropy of the surroundings. Hence, $-\frac{\Delta G_{sys}}{T}$ is equal to the change in the entropy of the universe when the process occurs with the enthalpy change of the system going into heat in the surroundings:

$$-\frac{\Delta G_{sys}}{T} = \Delta S_{surr} + \Delta S_{sys}.$$

Therefore, this term determines if the process proceeds spontaneously in the forward direction or in the reverse direction. We have encountered this in the discussion after Example 7.3 on hydrogen explosion. This is the reason the term $\frac{\Delta G}{T}$ occurs frequently in thermodynamic analysis of phase and chemical equilibria. Also note that both ΔH and ΔS are system properties, so that these are determined solely by examining the system. No consideration of the details of the surroundings is needed apart from indicating the temperature and pressure which it shares with the system in equilibrium states. This is indeed why it is convenient to set up the framework of thermodynamic potentials.

Consider taking the constant pressure derivative of $\frac{G}{T}$ with respect to the temperature:

$$\left[\frac{\partial}{\partial T}\left(\frac{G}{T}\right)\right]_P = \frac{1}{T}\left(\frac{\partial G}{\partial T}\right)_P - \frac{G}{T^2}$$

$$= -\frac{S}{T} - \frac{G}{T^2} = -\frac{H}{T^2}.$$

In the second line we have used $\left(\frac{\partial G}{\partial T}\right)_P = -S$, and the definition for the Gibbs free energy $G = H - TS$. Hence,

$$\left[\frac{\partial}{\partial T}\left(\frac{G}{T}\right)\right]_P = -\frac{H}{T^2}.$$

where G and H are the Gibbs free energy and the enthalpy of any equilibrium state. Then, considering the change between two equilibrium states, we have:

$$\left[\frac{\partial}{\partial T}\left(\frac{\Delta G}{T}\right)\right]_P = -\frac{\Delta H}{T^2}$$

which is the differential form of the Gibbs-Helmholtz equation. This equation is exact. Assuming that the enthalpy change is independent of temperature, this can be easily integrated over temperature from T_0 to T giving the approximate:

$$\frac{\Delta G(T,P)}{T} - \frac{\Delta G(T_0,P)}{T_0} = \Delta H\left(\frac{1}{T} - \frac{1}{T_0}\right).$$

This is the integrated form of the Gibbs-Helmholtz equation which gives an estimate of how $\Delta G(T,P)$ depends upon temperature T, given the value of $\Delta G(T_0,P)$ at some arbitrary reference temperature T_0. This assumes a temperature independent ΔH, which is not a bad assumption if the temperature interval from T_0 to T is small. It is worthwhile to expand on this.

Using the differential form of the Gibbs-Helmholtz equation, we can also see that:

$$\left[\frac{\partial(\Delta G/T)}{\partial(1/T)}\right]_P = \frac{dT}{d(1/T)}\left[\frac{\partial}{\partial T}\left(\frac{\Delta G}{T}\right)\right]_P = (-T^2)\left(-\frac{\Delta H}{T^2}\right)$$

$$= \Delta H.$$

That is, in any process or chemical reaction, if you plot the graph of $\frac{\Delta G}{T}$ against $\frac{1}{T}$, the slope at any temperature is equal to the value of ΔH at that temperature. We also see this from the approximate integrated form, although that assumes that this graph is a straight-line, which it is not in general.

§   *Example 7.14 Estimating the boiling point of water*

Water has an enthalpy of vaporization $\Delta_{vap}H^{\emptyset}$ equal to 44 kJ mol^{-1} and an entropy of vaporization $\Delta_{vap}S^{\emptyset}$ equal to 118.92 J K^{-1} mol^{-1} at standard temperature and pressure.

Is liquid water or water vapour more stable at standard temperature and pressure? Estimate the boiling point of water at standard pressure using the Gibbs-Helmholtz equation.

The standard Gibbs free energy of vaporization is:

$$\Delta_{vap}G^{\emptyset} = \Delta_{vap}H^{\emptyset} - T\Delta_{vap}S^{\emptyset} = 8.57 \text{ kJ mol}^{-1}$$

Since this is positive, liquid water is more stable than water vapour at 298.15 K and 1 bar. That is, 298.15 K is below the boiling temperature in Figure 7.1, in the region where $G_{liq} < G_{gas}$.

Let us estimate the boiling temperature at standard pressure using the Gibbs-Helmholtz equation. At the boiling point, indicated by the dotted line in Figure 7.1, $G_{liq} = G_{gas}$, so that $\Delta_{vap}G$ is equal to zero. Taking T_0 in the Gibbs-Helmholtz equation above to be the standard temperature, i.e., 298.15 K, and assuming that the enthalpy of vaporization at any other temperature remains equal to $\Delta_{vap}H^{\emptyset}$, we have

$$\frac{\Delta_{vap}G(T)}{T} - \frac{\Delta_{vap}G^{\emptyset}}{298} = \Delta_{vap}H^{\emptyset}\left(\frac{1}{T} - \frac{1}{298.15}\right).$$

Setting $\Delta_{vap}G(T) = 0$, this gives an estimate of the boiling point of water at standard pressure equal to 370 K. A better estimate can be obtained by treating $\Delta_{vap}H$ more accurately as temperature dependent.

7.13 ΔG and ΔH at low temperatures

We wrap this chapter up by considering the changes in Gibbs free energy ΔG and enthalpy ΔH of any substance at temperatures close to absolute zero. Thermodynamic measurements in various systems suggest that these quantities behave as illustrated schematically in Figure 7.3. That is, as the temperature goes to zero, the difference between the enthalpy change and the Gibbs free energy change of an isothermal process goes to zero. This difference is, of course, simply $T\Delta S$ as indicated in the Figure because

$$G \equiv H - TS.$$

That ΔS goes to zero for processes at temperatures close to absolute zero is the Nernst heat theorem. Indeed, it is not just the difference $\Delta H - \Delta G$ that approaches zero. Additionally, the slopes of ΔH and ΔG as functions of temperature also approach zero, so that ΔH and ΔG approach each other tangentially at absolute zero.

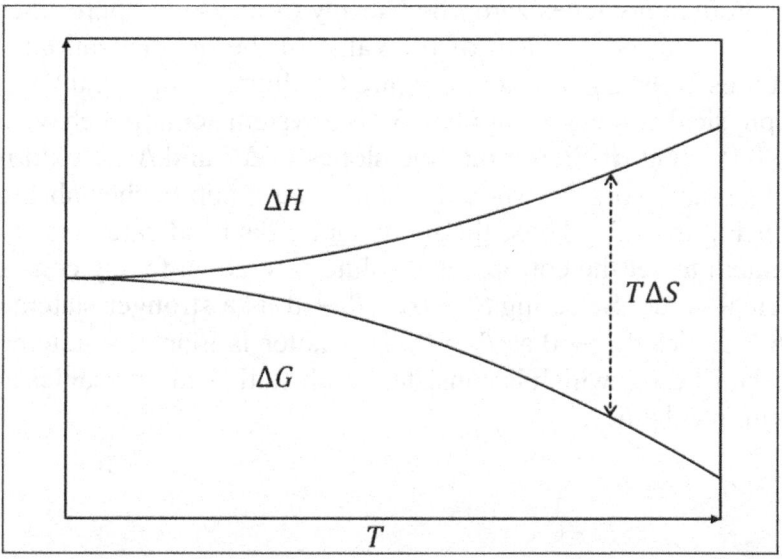

Figure 7.3: The temperature dependence of ΔH (upper curve) and ΔG (lower curve). The temperature axis starts at $T = 0$ at the left, so that $T\Delta S = \Delta H - \Delta G = 0$ at absolute zero. The slopes of ΔH and ΔG also go to zero at absolute zero.

Therefore, the experimental observations illustrated in Figure 7.3 for the dependence of ΔG and ΔH means:

$$\left(\frac{\partial(\Delta H)}{\partial T}\right)_P = \Delta C_P = 0$$

$$\left(\frac{\partial(\Delta G)}{\partial T}\right)_P = -\Delta S = 0.$$

Hence, the heat capacities and entropies of the initial and final thermodynamic equilibrium states of any thermodynamic transformation approach the same value as the absolute temperature goes to zero. For example, the process could be an isothermal magnetization, which we have encountered in Section 6.2. As the temperature approaches zero, the entropy of a system approaches a constant value independent of the value of the magnetization; the two curves in the right panel of Figure 6.6 illustrate this, and has the same physical content as Figure 7.3. If a system actually behaves as in the left panel of Figure 6.6, the slopes of ΔG and ΔH would not go tangentially to zero when the temperature approaches absolute zero in Figure 7.3. Thus, thermodynamic data indicates that it is convenient to set the entropy at absolute zero equal to any constant numerical value. Selecting $S \to 0$ as $T \to 0$, is a stronger statement which includes $\Delta S \to 0$ as $T \to 0$. This latter is Planck's statement of the Third Law, which is consistent with statistical mechanics and quantum mechanics.

Phase Equilibrium

8.1 Scaling of Gibbs free energy with system size

The business of Chemistry is the transformation of matter. Thus, it is necessary that we are able to handle systems where the number of moles of one or more chemical species can change, for example through a chemical reaction or a phase transformation. We here extend the discussion of the fundamental equations in the previous chapter so that we can deal with such variable compositions or open systems.

We first quickly review how we know that U, S, H, A, G are extensive quantities. Although this is a simple point, it is the foundation for the scaling arguments in this section. See Sections 2.1 and 7.5 for our previous discussions of extensive and intensive variables. Knowing that matter consists of atoms and that the internal energy is the sum of the kinetic energy of random atomic motion plus the *finite* range interatomic/intermolecular interactions, we can understand that U is proportional to the number of atoms in the system. That is, the internal energy is an extensive quantity. If we do not have recourse to this microscopic picture, or do not know that matter consists of atoms, we could empirically discover that the internal energy is extensive by performing the Joule experiments of Chapter 2 using different amounts of water in the container. An experiment which uses ten times as much water will require inputs of heat and work that are ten times larger for the same change of state, for the same rise in temperature. That is, the internal energy is extensive. With U extensive, then we use:

$$dU = TdS - PdV$$

to establish that S is extensive. Each term in this equation has to be extensive since dU is extensive. PdV is extensive since P is intensive and dV extensive. Similarly, dS is extensive since T is intensive. Then using the definitions:

$$H = U + PV$$
$$A = U - TS$$
$$G = H - TS$$

we can see that H, A, G are extensive.

Let us examine the effect of changing system size by working with G. Consider increasing the number of moles in a one-component system from n to $N = \lambda n$. Because the Gibbs free energy is extensive, increasing the size of the system by a factor of λ increases its Gibbs free energy by the same factor of λ. Therefore,

$$G(T, P, N) = \lambda G(T, P, n).$$

Take the derivative with respect to λ. The right-hand side gives just

$$\left[\frac{\partial}{\partial \lambda} \left(\lambda G(T, P, n) \right) \right]_{T,P} = G(T, P, n).$$

The left-hand side leads to

$$\left[\frac{\partial G(T, P, N)}{\partial \lambda} \right]_{T,P} = \left(\frac{dN}{d\lambda} \right) \left[\frac{\partial G(T, P, N)}{\partial N} \right]_{T,P} = n \left(\frac{\partial G(T, P, N)}{\partial N} \right)_{T,P}$$

using the product rule for derivatives and $N = \lambda n$. Thus, by using n instead of N as the size argument in the last derivative, we obtain

$$n \left(\frac{\partial G(T, P, N)}{\partial N} \right)_{T,P} = n \left(\frac{\partial G(T, P, n)}{\partial n} \right)_{T,P}.$$

Hence, equating the results we have:

$$n \left(\frac{\partial G(T, P, n)}{\partial n} \right)_{T,P} = G(T, P, n).$$

8.2 μ is the molar Gibbs free energy

If we then define the chemical potential μ by:

$$\mu = \left(\frac{\partial G(T,P,n)}{\partial n}\right)_{T,P}$$

then we will have:

$$n\mu = G(T,P,n)$$

from which we recognize that the chemical potential μ is the molar Gibbs free energy, that is

$$\mu = \frac{G(T,P,n)}{n} = \mu(T,P)$$

where μ is not dependent upon n; it is an intensive quantity because $G(T,P,n)$ is extensive and thus directly proportional to n so that the dependence upon n in $G(T,P,n)$ cancels the n in the denominator.

Furthermore, treating $G(T,P,n)$ as a function of three variables T, P, and n we have:

$$\begin{aligned}
dG &= \left(\frac{\partial G}{\partial T}\right)_{P,n} dT + \left(\frac{\partial G}{\partial P}\right)_{T,n} dP + \left(\frac{\partial G}{\partial n}\right)_{T,P} dn \\
&= \left(\frac{\partial G}{\partial T}\right)_{P,n} dT + \left(\frac{\partial G}{\partial P}\right)_{T,n} dP + \mu dn.
\end{aligned}$$

Note that keeping the number of moles n constant in $\left(\frac{\partial G}{\partial T}\right)_{P,n}$ and $\left(\frac{\partial G}{\partial P}\right)_{T,n}$ means that these quantities are really the same as $\left(\frac{\partial G}{\partial T}\right)_{P}$ and $\left(\frac{\partial G}{\partial P}\right)_{T}$, respectively, in the closed systems of Chapter 7 which have a constant amount of matter. Hence,

$$\left(\frac{\partial G}{\partial T}\right)_{P,n} = -S(T,P,n) = -ns(T,P)$$

$$\left(\frac{\partial G}{\partial P}\right)_{T,n} = V(T,P,n) = nv(T,P)$$

where $s(T,P)$ and $v(T,P)$ are molar entropy and volume. Therefore, we have:

$$dG = -SdT + VdP + \mu dn.$$

This is the fundamental equation for the Gibbs free energy of open systems with one chemical species.

8.3 Chemical potentials in multicomponent systems $G = \sum_i \mu_i n_i$

What if we have more than one chemical species? We have $G(T,P,n_1,n_2)$ for a two-component system with n_1 and n_2 moles of each of the two components, respectively. Similarly to the one-component system above, we can figure out the dependence of the Gibbs free energy upon the system size by examining

$$G(T,P,N_1,N_2) = \lambda G(T,P,n_1,n_2)$$

where $N_1 = \lambda n_1$ and $N_2 = \lambda n_2$. This gives

$$G(T,P,n_1,n_2) = n_1\mu_1 + n_2\mu_2$$

with the molar Gibbs free energy of the two-component system equal to

$$\frac{G(T,P,n_1,n_2)}{n_1 + n_2} = x_1\mu_1 + x_2\mu_2$$

where $x_1 = \frac{n_1}{n_1+n_2}$ and $x_2 = \frac{n_2}{n_1+n_2}$ are the mole fractions of species 1 and 2. This gives the Gibbs free energy for a system with $n_1 + n_2$ total number moles. The chemical potentials μ_1 and μ_2 of the components each depend upon the temperature, pressure *and* the composition. Now, treating G for a two-component system as a function of the four variables T, P, n_1 and n_2, we have:

$$dG = \left(\frac{\partial G}{\partial T}\right)_{P,n_1,n_2} dT + \left(\frac{\partial G}{\partial P}\right)_{T,n_1,n_2} dP + \sum_{i=1}^{2} \left(\frac{\partial G}{\partial n_i}\right)_{T,P,n_{j\neq i}} dn_i.$$

Again, identifying the partial derivatives of the Gibbs free energy with respect of each of the four variables, we have

$$\left(\frac{\partial G}{\partial T}\right)_{P,n_1,n_2} = -S$$

$$\left(\frac{\partial G}{\partial P}\right)_{T,n_1,n_2} = V$$

$$\left(\frac{\partial G}{\partial n_1}\right)_{T,P,n_2} = \mu_1$$

$$\left(\frac{\partial G}{\partial n_2}\right)_{T,P,n_1} = \mu_2.$$

Altogether, we have

$$dG = -SdT + VdP + \mu_1 dn_1 + \mu_2 dn_2.$$

With dG extensive, each of the terms on the right-hand side must be extensive. In particular, $\mu_1 dn_1$ and $\mu_2 dn_2$ are extensive quantities; with the extensive dn factors in these terms, μ_1 and μ_2 are thus intensive variables. Since the sum of the mole fractions $x_1 + x_2$ is equal to 1, only one of x_1 or x_2 fully specifies the composition of a two-component system. Hence,

$$\mu_1 = \mu_1(T, P, x_1)$$

$$\mu_2 = \mu_2(T, P, x_1)$$

arbitrarily selecting the mole fraction of 1 to indicate the composition. The chemical potential $\mu_1(T, P, x_1)$ of component 1 in the two-component system is related to but generally not equal to the chemical potential of the pure component 1 because we expect molecules of the two components to interact differently than they do in their respective pure systems. The chemical potential of a pure system $\mu(T, P)$ is a function of only the temperature and the pressure, whereas $\mu_1(T, P, x_1)$ depends upon T, P *and* the composition of the two-component system.

Extending the above to systems consisting of m chemical species:

$$dG = \left(\frac{\partial G}{\partial T}\right)_{P,n_1,\dots,n_m} dT + \left(\frac{\partial G}{\partial P}\right)_{T,n_1,\dots,n_m} dP + \sum_{i=1}^{m} \left(\frac{\partial G}{\partial n_i}\right)_{T,P,n_{j\neq i}} dn_i$$

where the derivative $\left(\frac{\partial G}{\partial n_i}\right)_{T,P,n_{j\neq i}}$ is taken with respect to n_i while keeping constant the number of moles of all other species $j \neq i$. Therefore,

$$\mu_i = \left(\frac{\partial G}{\partial n_i}\right)_{T,P,n_{j\neq i}}$$

gives the chemical potential of each component species i in a multicomponent system. Just as in the one- and two-component systems, we have:

$$G(T, P, n_1, \dots, n_m) = \sum_{i=1}^{m} n_i \mu_i (T, P, x_1, \dots, x_{m-1})$$

where the chemical potential of each component is a function of T, P and the composition of the system, now determined by $m - 1$ mole fractions since $\sum_{i=1}^{m} x_i = 1$; here we are arbitrarily using the mole fractions $(x_1, \dots, x_{m-1})$, but any subset of $m - 1$ mole fractions will work.

8.4 The fundamental equations for an open system

We thus have the fundamental equation for the Gibbs free energy for an open system:

$$dG = -SdT + VdP + \sum_{i=1}^{m} \mu_i \, dn_i$$

where the new terms with the chemical potentials extend the equations in Chapter 7 to allow us to deal with variable composition and open systems. Collecting the thermodynamic potentials and using the definitions for H, A and G, we have:

$$dU = TdS - PdV + \sum_{i=1}^{m} \mu_i \, dn_i$$

$$dH = TdS + VdP + \sum_{i=1}^{m} \mu_i \, dn_i$$

$$dA = -SdT - PdV + \sum_{i=1}^{m} \mu_i \, dn_i$$

$$dG = -SdT + VdP + \sum_{i=1}^{m} \mu_i \, dn_i$$

and,

$$U = U(S, V, n_1, \dots, n_m)$$

$$H = H(S, P, n_1, \dots, n_m)$$

$$A = A(T, V, n_1, \dots, n_m)$$

$$G = G(T, P, n_1, \dots, n_m)$$

where we have listed the natural variables for each of the thermodynamic potentials.

8.5 Partial molar quantities

Consider the internal energy $U = U(S, V, n_1, \ldots, n_m)$ in an m-component system. In the fundamental equation for dU, we have

$$\mu_i = \left(\frac{\partial U}{\partial n_i}\right)_{S,V,n_{j\neq i}}.$$

While this is the same physical quantity as

$$\mu_i = \left(\frac{\partial G}{\partial n_i}\right)_{T,P,n_{j\neq i}},$$

the function μ_i obtained from taking the derivative of the internal energy with respect to n_i is a function of variables $S, V, n_1, \ldots, n_m$ while the function μ_i obtained by taking the derivative of the Gibbs free energy with respect to n_i is a function of variables $T, P, n_1, \ldots, n_m$. These are two different mathematical functions describing the same physical quantity, i.e., the chemical potential of component species i in the multicomponent system.

We have, in general, $\mu_i = \left(\frac{\partial G}{\partial n_i}\right)_{T,P,n_{j\neq i}}$ and $G = \sum_{i=1}^{m} n_i \mu_i$ for a multicomponent system. From these we see that μ_i is the contribution per mole of species i to the Gibbs free energy of a multicomponent system at constant temperature and pressure. As a result of this relationship to the number of moles of i in the multicomponent system, μ_i is also referred to as the partial molar Gibbs free energy of species i in the mixture. This is sometimes given a separate symbol $\bar{g}_i$ with a bar at the top of the symbol:

$$\mu_i = \left(\frac{\partial G}{\partial n_i}\right)_{T,P,n_{j\neq i}} \equiv \bar{g}_i.$$

The partial molar quantity corresponding to any extensive quantity Z is similarly defined by:

$$\bar{z}_i \equiv \left(\frac{\partial Z}{\partial n_i}\right)_{T,P,n_{j\neq i}}.$$

Thus, the partial molar volume for species i is

$$\bar{v}_i \equiv \left(\frac{\partial V}{\partial n_i}\right)_{T,P,n_{j\neq i}}$$

where $V = V(T, P, n_1, ..., n_m)$ is the volume of a multicomponent system at T, P and consisting of $n_1, n_2, ..., n_m$ moles of each component. The partial molar volume is the rate of change of the volume with respect to the number of moles of component i, holding temperature, pressure and the number of moles of all other species constant. Therefore, any quantity Z can be written in terms of the corresponding partial molar quantities $\bar{z}_i$ for each component:

$$Z = \sum_{i=1}^{m} n_i \bar{z}_i.$$

In general, the partial molar quantity $\bar{z}_i$ for each component depends on the concentrations of all the components present. That is, in a system with m components

$$\bar{z}_i = \bar{z}_i \left(x_1, x_2, ... x_{m-1}\right).$$

Consider the case of the partial molar volume for a binary solution

$$\bar{v}_1 = \bar{v}_1(x_1)$$
$$\bar{v}_2 = \bar{v}_2(x_1)$$

since either of the two mole fractions completely specify the concentration; we have arbitrarily used x_1. Therefore, the volume of a binary solution consisting of, respectively, n_1 and n_2 moles of components 1 and 2 is

$$V = n_1 \bar{v}_1 + n_2 \bar{v}_2.$$

This gives a molar volume for the solution of

$$v = \frac{V}{(n_1 + n_2)} = x_1 \bar{v}_1 + (1 - x_1)\bar{v}_2.$$

In general, for any property Z for a binary system, the molar quantity is given by

$$z = x_1 \bar{z}_1 + x_2 \bar{z}_2.$$

The molar volume of the binary solution can, of course, be measured as a function of the mole fraction. Thus, we know $v = v(x_1)$. Hence, the partial molar volumes of the two components are related by

$$\bar{v}_2(x_1) = \frac{v(x_1) - x_1 \bar{v}_1(x_1)}{(1 - x_1)}.$$

Once the dependence of $\bar{v}_1$ upon concentration is known, the dependence of $\bar{v}_2$ upon concentration is completely determined. We will extend this property to systems consisting of more than two components in Section 8.8.2 below using the Gibbs-Duhem equation.

As noted above $\mu_i = \left(\dfrac{\partial U}{\partial n_i}\right)_{S,V,n_{j \neq i}}$ is the same physical quantity as $\left(\dfrac{\partial G}{\partial n_i}\right)_{T,P,n_{j \neq i}}$ but the former is not a partial molar quantity because the derivative is not taken holding temperature and pressure constant. The partial molar internal energy would be defined by $\bar{U}_i \equiv \left(\dfrac{\partial U}{\partial n_i}\right)_{T,P,n_{j \neq i}}$ and that is not equal to μ_i.

§   *Example 8.1 Partial molar volumes in a binary solution*

At constant temperature and pressure, the volume of a binary solution depends upon the number of moles of its components as follows:

$$V = n_1 v_1^* + n_2 v_2^* + \frac{a}{n} n_1 n_2$$

where n_1 and n_2 are the number of moles of the two components and $n = n_1 + n_2$ is the total number of moles.

i)   What physical meanings do the quantities v_1^*, v_2^* and a have? What are the units for each of them?

ii)   The mole fractions are

$$x_1 = n_1/n$$
$$x_2 = n_2/n.$$

Using these, show that the molar volume of the binary solution is

$$v = \frac{V}{n} = xv_1^* + x_2v_2^* + ax_1x_2.$$

iii) Using the definition of partial molar quantities,

$$\bar{v}_i \equiv \left(\frac{\partial V}{\partial n_i}\right)_{T,P,n_{j\neq i}}$$

obtain expressions for the partial molar volume of the two components in terms of n_1 and n_2. Pay attention when you take the partial derivatives.

iv)   Then write these expressions in terms of x_1 and x_2.

v)   Substitute your results in iv) for the partial molar volumes into

$$v = x_1\bar{v}_1 + x_2\bar{v}_2.$$

You should obtain the same expression for v as in part ii) above.

With the number of moles n_1 equal to the total number of moles and n_2 equal to zero, the volume of the system is $V = n_1v_1^* = nv_1^*$. Hence, the quantity v_1^* is the molar volume for the pure component 1. Similarly for v_2^*. Both v_1^* and v_2^* have units of volume per mole. The term $\frac{a}{n}n_1n_2$ is equal to the difference between the volume of the binary solution and the sum of the volumes of the pure components mixed to form the solution. The parameter a accounts for the effect of intermolecular interactions on the volume of the mixture. From the form of the equation for V, it has units of volume per mole.

The expression for molar volume can be written in terms of the mole fractions as

$$v = \frac{V}{n} = \frac{n_1}{n} v_1^* + \frac{n_2}{n} v_2^* + a \frac{n_1}{n} \frac{n_2}{n}$$

$$= x_1 v_1^* + x_2 v_2^* + a x_1 x_2.$$

The partial molar volume of component 1 is

$$\bar{v}_1 = \left(\frac{\partial V}{\partial n_1} \right)_{T,P,n_2}$$

$$= v_1^* + \frac{a n_2}{(n_1 + n_2)} - \frac{a n_1 n_2}{(n_1 + n_2)^2}$$

$$= v_1^* + a x_2^2.$$

Similarly,

$$\bar{v}_2 = v_2^* + a x_1^2.$$

The partial molar volumes tend to the respective molar volumes of the pure components when x_1 or x_2 tends to one. Using these expressions for the partial molar volumes, we get

$$v = x_1 \bar{v}_1 + x_2 \bar{v}_2$$

$$= x_1 v_1^* + a x_1 x_2 (x_1 + x_2) + x_2 v_2^*$$

$$= x_1 v_1^* + x_2 v_2^* + a x_1 x_2,$$

as we have above.

An important point to note here is that the partial molar volume depends only upon the composition of the mixture and not on how many total moles of the mixture you have. That is, $\bar{v}_1$ and $\bar{v}_2$ each depends on the ratios $\frac{n_1}{n}$ and $\frac{n_2}{n}$ but not upon n. Partial molar quantities are intensive, not extensive, properties of a system. §

8.6 Equilibrium condition: minimizing G in a multiphase system

As we have seen in Chapter 7, for a system at constant temperature and pressure, equilibrium is achieved when the Gibbs free energy of the system is minimized. We examine what this leads to in phase equilibrium. Consider a system consisting of one chemical species in two phases in equilibrium. For example, a system consisting of water vapour in equilibrium with liquid water. The fundamental equation is

$$dG = -SdT + VdP + \mu_{vap}dn_{vap} + \mu_{liq}dn_{liq}$$

where the subscripts vap and liq denote the vapour and liquid phases. If the system has not yet reached phase equilibrium, water molecules will transfer from the liquid phase to the vapour phase, or vice versa, so that an increase in the number of moles in one phase is equal to the decrease in the number of moles in the other phase:

$$dn_{vap} = -dn_{liq}.$$

Hence, at constant T and P, we have the following expression for the change in the Gibbs free energy:

$$dG = (\mu_{vap} - \mu_{liq})dn_{vap}.$$

The Second Law tells us that spontaneous processes bring a system toward equilibrium at constant T and P by decreasing the Gibbs free energy of the system toward its minimum. Thus, if $\mu_{vap} > \mu_{liq}$, then dn_{vap} must be negative: water is transferred spontaneously from the vapour phase which has a higher chemical potential to the liquid phase which has a lower chemical potential. Conversely if $\mu_{vap} < \mu_{liq}$, water molecules transfer spontaneously in the opposite direction from the liquid phase to the vapour phase. Since all spontaneous processes stop at equilibrium, the condition for vapour-liquid equilibrium in a one-component system is:

$$\mu_{vap} = \mu_{liq}.$$

This same idea can be simply extended to systems of any number of phases and any number of components. Equilibrium is reached when the chemical potential of each species is the same in all the co-existing phases:

$$\mu_i^1 = \mu_i^2 = \cdots = \mu_i^p$$

where i denotes the chemical species and the superscript denotes the phase the component is in. For a system with p phases, the chemical potential of each species is constrained by these $(p - 1)$ equations. We next use this equilibrium condition to deduce the number of independent intensive variables in a multicomponent multiphase system.

8.7 The Gibbs Phase Rule

In Section 2.2, we argued, simply by considering the equation of state of an ideal gas, that the amount of matter plus two variables such as P and V completely specify the state of a single-phase (homogeneous) single-component system. If we divide V by the number of moles in the system, we have the molar volume v which is an intensive variable. Specifying the pressure and the molar volume completely specifies the state of a homogeneous one-component system. Any other pair of intensive variables would also completely specify the state. The framework we have set up above provides a rigorous approach to count the minimum number of intensive variables needed to completely specify the state of a multicomponent, multiphase system. Thus, we can generalize what we had in Section 2.2 to such systems.

Consider a system with p phases and c chemical species. In each phase, the composition is completely specified by $(c - 1)$ mole fractions because $x_1^q + x_2^q + \cdots + x_c^q = 1$ in any phase q. Therefore, the number of mole fractions needed to completely specify the compositions of all p phases in a system is equal to:

$$p(c - 1) = cp - p.$$

To these composition variables, we add the variables T and P, giving a total number of intensive variables for the system that is equal to $(cp - p + 2)$. These $(cp - p + 2)$ intensive variables are not all independent. This is because at equilibrium the chemical potential of any one chemical species is the same in all phases. As seen in the previous section, this gives the $(p - 1)$ equations

$$\mu_i^1 = \mu_i^2 = \cdots = \mu_i^p$$

relating the temperature, pressure and the composition variables in the system since the chemical potentials are functions of these intensive variables. Hence, requiring that the system is in equilibrium imposes

$$c(p - 1) = cp - c$$

constraints on the system. With the $(cp - p + 2)$ intensive variables and the $(cp - c)$ equilibrium constraints relating these intensive variables, the total number of intensive variables that can be freely varied is:

$$f = (cp - p + 2) - (cp - c) = c - p + 2.$$

We refer to f as the number of degrees of freedom. This relationship between the number of degrees of freedom available to independently specify the value of intensive variables and the number of components c and phases p is known as the Gibbs Phase Rule.

8.8 The Gibbs-Duhem equation

8.8.1 How Gibbs derived his Phase Rule

We have seen in the previous section a derivation of the Gibbs Phase Rule from a consideration of the equilibrium condition for phase equilibrium. Here we examine a more general approach actually used by Gibbs. He considered that the thermodynamic state for each phase is specified by the temperature, pressure and the chemical potentials of the c components, namely $(c + 2)$ variables, taking the chemical potentials as the intensive variables specifying the state, rather than using the mole fractions as we have done above to specify the composition. Since at equilibrium each of these $(c + 2)$ variables is equal for all the phases present in the system, the state of the entire system is also completely specified by $(c + 2)$ intensive variables. The question is: how many of these $(c + 2)$ variables is independent in each phase. The discussion above on chemical potentials in a multicomponent system provides an elegant way to count the number of independent intensive variables. We have seen that for a multicomponent system:

$$G = \sum_{i=1}^{m} n_i \mu_i.$$

This equation holds for any one of the phases in a multiphase system. Thus, we have:

$$dG = \sum_{i=1}^{m} n_i \, d\mu_i + \sum_{i=1}^{m} \mu_i \, dn_i$$

simply by applying the product rule for derivatives. But we also have the fundamental equation for the Gibbs free energy:

$$dG = -S dT + V dP + \sum_{i=1}^{m} \mu_i \, dn_i.$$

Subtracting the latter of these two equations from the former gives us the Gibbs-Duhem equation:

$$0 = S dT - V dP + \sum_{i=1}^{m} n_i \, d\mu_i.$$

For all possible changes in temperature, pressure and the chemical potential of each component, this equation relates the differential quantities dT, dP and all the $d\mu_i$. That is, this equation provides a constraint on the variations in the $(c + 2)$ intensive variables of the system. For each phase in the system, there is one Gibbs-Duhem equation constraining the intensive variables. Thus, for a system with p phases there will be p constraints to be applied to the changes in these variables. Since there is a total of $(c + 2)$ variables, we are left with $(c + 2 - p)$ independent variables after applying these constraints. This is Gibbs original derivation of his Phase Rule. The Gibbs-Duhem equation is a general relationship between the intensive variables in any homogeneous phase of a multicomponent system. We will encounter it again when we consider specific examples of phase equilibria.

8.8.2 The Gibbs-Duhem equation for other partial molar quantities

Any extensive property Z depends upon temperature, pressure and the number of moles of the m components in the system. That is,

$$Z = Z(T, P, n_1, \dots n_m).$$

Therefore, simply from mathematics (not Thermodynamics) we can write

$$dZ = \left(\frac{\partial Z}{\partial P}\right)_{T,n_j} dP + \left(\frac{\partial Z}{\partial T}\right)_{P,n_j} dT + \sum_{i=1}^{m} \left(\frac{\partial Z}{\partial n_i}\right)_{T,P,n_{j \neq i}} dn_i.$$

If we consider systems at constant temperature and pressure, then we can drop the first two terms, leaving

$$dZ = \sum_{i=1}^{m} \left(\frac{\partial Z}{\partial n_i}\right)_{T,P,n_{j \neq i}} dn_i$$

$$= \sum_{i=1}^{m} \bar{z}_i dn_i$$

making use of the definition of the partial molar quantities $\bar{z}_i$. But we have already seen in Section 8.5 that

$$Z = \sum_{i=1}^{m} n_i \bar{z}_i.$$

Thus, taking the differential of this, gives us

$$dZ = \sum_{i=1}^{m} n_i d\bar{z}_i + \sum_{i=1}^{m} \bar{z}_i dn_i.$$

Comparing the two differentials dZ, we see that at constant temperature and pressure, we have

$$\sum_{i=1}^{m} n_i d\bar{z}_i = 0.$$

This is the general case for the Gibbs-Duhem equation which we encountered in Section 8.8.1 for the chemical potential which is

$$\sum_{i=1}^{m} n_i \, d\mu_i = 0.$$

Any physically allowed change in any set of partial molar properties of a system at constant temperature and pressure must satisfy its Gibbs-Duhem equation. As we have seen in the previous section, this places a constraint on the changes of, for example, the chemical potential leading to the Gibbs Phase Rule.

Now, consider changing the composition of a system at constant temperature and pressure by adding dn_1 moles to it. The rate of change of Z with respect to this change of n_1 would be

$$\left(\frac{\partial Z}{\partial n_1}\right)_{T,P} = \sum_{i=1}^{m} n_i \left(\frac{\partial \bar{z}_i}{\partial n_1}\right)_{T,P}$$

$$= \sum_{i=1}^{m} x_i \left(\frac{\partial \bar{z}_i}{\partial x_1}\right)_{T,P}$$

which is equal to zero from the Gibbs-Duhem equation. Thus, the changes in the partial molar quantities for the components in the system are related by

$$\sum_{i=1}^{m} x_i \left(\frac{\partial \bar{z}_i}{\partial x_1}\right)_{T,P} = 0.$$

For example, for a binary system, we have

$$x_1 \left(\frac{\partial \bar{z}_1}{\partial x_1}\right)_{T,P} + x_2 \left(\frac{\partial \bar{z}_2}{\partial x_1}\right)_{T,P} = 0,$$

noting that the derivatives are with respect to x_1 for both terms. In general, we also have the molar quantity z given by

$$z = x_1 \bar{z}_1 + x_2 \bar{z}_2,$$

for which the derivative with respect to x_1 gives

$$\left(\frac{\partial z}{\partial x_1}\right)_{T,P} = \bar{z}_1 - \bar{z}_2 + x_1 \left(\frac{\partial \bar{z}_1}{\partial x_1}\right)_{T,P} + x_2 \left(\frac{\partial \bar{z}_2}{\partial x_1}\right)_{T,P}.$$

From the Gibbs-Duhem equation the last two terms add up to zero, so that

$$\left(\frac{\partial z}{\partial x_1}\right)_{T,P} = \bar{z}_1 - \bar{z}_2.$$

Together with

$$z = x_1 \bar{z}_1 + x_2 \bar{z}_2.$$

We thus have two equations relating the partial molar quantities to the molar quantity z and its derivative $\left(\frac{\partial z}{\partial x_1}\right)_{T,P}$. Solving these for the partial molar quantities gives

$$\bar{z}_1 = z + x_2 \left(\frac{\partial z}{\partial x_1}\right)_{T,P}.$$

$$\bar{z}_2 = z - x_1 \left(\frac{\partial z}{\partial x_1}\right)_{T,P}.$$

These are rather useful relations for reading off partial molar quantities from experimental measurements of the dependence of the molar quantity z upon the mole fraction x_1.

To be concrete, let us consider the dependence of the molar volume upon mole fraction. The experimental data would consist of measurements of molar volume as a function of the mole fraction x_1 as illustrated by the solid line in Figure 8.1.

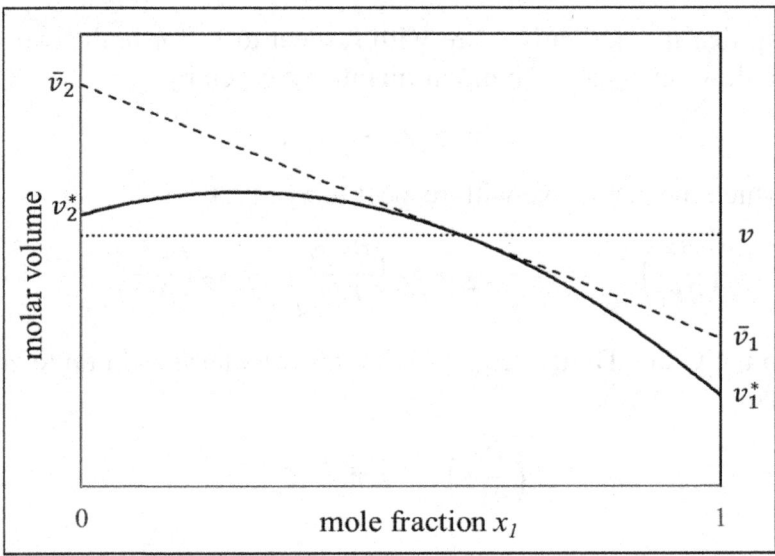

Figure 8.1 Illustration of the Gibbs-Duhem equation. The solid line is the molar volume of a binary system; the dashed tangent at $x_1 = 0.6$ intersects the vertical axes at the values of $\bar{v}_1$ and $\bar{v}_2$ for that mole fraction.

In this figure, we use an arbitrary model of the molar volume

$$v = x_1 v_1^* + x_2 v_2^* + a x_1 x_2$$

with the molar volumes of the pure components v_1^* and v_2^* illustrated, along with a positive value of the interaction parameter a. The Gibbs-Duhem analysis above gives the following equations for the partial molar volume at any specific mole fraction:

$$\bar{v}_1 = v + x_2 \left(\frac{\partial v}{\partial x_1}\right)_{T,P}$$

$$\bar{v}_2 = v - x_1 \left(\frac{\partial v}{\partial x_1}\right)_{T,P}$$

which are equations of the same straight line for v as a function of either x_1 or x_2, with a slope equal to $\pm \left(\frac{\partial v}{\partial x_1}\right)_{T,P}$ and intercepts of $\bar{v}_2$ or $\bar{v}_1$, respectively. This analysis shows that the values of the partial molar volumes for the two components are not independent of each other, and are readily obtained at any concentration from measurements of the molar volume of the mixture.

If the interaction parameter is zero, then the solution is ideal. The molar volume of the solution would then be linear with mole fraction. Then the partial molar volumes are independent of concentration and simply equal to the molar volume of the respective pure liquid:

$$\bar{v}_1 = v_1^*$$

$$\bar{v}_2 = v_2^*.$$

8.9 Phase diagrams for single-component systems

Consider a system of p phases containing only one chemical species. With $c = 1$, we have

$$f = c - p + 2 = 3 - p$$

giving the number of independent intensive variables. Since a one-component system has no composition variables, the total number of intensive variables is 2, i.e., just T and P. Armed with this we examine the phase diagram of a one-component system. With two phases, for example vapour and liquid, we have at equilibrium:

$$\mu_{vap}(T,P) = \mu_{liq}(T,P)$$

from the general condition for phase equilibrium. Since the chemical potentials are simply the molar Gibbs free energy, you will realise that we have already encountered this in our discussion in Section 7.10 of the role of the Gibbs free energy in determining the relative stability of vapour and liquid phases as temperature and pressure varies. For a one-component system with two phases, $f = 1$ leaves only one intensive variable that can be independently specified at equilibrium. Thus, if we specify T, then the pressure P is completely determined, and vice versa, specifying P completely determines the equilibrium temperature T. Hence, on a pressure-temperature phase diagram, a system in vapour-liquid equilibrium is represented by a line relating pressure to temperature. This line is the vapour-liquid equilibrium curve relating the boiling temperature to the pressure. Similarly, if the system has liquid in equilibrium with solid, a solid-liquid equilibrium curve relates the melting temperature to the pressure. A sublimation temperature versus pressure curve gives the pressure dependence of temperature for solid-gas equilibrium. These are, respectively, the vaporization, fusion and sublimation curves.

When three phases are present at equilibrium, the number of degrees of freedom is $f = 3 - p = 0$. This means that it is not possible to arbitrarily set the temperature and pressure of the triple point where solid, liquid and vapour co-exist. The triple point temperature and pressure are completely determined by the intermolecular interactions of the system. A corollary of this is that the three vaporization, fusion and sublimation curves all intersect at the triple point. Once the chemical species is specified, the temperature and pressure of the triple point is fully determined. For water, for example, this point is at $T = 273.16$ K and $P = 611.7$ N m^{-2} $\cong 0.006$ bar. We draw an illustrative generic phase diagram in Figure 8.2 for a one-component system.

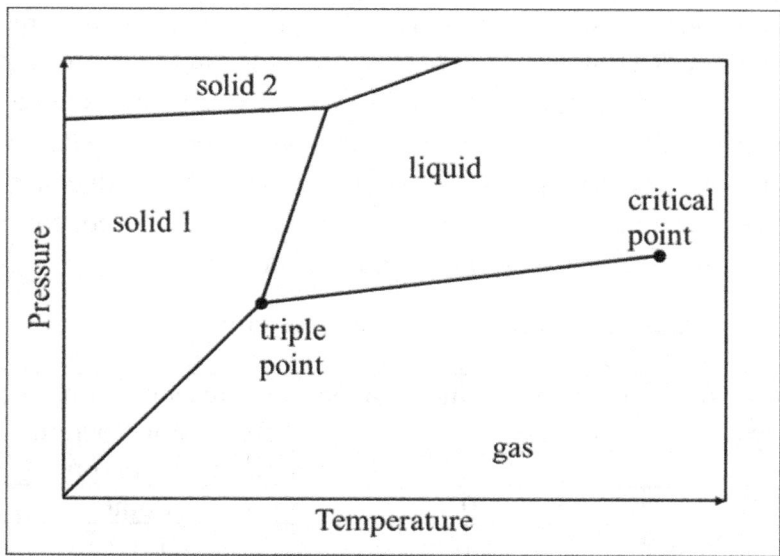

Figure 8.2: Schematic illustration of a one-component system phase diagram. Two different solid phases, the liquid and the gas phase are indicated. The vaporization line ends at the critical point.

The two-phase equilibrium curves are schematically drawn using straight-lines. We will discuss below how these curves depend upon temperature and pressure. In general, with different possible crystal structures for a substance, there will be equilibrium curves describing the pressure-temperature relationship for equilibrium between pairs of solid phases. We illustrate in the diagram two possible solid phases.

The vaporization curve, which divides the liquid from the gas, ends at the critical point. For temperatures and pressures above the critical point, there is no distinction between a liquid and its gas; this region of the phase diagram is sometimes labelled as supercritical fluid since there is no distinction between liquid and gas phases in this region. For water, the critical temperature and pressure are $T_c = 647.4$ K and $P_c = 218.3$ atm. In the regions of the phase diagram *solid* 1, *solid* 2, *liquid* and *gas*, only solid, liquid or gas exists, respectively. These are areas rather than lines because for a single-phase one-component system, the number of degrees of freedom from the Gibbs Phase Rule is $f = 2$. Thus, the pressure and the temperature can both be independently set to arbitrary values as long as these are within each of these single-phase regions. The Gibbs Phase Rule also tells us that there is no possibility of $p = 4$ giving a negative number of independent variables; in a one-component system it is not possible to have four phases co-existing in equilibrium.

To summarize, for a one-component system:

number of phases	number of independent variables f	Represented on the pressure-temperature phase diagram by:
3	0	Point
2	1	Line
1	2	Area

Thus, we can understand from Thermodynamic laws why all one-component phase diagrams have a general structure as illustrated in Figure 8.2. Water, for example has several crystalline phases at sufficiently high pressures so that the phase diagram has considerably more structure at high pressure than what we have schematically drawn above. However, the structure of the phase diagram and the basic principles that determine this structure are as illustrated in Figure 8.2 and is the same for all one-component systems.

8.10 The Clapeyron equation: equilibrium between two phases

We have seen the general structure of the phase diagram of a one-component system. What about the two-phase equilibria curves separating the single-phase regions of that phase diagram? How do we determine the equations of the fusion, vaporization and sublimation curves? For vapour-liquid equilibrium we have

$$\mu_{vap}(T, P) = \mu_{liq}(T, P)$$

along the vaporization curve. As we move along this curve, the equilibrium condition must remain satisfied. From one point to the next point along the vaporization curve the temperature and pressure changes by dT and dP. Using the fundamental equation for the Gibbs free energy, the differential change in the chemical potential of the vapour phase is:

$$d\mu_{vap} = -s_{vap}dT + v_{vap}dP$$

with s_{vap} and v_{vap} being the molar entropy and molar volume, respectively, of the vapour phase. Similarly, for the same dT and dP, the differential change in the chemical potential of the liquid phase is

$$d\mu_{liq} = -s_{liq}dT + v_{liq}dP.$$

In order to maintain the equilibrium condition of equal chemical potential for the two phases, we must have

$$d\mu_{vap} = d\mu_{liq}.$$

Hence, we have the following relationship between the molar entropies and molar volumes of the two phases:

$$-s_{vap}dT + v_{vap}dP = -s_{liq}dT + v_{liq}dP$$

giving,

$$dP = \frac{\left(s_{vap} - s_{liq}\right)}{\left(v_{vap} - v_{liq}\right)}dT$$

for any variation in the temperature and pressure along the vaporization curve. Therefore, the rate of change of pressure with temperature along the vaporization curve is given by

$$\left(\frac{dP}{dT}\right)_{\substack{vaporization \\ curve}} = \frac{\Delta_{vap}s}{\Delta_{vap}v}$$

where $\Delta_{vap}s$ and $\Delta_{vap}v$ are the molar entropy and the molar volume of vaporization, that is, the *change* in the entropy and volume upon vaporization of one mole of the substance. The subscript for the derivative indicates that it is the change of pressure with temperature *along* the vapour-liquid equilibrium curve. Using what we have learned about entropy and enthalpy we thus have:

$$\left(\frac{dP}{dT}\right)_{\substack{vaporization \\ curve}} = \frac{\Delta_{vap}s}{\Delta_{vap}v} = \frac{\Delta_{vap}q}{T\Delta_{vap}v} = \frac{\Delta_{vap}h}{T\Delta_{vap}v}$$

where $\Delta_{vap}h$ is the molar enthalpy of vaporization at a phase equilibrium pressure and temperature. This equation is the Clapeyron equation, and is exact. In the same way, the fusion and sublimation curves are given by:

$$\left(\frac{dP}{dT}\right)_{\substack{fusion \\ curve}} = \frac{\Delta_{fus}h}{T\Delta_{fus}v}$$

and,

$$\left(\frac{dP}{dT}\right)_{\substack{sublimation \\ curve}} = \frac{\Delta_{sub}h}{T\Delta_{sub}v}.$$

We illustrate these equilibrium curves in the vicinity of the triple point in the Figure 8.3 below. The left panel shows the typical positive slope $\left(\frac{dP}{dT}\right)$ for the fusion curve for most substances while the right panel shows a fusion curve with a negative slope $\left(\frac{dP}{dT}\right)$. The latter is the case for water which has a denser liquid than solid phase. For the vaporization curve $\Delta_{vap}h > 0$ since latent heat is always needed to vaporize a liquid, and $\Delta_{vap}v = v_{vap} - v_{liq} \cong v_{vap}$. Typically, v_{vap} is approximately three orders of magnitude larger than v_{liq}. Thus, we expect the slope of the vaporization curve to be positive and small.

For the fusion curve $\Delta_{fus}h > 0$ and $\Delta_{fus}v = v_{liq} - v_{solid} \gtrsim 0$. Typically, v_{liq} is just slightly larger than v_{solid}. Thus, where they meet at the triple point, we expect the slope of the fusion curve to be positive and significantly larger than the slope of the vaporization curve. One important exception is water, for which $\Delta_{fus}v \lesssim 0$ because ice is less dense than liquid water. Hence, for water the fusion curve has a negative slope, and the slope is significantly larger than that of its vaporization curve. For the sublimation curve $\Delta_{sub}h > 0$, and it is typically larger than $\Delta_{vap}h$. Similarly to the case of vaporization, the change in the volume upon sublimation $\Delta_{sub}v = v_{vap} - v_{solid} \cong v_{vap}$. Typically, v_{vap} is approximately three orders of magnitude larger than v_{solid}. Thus, we expect the slope of the sublimation curve to be positive and larger than the slope of the vaporization curve close to the triple point

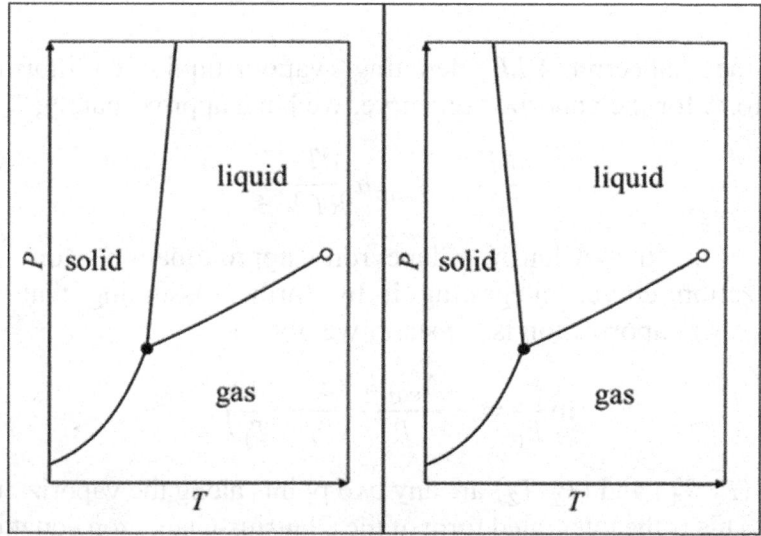

Figure 8.3: Vaporization, fusion and sublimation curves in the vicinity of the triple point indicated by the dot. The critical point is indicate by a circle. On the left for typical substances where $v_{liq} > v_{solid}$, on the right for substances such as water, silica, germanium and silicon where $v_{liq} < v_{solid}$.

8.11 Using the Clapeyron equation

8.11.1 The Clausius-Clapeyron equation

We can get simple approximate equations for the vaporization and sublimation curves if we apply a couple of approximations to the Clapeyron equation. For the vaporization curve, we can approximate the molar volume of vaporization to the molar volume of the vapour itself: $\Delta_{vap}v = v_{vap} - v_{liq} \cong v_{vap}$. This is because the vapour phase of a substance is typically much less dense than its liquid phase. We can then treat the vapour as an ideal gas so that a simple approximation for the volume of the vapour can be explicitly written $v_{vap} \cong \frac{RT}{P}$. Hence, the Clapeyron equation is approximated by

$$\left(\frac{dP}{dT}\right)_{VLE} = \frac{\Delta_{vap}h}{T\Delta_{vap}v} \cong \frac{\Delta_{vap}h}{\frac{RT^2}{P}}$$

with the subscript VLE denoting vapour-liquid equilibrium. Therefore, for the vaporization curve, we have approximately,

$$\frac{dP}{P} = \Delta_{vap}h\frac{dT}{RT^2}.$$

This is the Clausius-Clapeyron approximation for the vaporization curve. Integrating it by further assuming that the enthalpy of vaporization is constant, we obtain:

$$\ln\frac{P_2}{P_1} = -\frac{\Delta_{vap}h}{R}\left(\frac{1}{T_2} - \frac{1}{T_1}\right)$$

where (P_1, T_1) and (P_2, T_2) are any two points along the vaporization curve. This is the integrated form of the Clausius-Clapeyron equation. The same approximations, that is, $\Delta_{sub}v = v_{vap} - v_{solid} \cong v_{vap}$

and $v_{vap} \cong \frac{RT}{P}$, can be used for the sublimation curve, to give the Clausius-Clapeyron equation for sublimation:

$$\ln\frac{P_2}{P_1} = -\frac{\Delta_{sub}h}{R}\left(\frac{1}{T_2} - \frac{1}{T_1}\right).$$

The Clausius-Clapeyron equation is used to sketch the vaporization and sublimation curves in Figure 8.3. We will encounter it again when we learn more about vapour-liquid equilibrium in Chapter 11.

§   ***Example 8.2 How to boil water at room temperature***

At one atmosphere pressure, the boiling point of water is $100°C = 373.15\ K$. How low must the pressure be in order for water to boil at room temperature of 298 K?

The standard enthalpy of vaporization of water $\Delta_{vap}H^{\emptyset}$ is equal to 43.988 kJ mol^{-1}. This is approximately constant with pressure.

Hence, we have

$$T_1 = 373.15\ K$$

$$P_1 = 101325\ N\ m^{-2}$$

$T_2 = 298$ K. We use these values to calculate P_2 with the Clausius-Clapeyron equation:

$$\ln\frac{P_2}{101325} = -\frac{43988}{8.314}\left(\frac{1}{298} - \frac{1}{373.15}\right)$$

$\therefore P_2 = 2837\ N\ m^{-2} = 0.028$ atm

At a low pressure of about 3% of atmospheric pressure, water boils at room temperature of 25°C.

§   ***Example 8.3 Boiling water at the Everest base camp***

The Everest base camp is at an altitude of 5364 m, at which height the pressure is equal to 401 mmHg. One atmosphere pressure is 760 mmHg. Calculate the temperature of boiling water when you make your tea at the Everest base camp.

$$T_1 = 373.15 \text{ K}$$
$$P_1 = 760 \text{ mmHg}$$
$$P_2 = 401 \text{ mmHg}.$$

With the same enthalpy of vaporization as Example 8.2, we use the Clausius-Clapeyron equation to estimate T_2

$$\ln\left(\frac{401}{760}\right) = -\frac{43988}{8.314}\left(\frac{1}{T_2} - \frac{1}{373.15}\right).$$

Hence,

$$T_2 \cong 357 \text{ K} = 84°\text{C}.$$

This is several degrees lower than the optimal temperature for brewing Oolong. $\S$

8.11.2 Temperature-dependent $\Delta_{vap}H$

In arriving at the Clausius-Clapeyron equation, we have assumed that enthalpy of vaporization is a constant. How is the change in the enthalpy of vaporization with temperature considered, if necessary? We start with the fundamental equation for enthalpy

$$dH = TdS + VdP$$
$$= T\left(\frac{\partial S}{\partial T}\right)_P dT + \left[T\left(\frac{\partial S}{\partial P}\right)_T + V\right]dP$$
$$= T\left(\frac{\partial S}{\partial T}\right)_P dT + \left[-T\left(\frac{\partial V}{\partial T}\right)_P + V\right]dP$$

where we have expanded the entropy S as a function of T and P in the second line and used the Maxwell relation $\left(\frac{\partial S}{\partial P}\right)_T = -\left(\frac{\partial V}{\partial T}\right)_P$ in the third line. Applying this expression to the difference in the enthalpies of the vapour and the liquid phases, we get the enthalpy of vaporization in terms of the molar entropy and molar volume of vaporization:

$$d(\Delta_{vap}h) = T\left(\frac{\partial \Delta_{vap}s}{\partial T}\right)_P dT + \left[-T\left(\frac{\partial \Delta_{vap}v}{\partial T}\right)_P + \Delta_{vap}v\right]dP.$$

The quantity $\Delta_{vap}v$ is the difference between the molar volume of the vapour and the molar volume of the liquid.

Assuming that the vapour phase is an ideal gas, only the terms involving the molar volume of the liquid are left in the square brackets because the two terms involving the molar volume of the ideal gas cancel. Hence, we have:

$$d(\Delta_{vap}h) = T\left(\frac{\partial \Delta_{vap}s}{\partial T}\right)_P dT - \left[-T\left(\frac{\partial v_{liq}}{\partial T}\right)_P + v_{liq}\right]dP$$

$$= T\left(\frac{\partial \Delta_{vap}s}{\partial T}\right)_P dT - v_{liq}\left[-\frac{T}{v_{liq}}\left(\frac{\partial v_{liq}}{\partial T}\right)_P + 1\right]dP$$

$$\cong T\left(\frac{\partial \Delta_{vap}s}{\partial T}\right)_P dT$$

$$= (C_P^{vap} - C_P^{liq})dT$$

where we have dropped the term in dP assuming that the liquid molar volume and its isobaric expansion coefficient are small. For water, for example, $\frac{1}{v_{liq}}\left(\frac{\partial v_{liq}}{\partial T}\right)_P$ is approximately 2×10^{-4} at room temperature.

We can thus use Hess's law to calculate $\Delta_{vap}h(T)$. Taking the heat capacities to be independent of temperature, we get:

$$\Delta_{vap}h(T) - \Delta_{vap}h(T_0) = \int_T^{T_0} C_P^{liq} dT + + \int_{T_0}^T C_P^{vap} dT$$

$$= \int_{T_0}^T (C_P^{vap} - C_P^{liq})dT$$

$$= (C_P^{vap} - C_P^{liq})(T - T_0)$$

$$= \Delta_{vap}C_P(T - T_0)$$

where $\Delta_{vap}C_P \equiv (C_P^{vap} - C_P^{liq})$, T_0 is the standard temperature and $\Delta_{vap}h(T_0)$ is the molar enthalpy of vaporization at the standard temperature and pressure.

If the heat capacities are constant at $C_P^{vap} = 33.58$ J K^{-1} mol^{-1} and $C_P^{liq} = 75.29$ J K^{-1} mol^{-1}, then the enthalpy of vaporization of water at 100°C is 40.860 kJ mol^{-1}, compared to 43.988 kJ mol^{-1} at 25 °C. Using our expression for $\Delta_{vap}h(T)$ in the differential Clausius-Clapeyron equation, we have

$$\frac{dP}{P} = \Delta_{vap}h \frac{dT}{RT^2}$$

$$= \left[\frac{\Delta_{vap}h(T_0)}{RT^2} + \frac{\Delta_{vap}C_P}{R}\left(\frac{1}{T} - \frac{T_0}{T^2}\right)\right]dT.$$

Integrating this from (T_1, P_1) to (T_2, P_2) we obtain a slightly better approximation for the vaporization curve:

$$\ln\left(\frac{P_2}{P_1}\right) = -\frac{1}{R}[\Delta_{vap}h(T_0) - T_0\Delta_{vap}C_P]\left(\frac{1}{T_2} - \frac{1}{T_1}\right)$$
$$+ \frac{\Delta_{vap}C_P}{R}\ln\left(\frac{T_2}{T_1}\right).$$

§   *Example 8.4 A better estimate of the vapour pressure of water*

Using this approximation, estimate the pressure required to have water boil at room temperature.

The normal boiling point of water is $T_1 = 373.15$ K at a pressure of $P_1 = 101325$ N m^{-2}. The difference in the heat capacity is

$$\Delta_{vap}C_P = C_P^{vap} - C_P^{liq} = -41.71 \text{ J K}^{-1} \text{ mol}^{-1}.$$

Therefore, we have

$$\ln\left(\frac{P_2}{101325}\right) = -\frac{1}{R}[43988 + (298 \times 41.71)]\left(\frac{1}{298} - \frac{1}{373.15}\right)$$
$$-\frac{41.71}{8.314}\ln\left(\frac{298}{373.15}\right),$$

giving

$$P_2 = 3192 \text{ N m}^{-2}$$

compared to the measured value of 3173 N m^{-2}. Similarly, a better estimate can be made of the temperature of the temperature of boiling water at the Everest base camp; however, a numerical method of solution might be needed in that case. §

8.11.3 A non-ideal vapour phase

We can treat vapour-liquid equilibrium by numerically integrating the Clapeyron equation

$$\left(\frac{dP}{dT}\right)_{\substack{vaporization \\ curve}} = \frac{\Delta_{vap}h}{T\Delta_{vap}v}$$

even if the vaporization enthalpy varies with temperature and the vapour phase is not well approximated by an ideal gas. For this, we need better estimates of $\Delta_{vap}h$ and $\Delta_{vap}v$ as the temperature and pressure vary along the vapour-liquid equilibrium curve.

Consider the equilibrium illustrated by the solid curve in Figure 8.4. For pressures in the region above the curve the equilibrium state is a liquid, while for pressures below the curve the equilibrium state is a vapour. Given the enthalpies of the liquid and the vapour phases at point A with temperature and pressure (T_1, P_1), let us estimate the enthalpies of each phase at the point C, also on the equilibrium curve but with temperature and pressure equal to (T_2, P_2). The liquid is in a well-defined thermodynamic equilibrium state at every point above the equilibrium curve. Similarly for the vapour below the equilibrium curve.

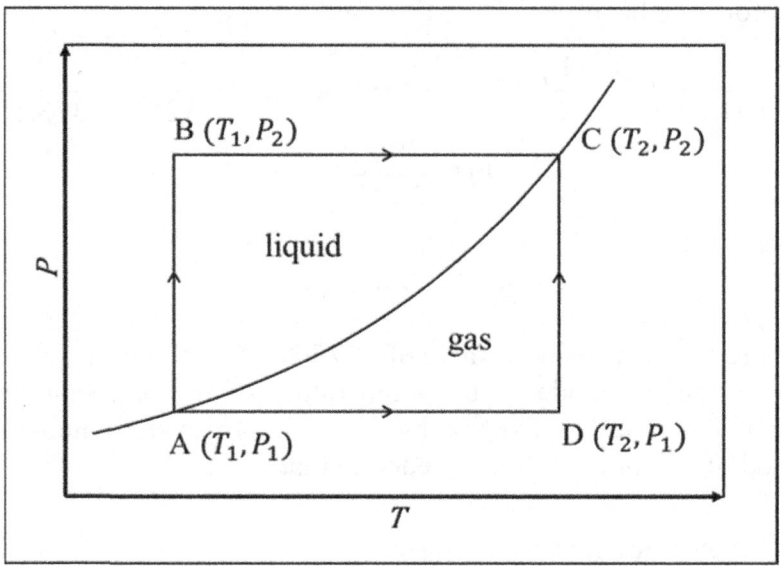

Figure 8.4: Paths used in integrating the fundamental equation for enthalpy to determine the enthalpy change for vaporization.

At every point along the path ABC, the liquid phase is in a well-defined thermodynamic equilibrium state. Therefore, we can integrate the changes in enthalpy of the liquid phase along this path from state A through state B to state C as follows:

$$h_{liq}(T_2, P_2) - h_{liq}(T_1, P_1)$$

$$= \int_{(T_1, P_1)}^{(T_1, P_2)} v_{liq}(T_1, P)\, dP + \int_{(T_1, P_2)}^{(T_2, P_2)} C_P^{liq}(T, P_2)\, dT.$$

The first term is an integral along AB over the pressure from P_1 to P_2 holding the temperature constant at T_1. The second term is an integral along BC over the temperature from T_1 to T_2 holding the pressure constant at P_2. The quantities $v_{liq}(T_1, P)$ and $C_P^{liq}(T, P_2)$ are well-defined equilibrium properties along AB and BC, respectively. Similarly, for the vapour phase along path ADC:

$$h_{vap}(T_2, P_2) - h_{vap}(T_1, P_1)$$

$$= \int_{(T_1,P_1)}^{(T_2,P_1)} C_P^{vap}(T, P_1)dT + \int_{(T_2,P_1)}^{(T_2,P_2)} v_{vap}(T_2, P) \, dP.$$

The first term is an integral over the temperature from T_1 to T_2 holding the pressure constant at P_1. The second term is an integral over the pressure from P_1 to P_2 holding the temperature constant at T_2.

Both of these equations for h_{liq} and h_{vap} follow directly from the fundamental equation for enthalpy

$$dH = TdS + VdP$$

$$= C_P dT + VdP$$

with no approximations. From these equations, the enthalpy of vaporization at any point along the equilibrium curve can be determined, albeit numerically, from calorimetric and PVT data, together with the enthalpy of vaporization at state A. Thus, the numerical value of $\Delta_{vap}h(T, P) = h_{vap}(T, P) - h_{liq}(T, P)$ can be determined at an appropriate number of points along the equilibrium curve AC.

Similarly, from

$$v_{vap}(T_2, P_2) - v_{vap}(T_1, P_1)$$

$$= \int_{(T_1,P_1)}^{(T_2,P_1)} v_{vap}(T, P_1)\beta_{vap}(T, P_1)dT$$

$$+ \int_{(T_2,P_1)}^{(T_2,P_2)} v_{vap}(T_2, P)\kappa_{vap}(T_2, P) \, dP$$

and

$$v_{liq}(T_2, P_2) - v_{liq}(T_1, P_1)$$

$$= \int_{(T_1,P_1)}^{(T_1,P_2)} v_{liq}(T_1, P)\kappa_{liq}(T_1, P)\, dP$$

$$+ \int_{(T_1,P_2)}^{(T_2,P_2)} v_{liq}(T, P_2)\beta_{liq}(T, P_2)dT$$

the volume change upon vaporization $\Delta_{vap}v(T, P) = v_{vap}(T, P) - v_{liq}(T, P)$ can be obtained. Here $\beta = \frac{1}{V}\left(\frac{\partial V}{\partial T}\right)_P$ and $\kappa = \frac{1}{V}\left(\frac{\partial V}{\partial P}\right)_T$ are the isobaric expansion coefficient and the isothermal compressibility, respectively.

Hence, both factors $\Delta_{vap}h(T, P)$ and $\Delta_{vap}v(T, P)$ on the right-hand side of the Clapeyron equation can be determined directly from thermodynamic data at any point along the vapour-liquid equilibrium curve. With these quantities the vaporization curve can be calculated by numerically integrating the Clapeyron equation. This gives a more accurate vapour pressure dependence upon temperature than our approximations from the Clausius-Clapeyron equation, although more work is needed. As this discussion of phase equilibrium illustrates, Thermodynamics provides the general framework for relating equilibrium macroscopic properties to each other. In applying this general framework to any material, the dependence of C_P^{vap}, C_P^{liq}, v_{vap}, v_{liq}, β_{vap}, β_{liq}, κ_{vap}, κ_{liq} upon the temperature and pressure for example, has to be measured experimentally or obtained from theoretical modelling of the specific material.

9

Simple Models

In learning a new physical theory, it is always easier to have accessible concrete models to apply the theory to. We offer some such models in decreasing levels of familiarity in this chapter. Knowing the Gibbs free energy as a function of temperature and pressure provides all thermodynamic information. Thus, our approach is to get to expressions for the chemical potential of various systems. In Chapter 1 we introduced the ideal gas and the van der Waals gas as model substances used to illustrate the calculation of thermodynamic properties. We have used the equations of state of these substances in various examples along the way. At this point we have a more sophisticated framework for thermodynamics than can be usefully illustrated with only the ideal gas and the van der Waals gas. Therefore, we apply this framework to a few model substances/systems which are slightly more sophisticated than the ideal gas. These model systems will be used in further applications in later chapters.

9.1 μ for ideal gas mixtures

Since there are no intermolecular interactions in ideal gases, one might think that the thermodynamic properties of a mixture of ideal gases are the same as those of a single-component ideal gas. However, a mixture of ideal gases has a composition variable that distinguishes a particular mixture from other mixtures at the same T and P. Indeed, the Gibbs Phase rule tells us that for a two-component mixture in the gas phase, the number of independent intensive variables specifying the thermodynamic state is:

$$f = c - p + 2 = 2 - 1 + 2 = 3.$$

That is, aside from temperature and pressure, the mole fraction of one of the components, for example, is needed to identify the thermodynamic state. But what thermodynamic properties of a mixture are dependent upon its composition? To address this question, let us examine a mixture of two ideal gases A and B. Consider the system shown in Figure 9.1 consisting of two compartments of gases separated by a rigid, immovable partition which is permeable to only gas A.

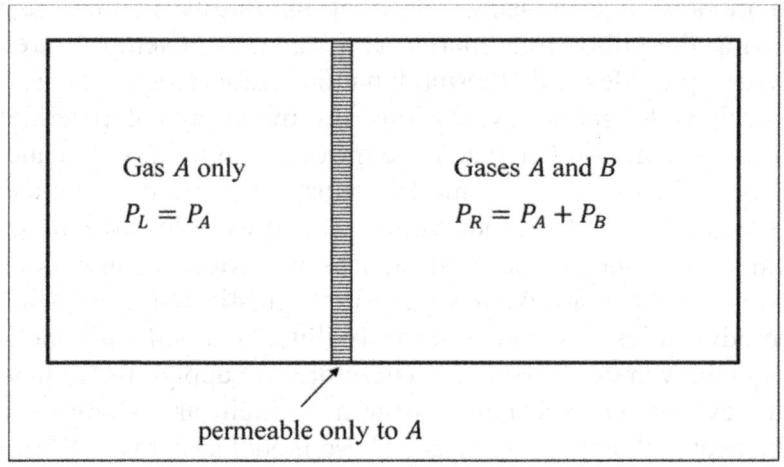

Figure 9.1 Left compartment contains only gas A, while right compartment contains gases A and B. The rigid semi-permeable membrane separating the two compartments is permeable only to gas A.

The temperature T is the same for both compartments. In the left compartment there is only gas A, while in the right compartment both gases A and B are present with mole fraction of A equal to y_A. Since the partition is rigid but permeable to only A, the pressures in the two compartments need not be equal when the flow of gases through the partition is allowed to equilibrate. The pressure in the right compartment P_R is given by the sum of the partial pressures of A and B,

$$P_R = P_A + P_B,$$

where the partial pressures are

$$P_A = y_A P_R$$
$$P_B = y_B P_R = (1 - y_A)P_R.$$

The pressure on the left is simply P_L, all due to gas A. When the system gets to equilibrium, the pressure of A in both compartments must be equal because gas A traverses the partition. Hence, we have

$$P_L = P_A.$$

We can consider this equilibrium system as consisting of two phases: in the left compartment, the pure gas A, and in the right compartment a different phase consisting of the mixture. When this system gets to equilibrium, we have from Chapter 7 that the chemical potential of A must be the same in the two compartments:

$$\mu_A^L(T, P_L) = \mu_A^R(T, P_R, y_A)$$

where the superscripts R and L denote the right and left compartments. We have explicitly indicated the total pressures and temperatures for the two compartments, and the composition for the right compartment.

Since the left compartment contains only A, the chemical potential $\mu_A^L(T, P_L)$ is really just the molar Gibbs free energy of pure A at temperature T and pressure $P_L = P_A$. The pressure dependence of the chemical potential of pure A is thus given by:

$$\mu_A^L(T, P_L) = \mu_A^L(T, P_0) + \int_{P_0}^{P_L} v\, dP$$
$$= \mu_A^L(T, P_0) + RT \ln\left(\frac{P_L}{P_0}\right),$$

where in the first line, we are integrating over pressure, from an arbitrary standard pressure P_0 to the pressure of the system P_L, making use of $\left(\frac{\partial G}{\partial P}\right)_T = V$ from the fundamental equation for the Gibbs free energy; $\mu_A^L(T, P_0)$ is the chemical potential of the system at pressure P_0. In the second line, we have used the ideal gas equation of state for the molar volume $v = \frac{RT}{P}$. We rewrite this equation with a superscript $*$, to denote that this is the chemical potential for a system containing only pure A, at temperature T and pressure P, generally:

$$\mu_A^*(T, P) = \mu_A^*(T, P_0) + RT \ln\left(\frac{P}{P_0}\right).$$

Since the two compartments are in equilibrium, the chemical potential of A in the right compartment with the mixture must also be given by this same expression, that is:

$$
\begin{aligned}
\mu_A^R(T, P_R, y_A) &= \mu_A^*(T, P_0) + RT \ln\left(\frac{P_L}{P_0}\right) \\
&= \mu_A^*(T, P_0) + RT \ln\left(\frac{P_R}{P_0}\frac{P_L}{P_R}\right) \\
&= \mu_A^*(T, P_0) + RT \ln\left(\frac{P_R}{P_0}\right) + RT \ln\left(\frac{P_L}{P_R}\right) \\
&= \mu_A^*(T, P_0) + RT \ln\left(\frac{P_R}{P_0}\right) + RT \ln\left(\frac{P_A}{P_R}\right) \\
&= \mu_A^*(T, P_0) + RT \ln\left(\frac{P_R}{P_0}\right) + RT \ln y_A \\
&= \mu_A^*(T, P_R) + RT \ln y_A.
\end{aligned}
$$

That is, the chemical potential of component A in an ideal gas mixture at temperature T and total pressure P, is equal to the chemical potential of pure A at the same temperature and pressure, plus a term depending on the natural logarithm of the mole fraction of A. In line 4, we note that $P_L = P_A$; the pressure in the left compartment is equal to the partial pressure of A in the right compartment. In line 5, we used Dalton's law of partial pressures for the ideal gas mixture, and in line 6, we used the expression for the chemical potential of a pure ideal gas. Rewriting this, the chemical potential of component A in an ideal gas mixture where temperature is T, total pressure is P, and the mole fraction of A is y_A is:

$$\mu_A(gas, T, P, y_A) = \mu_A^*(gas, T, P) + RT \ln y_A$$
$$= \mu_A^*(gas, T, P) + RT \ln \frac{P_A}{P}$$

where we have added "gas" to explicitly indicate that this is for the gas phase. We will use this explicit notation when we need to be clear in systems consisting of both liquid and gas phases.

By expressing the chemical potential of the pure gas at pressure P in terms of its chemical potential at an arbitrary standard pressure P_0, this equation is frequently written as

$$\mu_A(gas, T, P, y_A) = \mu_A^*(gas, T, P_0) + RT \ln \frac{P}{P_0} + RT \ln y_A$$
$$= \mu_A^*(gas, T, P_0) + RT \ln \frac{P_A}{P_0}$$

using Dalton's law of partial pressures in the second line.

9.2 Entropy of mixing

As an important detour, we apply this to the mixing of two ideal gases. Indeed, through this application, we can begin to understand why the chemical potential of a component in an ideal mixture depends upon the logarithm of the mole fraction. The mixing process is illustrated in Figure 9.2. Initially, two ideal gases are separated by an impermeable partition. In the left compartment we have n_A moles of A occupying volume V_A, and in the right compartment n_B moles of B occupying volume V_B. The gases in both compartments are at the same temperature T and pressure P, so that:

$$PV_A = n_A RT$$
$$PV_B = n_B RT.$$

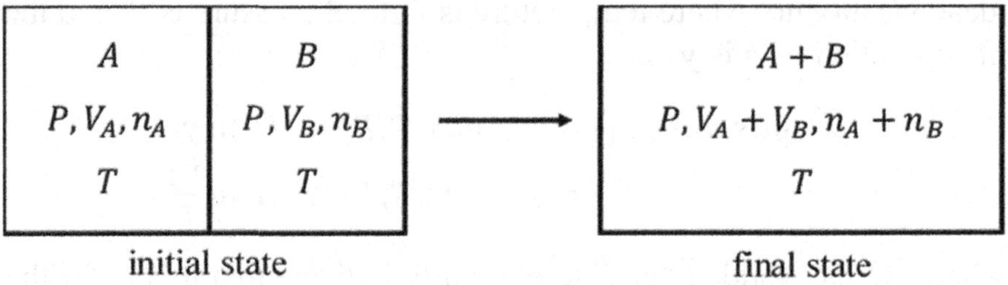

Figure 9.2: Mixing of ideal gases A and B upon removal of partition.

Upon removal of the partition, and the gases are mixed, we have $n = n_A + n_B$ moles of ideal gas occupying a volume $V = V_A + V_B$ with temperature T and P, since $P(V_A + V_B) = (n_A + n_B)RT$. What is the change in the Gibbs free energy of this process? Before mixing, the chemical potentials are:

$$\mu_A^*(T,P) = \mu_A^*(T,P_0) + RT \ln \left(\frac{P}{P_0}\right)$$

in the left compartment with n_A moles of pure A, and

$$\mu_B^*(T,P) = \mu_B^*(T,P_0) + RT \ln \left(\frac{P}{P_0}\right)$$

in the right compartment with n_B moles pure B. At the end of the mixing process, the chemical potentials of the components in the mixture are:

$$\mu_A(T,P,y_A) = \mu_A^*(T,P) + RT \ln y_A$$
$$\mu_B(T,P,y_B) = \mu_B^*(T,P) + RT \ln y_B.$$

Therefore, the change in the Gibbs free energy upon mixing is:

$$\Delta_{mix}G = [n_A\mu_A(T,P,y_A) + n_B\mu_B(T,P,y_B)]$$
$$-[n_A\mu_A^*(T,P) + n_B\mu_B^*(T,P)]$$
$$= n_A RT \ln y_A + n_B RT \ln y_B$$
$$= (n_A + n_B)RT(y_A \ln y_A + y_B \ln y_B).$$

Since mole fractions are always less than unity, the logarithms are negative so that the Gibbs free energy of mixing is always negative. Mixing of gases is therefore a spontaneous process. You will never observe the spontaneous unmixing of an ideal gas mixture into its pure component gases for any macroscopic system.

It is worthwhile to examine the expression for $\Delta_{mix}G$ a little bit more. From the definition of the Gibbs free energy, we have

$$\Delta_{mix}G = \Delta_{mix}(U + PV - TS)$$
$$= \Delta_{mix}U + P\Delta_{mix}V - T\Delta_{mix}S$$

since P and T are constant. We are dealing with ideal gases, so that $\Delta_{mix}U = 0$ because the temperature remains constant. We also have a constant total volume, $\Delta_{mix}V = 0$. Thus, we have, for mixing ideal gases at constant temperature and pressure

$$\Delta_{mix}G = -T\Delta_{mix}S.$$

With $\Delta_{mix}G$ negative, we have a positive mixing entropy. Hence, per mole of the mixture, the entropy of mixing is equal to

$$\Delta_{mix}S = -R(y_A \ln y_A + y_B \ln y_B).$$

And per molecule in the system:

$$\Delta_{mix}S = -k_B(y_A \ln y_A + y_B \ln y_B),$$

where k_B is the Boltzmann constant. This formula is known as the Gibbs formula for the entropy of a mixture. It connects the entropy with the mole fractions of its components. In Section 6.1, we encountered the Boltzmann formula $S = k_B \ln W$, which connects entropy to the number of microstates available to a system. As we stated in that section, the entropy is proportional to the logarithm of the number of possible microscopic arrangements of a system given its thermodynamic state. You will learn in statistical thermodynamics that the quantity $(-y_A \ln y_A - y_B \ln y_B)$ is equal to the logarithm of the number of ways of arranging A and B molecules in a random mixture with mole fractions y_A and y_B. Consider arranging N_R indistinguishable red and N_B indistinguishable black marbles in a row. The number of ways W of doing this is $\frac{(N_R+N_B)!}{N_R!N_B!}$ from combinatorics. If N_R and N_B are as large as the number of molecules in a macroscopic box of gas, then using Stirling's formula for the logarithm of a factorial,

$$\ln W = \ln\left[\frac{(N_R + N_B)!}{N_R!\,N_B!}\right]$$

$$\cong N_{tot}\left[-\frac{N_R}{N_{tot}}\ln\frac{N_R}{N_{tot}} - \frac{N_B}{N_{tot}}\ln\frac{N_B}{N_{tot}}\right],$$

to an excellent approximation, where $N_{tot} = N_R + N_B$ is the total number of marbles. Thus, per marble,

$$\ln W = -y_R \ln y_R - y_B \ln y_B,$$

with y_R and y_L being the fraction of red and black marbles. The number of ways of mixing indistinguishable marbles is the same as the number of ways of mixing non-interacting gas molecules. In an ideal mixture, the molecules are non-interacting and thus the arrangements are random. This is not the case for real gases where the number of arrangements will be modified by the intermolecular interactions. The equations for the Boltzmann and Gibbs entropy are fundamental in statistical thermodynamics, providing the link between the macroscopic and microscopic views of matter.

Because there are no intermolecular interactions in an ideal gas, the Gibbs formula for entropy is a general expression for the entropy arising solely from the fact that mixing increases the number of microstates for arranging the molecules randomly in a system. For a multicomponent system we sum over all the components, so that the entropy of mixing in terms of the mole fractions y_i is

$$\Delta_{mix}S = -k_B \sum_i y_i \ln y_i.$$

§   *Example 9.1 Entropy of mixing*

A system consists of a box of gas at room temperature T and pressure P with two compartments separated by an impermeable partition. One compartment is filled with one litre of oxygen. The other compartment is filled with two litres of nitrogen. When the partition is removed, the gases mix. Do the internal energy and the enthalpy of the system change? Treating the gases as ideal, calculate the change in the entropy of the system. Calculate the change in the entropy of the universe.

The gases in the two compartments are at the same temperature and pressure. Thus, using the ideal gas equation of state, mixing the gases do not change the temperature and pressure in the system. The internal energy and the enthalpy remain constant. The change in the entropy per molecule of the system is given by

$$\Delta_{mix}S = -k_B \left(\frac{1}{3}\ln\frac{1}{3} + \frac{2}{3}\ln\frac{2}{3}\right)$$

$$= 0.64k_B$$

since the mole fractions of oxygen and nitrogen are $\frac{1}{3}$ and $\frac{2}{3}$, respectively. The internal energy does not change, and the system does no work since volume is constant. Thus, no heat interactions occur with the surroundings. Thus, the entropy change of the surroundings is zero. Hence, the entropy change of the universe is equal to $0.64k_B$ per molecule of the gas, which is a positive quantity. The mixing of gases is spontaneous because this entropy of mixing dominates the weak intermolecular interactions in gases. §

9.3 μ for ideal solutions: Raoult's law

As a second model substance with multiple components, we discuss the ideal solution. "Ideal" here means that the molecules in the solution have the same intermolecular interactions each would have in their pure liquids; here we are thinking only of liquid solutions although the same idea can be applied to solids. The ideal solution is not the same as the ideal gas mixture where the intermolecular interactions are all zero. Since the interactions experienced by each molecule in an ideal solution is the same as those it would experience in its pure liquid, we expect that the thermodynamics of the ideal solution reflects the entropy, rather than the energy, of mixing of the component liquids. Since interactions are not affected by the mixing, $\Delta_{mix}U$ and $\Delta_{mix}V$ are zero, as in the ideal gas mixture discussed in the previous section: only the entropy of mixing plays a role when the composition of the solution changes. Therefore, we guess that the chemical potential of any component A in an ideal solution is given by:

$$\mu_A(liq, T, P, x_A) = \mu_A^*(liq, T, P) + RT \ln x_A,$$

where $\mu_A^*(liq, T, P)$ is the chemical potential of the pure liquid A, and the $\ln x_A$ term arises from the entropy of mixing, with x_A being the mole fraction of A in the solution. Let us deduce this expression by putting the ideal solution into equilibrium with its vapour as illustrated in Figure 9.3, with the vapour phase being a mixture of ideal gases.

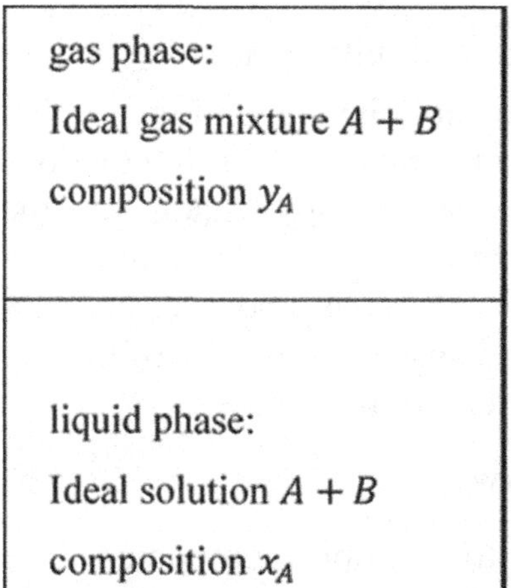

| gas phase: Ideal gas mixture $A + B$ composition y_A |
| liquid phase: Ideal solution $A + B$ composition x_A |

Figure 9.3: Equilibrium between an ideal liquid solution and an ideal gas mixture of components A and B.

The compositions of the gas and liquid phases in equilibrium at temperature T and pressure P are given by the mole fractions y_A and x_A, respectively. For each component in the system we have at equilibrium:

$$\mu_A(liq, T, P, x_A) = \mu_A(gas, T, P, y_A).$$

Writing $\mu_A(gas, T, P, y_A)$ using what we have deduced above for the chemical potential in an ideal gas mixture,

$$\mu_A(liq, T, P, x_A) = \mu_A^*(gas, T, P) + RT \ln\left(\frac{P_A}{P}\right).$$

Additionally, if we have a solution for which the equilibrium partial pressure of A is given by the product of its mole fraction and its vapour pressure P_A^*, then

$$x_A P_A^* = P_A.$$

Such solutions are said to obey Raoult's law and are called ideal solutions. This is an empirical law which is valid for some solutions. Thus, for ideal solutions

$$\mu_A(liq, T, P, x_A) = \mu_A^*(gas, T, P) + RT \ln\left(\frac{P_A^*}{P}\right) + RT \ln x_A.$$

This relationship also holds for a system where pure liquid A is in equilibrium with a mixed ideal gas which contains A along with other "non-condensable" components which are not present in the liquid phase. In that case, the mole fraction in the liquid phase x_A is equal to unity and the partial pressure P_A is equal to P_A^*. An example is liquid water in equilibrium with a mixture of water vapour and hydrogen gas, the latter of which is rather insoluble in water, which means the liquid phase is pure water to a good approximation. Then the pressure P acting on the liquid is the sum of the partial pressures of water vapour and hydrogen, that is $P = P_{H_2O} + P_{H_2}$. Setting the mole fraction x_A to unity, we have

$$\mu_A^*(liq, T, P) = \mu_A^*(gas, T, P) + RT \ln\left(\frac{P_A^*}{P}\right)$$

where the asterisk on the left-hand side emphasizes that it now describes the pure liquid. The right-hand side of this equation is the chemical potential of pure gas phase A at a pressure of P_A^*. The left-hand side is the chemical potential of the pure liquid phase A at any pressure P. Thus, using Raoult's law, we are approximating the chemical potential of the liquid phase to be independent of the pressure. We use this relationship to write the chemical potential of pure gas A in terms of the chemical potential of its pure liquid:

$$\mu_A^*(gas, T, P) = \mu_A^*(liq, T, P) - RT \ln\left(\frac{P_A^*}{P}\right).$$

Putting this back into the general expression for the chemical potential of A in an ideal solution, we have

$$\mu_A(liq, T, P, x_A) = \mu_A^*(gas, T, P) + RT \ln\left(\frac{P_A^*}{P}\right) + RT \ln x_A$$
$$= \mu_A^*(liq, T, P) + RT \ln x_A$$

giving the expression we guessed at the start of this section, and which is completely in terms of liquid phase properties.

We quickly run the above discussion in reverse, starting with the chemical potentials, and obtaining Raoult's law from them. We have the following chemical potentials for the ideal gas mixture and the ideal solution:

$$\mu_A(gas, T, P, y_A) = \mu_A^*(gas, T, P_A^*) + RT \ln\frac{P_A}{P_A^*}$$
$$\mu_A(liq, T, P, x_A) = \mu_A^*(liq, T, P) + RT \ln x_A.$$

For the gas phase chemical potential, we have used the reference pressure of P_A^* for later convenience. At any temperature and pressure for which there is gas-liquid equilibrium, we equate the chemical potentials on the left-hand side. This gives, upon some rearrangement of the terms on the right-hand side,

$$\frac{P_A}{x_A} = P_A^* \exp\left(\frac{\mu_A^*(liq, T, P) - \mu_A^*(gas, T, P_A^*)}{RT}\right).$$

The right-hand side is independent of the composition of either the liquid or the gas phase because the μ^* terms are the chemical potentials of pure A in gas and liquid phases. If the chemical potential of the pure liquid can be taken to be independent of pressure, then we can replace $\mu_A^*(liq, T, P)$ in the exponent by $\mu_A^*(liq, T, P_A^*)$. But the gas and liquid phases of pure A are in equilibrium at P_A^*, so that the exponent

$$\frac{\mu_A^*(gas, T, P_A^*) - \mu_A^*(liq, T, P_A^*)}{RT} = 0.$$

Thus, the exponential factor is equal to unity and we recover Raoult's law:

$$\frac{P_A}{x_A} = P_A^*.$$

Hence, the empirical Raoult's law applies for an ideal solution chemical potential of the form

$$\mu_A(liq, T, P, x_A) = \mu_A^*(liq, T, P) + RT \ln x_A$$

with a pressure independent pure liquid chemical potential.

§   *Example 9.2 Raoult's law: gas-liquid equilibrium of pentane and hexane*

You have an equimolar mixture of pentane and hexane in the liquid phase at room temperature. This is in equilibrium with a gas phase consisting only of pentane and hexane. Assuming both liquid and gas phases to be ideal, what is the total pressure of the gas phase, and its pentane mole fraction?

At room temperature the vapour pressure of pure pentane is 425 mmHg and that for hexane is 151 mmHg. With the higher vapour pressure, pentane is the more volatile component, and thus we expect that its mole fraction in the gas phase will be larger than that of hexane. Using Raoult's law, the partial pressure of pentane in the gas phase is

$$P_{pent} = x_{pent} P_{pent}^* = 0.5 \times 425 = 212.5 \text{ mmHg}.$$

Similarly, for hexane:

$$P_{hex} = x_{hex} P_{hex}^* = 0.5 \times 151 = 75.5 \text{ mmHg}.$$

Thus, the total pressure $\left(P_{pent} + P_{hex}\right)$ is equal to 288 mmHg, and the mole fraction of pentane in the gas phase is

$$y_{pent} = \frac{P_{pent}}{\left(P_{pent} + P_{hex}\right)} = 0.738.$$

As expected, the mole fraction of the more volatile component is larger in the gas phase in equilibrium with an equimolar liquid phase. In Chapter 11 we will see that this enrichment of the more volatile component in the gas phase is the basis for separation of the components in a liquid mixture using distillation. §

§   *Example 9.3 Pure liquid-mixed gas equilibrium*

Assume that nitrogen is not soluble in hexane. (Its equilibrium mole fraction in hexane is approximately 1.4×10^{-3} at 298.15 K.) What is the mole fraction of nitrogen in a gas phase mixture of nitrogen and hexane in equilibrium with liquid hexane at a total pressure of 1 atm which is equal to 760 mmHg at room temperature?

The vapour pressure of hexane is 151 mmHg, so that a pure hexane liquid has a partial pressure of hexane equal to

$$P_{hex} = x_{hex}P^*_{hex} = 1.0 \times 151 = 151 \text{ mmHg}.$$

Since the total pressure is 760 mmHg, then the partial pressure of nitrogen is $760 - 151 = 609$ mmHg. Hence, the mole fraction of nitrogen in the gas phase is equal to $\frac{609}{760} = 0.8013$. To increase the total pressure of the system at constant temperature, the partial pressure and mole fraction of nitrogen have to increase since the partial pressure of hexane stays at 151 mmHg assuming Raoult's law for the solution. §

   This is a system where we have a pure liquid phase in equilibrium with a mixed gas phase. With the liquid phase being pure A, we have $x_A = 1$ so that

$$\mu_A(liq, T, P, x_A) = \mu^*_A(liq, T, P) + RT \ln x_A$$

$$= \mu_A^*(liq, T, P)$$
$$= \mu_A(gas, T, P, y_A).$$

Note that the pressure P in this equation is not equal to the vapour pressure P_A^* because the gas phase has a non-condensable component, y_A being less than unity, which adds to the pressure of the gas phase. Indeed, the pressure of the gas phase may be much larger than the vapour pressure of pure A, depending upon how much of the non-condensable component is present. We have seen in Section 7.10 that the Gibbs free energy depends upon the pressure through

$$V = \left(\frac{\partial G}{\partial P}\right)_T.$$

Since volume is a positive quantity, the chemical potential increases with pressure. Thus, the chemical potential of a pure liquid in equilibrium with a gas phase with a non-condensable component actually increases with the pressure P according to

$$\mu_A^*(liq, T, P) = \mu_A^*(liq, T, P_0) + v_A(P - P_0)$$

where v_A is the molar volume of liquid A and P_0 is some standard pressure. We neglected the second, pressure-dependent term, in our discussion of the ideal solution above. The chemical potential of an ideal solution depends upon the mole fraction according to the term $RT \ln x_A$. We compare the magnitudes of these terms for the system in Example 9.2.

§   *Example 9.4 Pressure dependence of the chemical potential*

Revisit the pentane-hexane gas-liquid equilibrium system in Example 9.1 with the same total pressure of 288 mmHg. The molar volume of hexane is 130.7 cm³ mol⁻¹. How large is the contribution of the pressure-dependent term compared to the other contributions to the chemical potential of pure liquid hexane?

Including its pressure dependence, the chemical potential of hexane in the liquid phase is written as

$$\mu_{hex}(liq, T, P, x_{nex}) = \mu_{hex}^*(liq, T, P) + RT \ln x_{hex}$$
$$= \mu_{hex}^*(liq, T, P_{hex}^*) + v_{hex}(P - P_{hex}^*)$$
$$+ RT \ln x_{hex}.$$

The gas phase chemical potential is

$$\mu_{hex}(gas, T, P, x_{hex})$$
$$= \mu_{hex}^*(gas, T, P_{hex}^*) + RT \ln \frac{P}{P_{hex}^*} + RT \ln y_{hex}.$$

We use the vapour pressure of pure hexane at temperature T as the reference pressure for the pure liquid and gas hexane so that the terms $\mu_{hex}^*(liq, T, P_{hex}^*)$ and $\mu_{hex}^*(gas, T, P_{hex}^*)$ are equal and cancel. Let us consider the magnitudes of each of the remaining terms:

$$v_{hex}(P - P_{hex}^*) = 130.7 \times 10^{-6} \left(\frac{288 - 151}{760} \right) 101325 \text{ J}$$
$$= 2.4 \text{ J}.$$

You will notice that this first term is three orders of magnitude smaller than the following terms.

$$RT \ln x_{hex} = 8.314 \times 298.15 \times \ln 0.5 \text{ J}$$
$$= -1718.2 \text{ J}$$
$$RT \ln \frac{P}{P_{hex}^*} = 8.314 \times 298.15 \times \ln \frac{288}{151} \text{ J}$$
$$= 1600.5 \text{ J}$$
$$RT \ln y_{hex} = RT \ln \frac{x_{hex} P_{hex}^*}{P}$$

$$= 8.314 \times 298.15 \times \ln \left(\frac{0.5 \times 151}{288}\right)$$

$$= -3318.7 \text{ J.}$$

The last three terms exactly cancel out when Raoult's law is applied, that is

$$RT \ln x_{hex} = RT \ln \frac{P}{P_{hex}^*} + RT \ln y_{hex}.$$

This is because with Raoult's law, we have

$$P_{hex} = x_{hex} P_{hex}^* = y_{hex} P.$$

Hence, using Raoult's law, we make an approximation that neglects the 2.4 J from the first term. This term is of considerably smaller magnitude compared to the composition dependence of the liquid given by the $RT \ln x_{hex}$. This gives a sense of the (reasonably) small numerical error involved in neglecting the pressure dependence of the liquid phase chemical potential.

§ *Example 9.5 Composition versus pressure effects*

Calculate the change in the chemical potential of liquid ethanol at 298.15 K when its pressure changes from its vapour pressure to 1 atm. You have an equimolar solution of ethanol and water at 298.15 K and 1 atm. Assume it behaves as an ideal solution. Calculate the change in the chemical potential of ethanol in this solution relative to that of pure liquid ethanol at the same temperature and pressure.

At a temperature of 298.15 K, the vapour pressure of pure ethanol is 133.3 Nm^{-2} and its density is 0.788 g cm^{-3}. The molar mass of ethanol is 46.08 g mol^{-1}. Thus, the change in chemical potential due to the change in the pressure is

$$v_m (P - P_{eth}^*) = \frac{46.08}{0.788 \times 10^6} (101325 - 133.3) \text{ J mol}^{-1}$$

$$= 5.9 \text{ J mol}^{-1}.$$

The change in the chemical potential when the solution concentration decreases from pure ethanol to a mole fraction of 0.5 is equal to

$$RT \ln x_{eth} = 8.314 \times 298.15 \times \ln(0.5)$$
$$= -1718.2 \text{ J mol}^{-1}.$$

Typically, the chemical potential of a solution changes much more rapidly with composition than with the pressure. Compare this with Example 9.4 to see that the composition dependence is independent of the identity of the chemical species. §

To summarize this section, we can define the standard chemical potential of component A in an ideal solution to be equal to the chemical potential of the pure liquid at the pressure of the system

$$\mu_A^{\varnothing}(liq, T, P) = \mu_A^*(liq, T, P).$$

With Raoult's law, the general expression for the chemical potential of A in an ideal solution is approximated as

$$\mu_A(liq, T, P, x_A) = \mu_A^{\varnothing}(liq, T, P) + RT \ln x_A.$$

The pressure variation of the standard state chemical potential is neglected. For pressures on the order of a few atmospheres this is not an unreasonable approximation as we have seen through Examples 9.4 and 9.5.

The same definition is generally used for an ideal solid solution. For any component A in an ideal solid solution

$$\mu_A(solid, T, P, x_A) = \mu_A^{\varnothing}(solid, T, P) + RT \ln x_A$$

where x_A is the mole fraction of A in the solid solution and

$$\mu_A^{\varnothing}(solid, T, P) = \mu_A^*(solid, T, P)$$

is the chemical potential of the pure solid at T and P.

9.4 Fugacity: μ for non-ideal gases

At constant temperature, we integrate $\left(\frac{\partial \mu}{\partial P}\right)_T = v$ with respect to pressure, using the ideal gas equation of state $Pv = RT$ to obtain the dependence of its chemical potential upon pressure:

$$\mu^*(T, P) = \mu^*(T, P_0) + RT \ln\left(\frac{P}{P_0}\right).$$

Because non-ideal gases have a more complicated equation of state, the pressure dependence arising from the integral of the molar volume will not be this same logarithmic term in P. However, we can keep such a form for the pressure dependence of a non-ideal gas by defining a new quantity called the fugacity f which has the same units as pressure. The defining equation for this is:

$$\mu^*(T, P) = \mu^\emptyset(T, P_0) + RT \ln\left(\frac{f}{P_0}\right)$$

where $\mu^\emptyset(T, P_0)$ depends only upon temperature, and $\frac{f}{P} \to 1$ when $P \to 0$; note that the pressure dependence on the right-hand side is through f. Thus, a real gas is in its standard state when its fugacity, not its pressure, is equal to the standard pressure P_0. Because the fugacity of any gas is equal to its pressure at sufficiently low pressures when all gases behave ideally, at that limit $\mu^\emptyset(T, P_0) = \mu^*(T, P_0)$. The term fugacity shares the same etymological root as the word "fugitive", and thus fugacity refers, quite appropriately, to the tendency of the gas to escape. A fugacity smaller than the pressure indicates that the attractive intermolecular interactions are dominant over the repulsive interaction, giving the gas a smaller "tendency to escape" than if it were an ideal gas. The ratio $\frac{f}{P_0} = a$, is dimensionless and is called the activity. Rewriting this defining equation to include the pressure P, we have:

$$\mu^*(T, P) = \mu^\emptyset(T, P_0) + RT \ln a = \mu^\emptyset(T, P_0) + RT \ln\left(\frac{f}{P_0}\right)$$

$$= \mu^{\emptyset}(T, P_0) + RT \ln\left(\frac{P}{P_0}\right) + RT \ln\left(\frac{f}{P}\right)$$

$$= \mu^{\emptyset}(T, P_0) + RT \ln\left(\frac{P}{P_0}\right) + RT \ln \phi$$

where $\phi = \frac{f}{P}$ is the fugacity coefficient. If the gas is ideal, the first two terms on the right-hand side exactly describe its chemical potential. Thus, the third term with the fugacity coefficient captures all the non-ideal contributions to the chemical potential, accounting for the effects of intermolecular interactions in a non-ideal gas. When the pressure approaches zero the fugacity coefficient approaches unity so that this third term is equal to zero; real gases behave ideally as pressure tends to zero. The fugacity coefficient is a function of the temperature and pressure.

We can explicitly calculate the fugacity coefficient as follows. Take the derivative of μ^* with respect to pressure holding the temperature constant:

$$\left(\frac{\partial \mu^*}{\partial P}\right)_T = \frac{\partial}{\partial P}\left(\mu^{\emptyset}(T, P_0) + RT \ln\left(\frac{P}{P_0}\right) + RT \ln\left(\frac{f}{P}\right)\right)_T$$

$$= \frac{RT}{P} + RT \frac{\partial}{\partial P}\left(\ln\left(\frac{f}{P}\right)\right)_T.$$

But we know that $\left(\frac{\partial \mu}{\partial P}\right)_T = v$, from the fundamental equation for the Gibbs free energy. Therefore, equating our derivative to the molar volume v, and integrating over pressure (with dummy variable for pressure P'), we have:

$$RT \ln\left(\frac{f}{P}\right) = \int_0^P \left(v - \frac{RT}{P'}\right) dP',$$

where at the lower limit of zero pressure, we have $\frac{f}{P} = 1$ from the definition of the fugacity. This is the defining equation for the fugacity of any gas. It is clear from the integrand that the fugacity depends upon the difference between the pressure dependence of the actual molar volume v and the ideal gas molar volume $\frac{RT}{P}$. For an ideal gas, the molar volume is given by $v = \frac{RT}{P}$ so that the integrand is zero, and the fugacity is always equal to the pressure. This is not the case for real gases where intermolecular interactions may not always be ignored. The molar volume v is generally a function of the temperature and the pressure, and thus fugacity depends on T and P. This is how the fugacity, and then the chemical potential, of any gas can be calculated from experimental PVT-data. We illustrate this for the van der Waals equation of state.

§   ***Example 9.6 Fugacity of a van der Waals gas***

Derive an expression for the fugacity of a van der Waals gas in terms of its molar volume and temperature.

Doing an integration by parts for the molar volume term in the integrand,

$$\int_0^P \left(v - \frac{RT}{P'}\right) dP' = vP'|_0^P - \int_{v(0)}^{v(P)} P'\, dv - RT(\ln P')|_0^P$$

where the upper limit of the integral on the right-hand side is the molar volume at pressure P, and the lower limit of the integral is the molar volume as $P \to 0$. Using the van der Waals equation of state: $\left(P + \frac{a}{v^2}\right)(v - b) = RT$, the first term on the right-hand side can be written in terms of the molar volume as:

$$vP'|_0^P = v\left(\frac{RT}{v - b} - \frac{a}{v^2}\right) - RT$$
$$= RT\left(\frac{v}{v - b} - \frac{a}{v} - 1\right)$$

because in the lower limit as $P \to 0$, a van der Waals gas behaves ideally and $vP = RT$. The second term is:

$$-\int_{v(0)}^{v(P)} P'\, dv = -\int_{v(0)}^{v(P)} \left\{\left(\frac{RT}{v-b}\right) - \frac{a}{v^2}\right\} dv$$

$$= -RT\{\ln(v-b)\}\big|_{v(0)}^{v(P)} - \frac{a}{v}$$

where we have used $\frac{a}{v} = \frac{aP}{RT} = 0$ as $P \to 0$ for the lower limit. Adding to this the third term, we get

$$-RT\{\ln(v-b)\}\big|_{v(0)}^{v(P)} - \frac{a}{v} - RT(\ln P')\big|_0^P$$

$$= -RT \ln\left(\frac{P(v-b)}{RT}\right) - \frac{a}{v}$$

$$= -RT \ln\left(1 - \frac{a(v-b)}{v^2 RT}\right) - \frac{a}{v}$$

where in the second line we have combined terms from the corresponding integration limits of the two logarithmic terms, noting that $P(v-b) = Pv = RT$, when $P \to 0$. In the third line, we have used the van der Waals equation to write P in terms of v. Hence, the fugacity of the van der Waals gas is given by:

$$\ln\left(\frac{f}{P}\right) = \left(\frac{b}{v-b}\right) - \ln\left(1 - \frac{a(v-b)}{v^2 RT}\right) - \frac{2a}{v}.$$

The graph of this equation is the solid line in Figure 9.4 below for the van der Waals approximation of the equation of state of water, where $a = 5.464$ atm L^{-2} mol^{-2} and $b = 3.05 \times 10^{-2}$ L mol^{-1} at a temperature of 298.15 K. The fugacity coefficient $\phi = \frac{f}{P}$ tends to unity from below when the pressure approaches zero. For pressures of a few atmospheres, the attractive intermolecular interactions in van der Waals water dominates the repulsive interaction. Its chemical potential is less than what it would be if water were an ideal gas in this pressure range. This is the solid line in Figure 9.4. §

Conversely, the fugacity coefficient is greater than unity if repulsive intermolecular interactions dominate, for example for a gas with equation of state $P(v - b) = RT$, from which it is rather less tedious than the van der Waals equation of state to obtain the expression for the fugacity:

$$RT \ln\left(\frac{f}{P}\right) = \int_0^P \left(v - \frac{RT}{P'}\right) dP' = bP.$$

For a purely repulsive van der Waals model for water, we keep the value for b and set a to zero. Since bP is always positive, the fugacity is always larger than the pressure. From this equation, we obtain the dashed line in Figure 9.4.

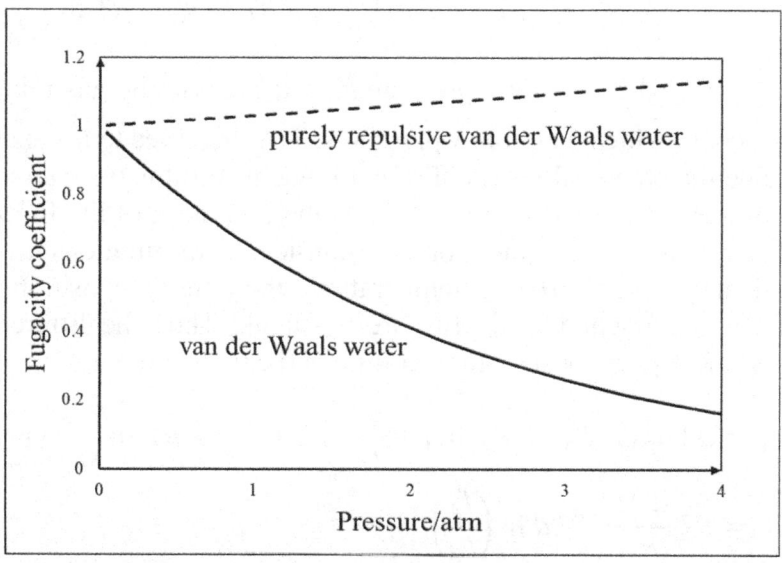

Figure 9.4: The solid line is the fugacity coefficient $\phi = f/P$ of the van der Waals approximation for water vapour at 298.15 K and pressures close to 1 atm. The dashed line is f/P for a model gas with equation of state $P(v - b) = RT$ with the same value of b as van der Waals water.

Helium has $a = 0.0341$ atm L^{-2} mol^{-2} and $b = 2.38$ L mol^{-1}. This means that the repulsive intermolecular interaction dominates in helium; its fugacity at normal temperatures is larger than unity like the dashed line in Figure 9.4. The graphs in Figure 9.4 are again consistent with the intuitive sense of fugacity as the "tendency to escape".

Following exactly the same approach, the equation for the chemical potential of component A in a non-ideal gas mixture is given by replacing its partial pressure by its fugacity:

$$\mu_A(T, P, y_A) = \mu_A^{\varnothing}(T, P_0) + RT \ln\left(\frac{f_A}{P_0}\right)$$

$$= \mu_A^{\varnothing}(T, P_0) + RT \ln\left(\frac{P_A}{P_0}\right) + RT \ln\left(\frac{f_A}{P_A}\right)$$

$$= \mu_A^{\varnothing}(T, P_0) + RT \ln\left(\frac{P}{P_0}\right) + RT \ln y_A + RT \ln\left(\frac{f_A}{P_A}\right)$$

with $\frac{f_A}{P_A} = 1$ when the total pressure $P \to 0$ because the gas mixture approaches an ideal gas at low pressures. The first three terms exactly describe the chemical potential of component A if this were an ideal gas mixture. The fourth term, which is the logarithm of the fugacity coefficient for A accounts for the non-ideal contributions to the chemical potential. If the temperature and the composition are constant, the first and third terms are constants. Thus, the differential of $\mu_A(T, P, y_A)$ at constant temperature and composition is:

$$d\mu_A = d\left(\mu_A^{\varnothing}(T, P_0) + RT \ln\left(\frac{P}{P_0}\right) + RT \ln y_A + RT \ln\left(\frac{f_A}{P_A}\right)\right)_{T, y_A}$$

$$= RT \frac{dP}{P} + RT d \ln\left(\frac{f_A}{P_A}\right),$$

remembering that the fugacity is a function of pressure. From Section 9.3, we know that the derivative of chemical potential at constant temperature and composition is the molar volume

$$\left(\frac{\partial \mu_A}{\partial P}\right)_{T, n_1, \dots, n_m} = v_A.$$

Therefore, we can determine f_A the fugacity of component A by integrating PVT-data:

$$RT \ln \left(\frac{f_A}{P_A}\right) = \int_0^P \left(d\mu_A - RT \frac{dP'}{P'}\right) = \int_0^P \left(v_A - \frac{RT}{P'}\right) dP'.$$

From this, the chemical potential of each component in a non-ideal gas mixture can be determined experimentally or from an equation of state as in Example 9.6.

9.5 Non-ideal solutions

9.5.1 Activity: μ for non-ideal solutions

Considering just the entropy of mixing $\Delta_{mix}S$ is not sufficient to describe real solutions. Due to significantly differing intermolecular interactions between the components in non-ideal solutions, the energy of mixing $\Delta_{mix}U$ and volume of mixing $\Delta_{mix}V$ cannot be neglected. Hence, the $RT \ln x$ term that exactly describes the entropy of mixing in the chemical potential of ideal solutions needs to be modified or augmented to account for non-ideal behaviour.

The way to proceed is motivated by noticing that even in non-ideal solutions the ideal solution description is a good approximation when the solution is dilute. Arbitrarily selecting A as the solvent, this is the limit where the mole fraction $x_A \to 1$ and the mole fraction of any solute $x_B \to 0$. Then, the chemical potential of each component is of the form

$$\mu(T, P, x) = \mu^{\emptyset}(T, P) + RT \ln x$$

where $\mu^{\emptyset}$ is the standard chemical potential of the component and x is its mole fraction. Given this, it makes sense to define the chemical potentials of non-ideal solutions with an expression of the same structure as that for ideal solutions, keeping the composition dependence through a term corresponding to $\ln x$. Hence, we write the chemical potential of component A in a non-ideal solution as

$$\mu_A(liq, T, P, \{x_i\}) = \mu_A^{\emptyset}(liq, T, P) + RT \ln a_A$$
$$= \mu_A^{\emptyset}(liq, T, P) + RT \ln \gamma_A x_A$$
$$= \mu_A^{\emptyset}(liq, T, P) + RT \ln x_A + RT \ln \gamma_A$$

where $a_A \equiv \gamma_A x_A$ is the activity of species A in the solution and γ_A is the activity coefficient. Both a_A and γ_A are dimensionless. Since $\ln a_A$ is zero when the activity is equal to unity, writing the chemical potential in this form means that we are setting the reference $\mu_A^{\emptyset}(liq, T, P)$ to be the chemical potential when $a_A = 1$. For ideal gases and solutions, the standard state is well-defined by specifying just the pressure and temperature of the pure gases and liquids. For non-ideal systems, we need to extend the definition of the standard state to specify a reference state with unit activity.

Although we include pressure as an argument of $\mu_A^{\emptyset}$, liquids are typically rather incompressible so that the pressure dependence is weak. Comparing the form of the equation in the first line to that for an ideal solution, you see that the activity a_A is a generalised mole fraction. The ideal solution is exactly described by the first two terms in the third line, so that the all the non-ideal contributions to the chemical potential are accounted for by the $RT \ln \gamma_A$ term. The activity coefficient γ_A is a function of the temperature, pressure and the concentrations of all the species present in the solution $\{x_i\}$, not just of x_A. That is,

$$a_A \equiv \gamma_A x_A = x_A \gamma_A(T, P, \{x_i\}).$$

In the same way that the entropy contributions arising from intermolecular interactions in a non-ideal gas mixture are accounted for by the logarithm of the fugacity coefficient, the logarithm of the activity coefficient completely captures all entropy contributions in a non-ideal solution. Thus, for the non-ideal gas we already have, from Section 9.4

$$\mu_A(T, P, y_A) = \mu_A^*(T, f_A) + RT \ln\left(\frac{f_A}{P_0}\right) = \mu_A^{\emptyset}(T, P_0) + RT \ln a_A.$$

Since we can readily determine the fugacity of a gas from PVT-data, we can quite straightforwardly determine the state of a real gas corresponding to unit activity, i.e. when $f_A = P_0$.

The general form of the chemical potential in terms of activity does not completely specify the meaning of each term in it. From the form of the equation, $\mu_A^\emptyset$ is for a standard state with an activity equal to unity. But, how do we measure the activity? Which physical state of the solution does $a_A = 1$ correspond to? For a non-ideal gas, the activity is equal to unity when its fugacity is equal to the standard pressure P_0. What about solutions? Different choices are made depending upon the system. We discuss some of these choices here.

9.5.2 Solvent standard state; Raoult's law

From our results in Sections 9.4 and 9.5.1 for non-ideal gas mixtures and solutions, equilibrium at T and P between two such phases is described by

$$\mu_A(gas, T, P, y_A) = \mu_A^\emptyset(gas, T, P_0) + RT \ln\left(\frac{f}{P_0}\right)$$
$$= \mu_A^\emptyset(liq, T, P) + RT \ln a_A$$
$$= \mu_A^\emptyset(liq, T, P) + RT \ln x_A + RT \ln \gamma_A.$$

In the limit as the solution and the gas phase goes to pure A, we then have

$$\mu_A^*(gas, T, P) = \mu_A^\emptyset(liq, T, P) + RT \ln \gamma_A = \mu_A^*(liq, T, P).$$

The molecular environment of any component A in a solution varies linearly with x_A in both limits when x_A approaches one or zero, so that in both limits, the solution behaves ideally in the sense that the activity is equal to x_A. Therefore, we can set the standard chemical potential for the solution to be

$$\mu_A^\emptyset(liq, T, P) = \mu_A^*(liq, T, P)$$

along with $\gamma_A \to 1$ in the limit when $x_A \to 1$. With this choice of $\mu_A^\emptyset(liq, T, P)$ the activity is $a_A = \gamma_A x_A$, so that

$$\mu_A^\emptyset(gas, T, P_0) + RT \ln\left(\frac{f_A}{P_0}\right) = \mu_A^\emptyset(liq, T, P) + RT \ln \gamma_A x_A.$$

Thus, at gas-liquid equilibrium

$$\frac{f_A/P_0}{\gamma_A x_A} = exp\left[-\frac{\left(\mu_A^\emptyset(gas, T, P_0) - \mu_A^\emptyset(liq, T, P)\right)}{RT}\right] \equiv \kappa$$

with κ independent of composition by the definitions of the standard state chemical potentials. Taking the limit where x_A and γ_A go to one, and f_A equal to P_A we have,

$$\kappa = \frac{f_A}{P_0} = \frac{P_A}{P_0} = \frac{P_A^*}{P_0}$$

since the partial pressure of A for the pure liquid A is equal to its vapour pressure P_A^*. Therefore, in general we have

$$f_A = \gamma_A x_A P_A^*.$$

f_A, x_A and P_A^* are directly measurable, so that this relationship can be used to determine activity coefficients of solutions. The activity is thus $a_A = \gamma_A x_A \equiv \frac{f_A}{P_A^*}$. Specifically, for ideal solution in equilibrium with an ideal gas phase,

$$P_A = x_A P_A^*$$

which is Raoult's law discussed in Section 9.3. In summary, one convention to describe the chemical potential in a non-ideal solution is

$$\mu_A(liq, T, P, x_A) = \mu_A^\emptyset(liq, T, P) + RT \ln \gamma_A x_A$$

where the activity coefficient $\gamma_A \to 1$ as $x_A \to 1$, and the standard state chemical potential is simply the pure liquid chemical potential:

$$\mu_A^\emptyset(liq, T, P) \equiv \mu_A^*(liq, T, P).$$

This convention is typically used when A is a liquid at the temperature and pressure of the solution. The vapour pressure of pure liquid A is then generally convenient to measure.

9.5.3 Solute standard state; Henry's law

If we have a solute B for which the pure substance is not a liquid at the temperature and pressure of the solution, then the solvent standard state is not appropriate. In the limit when $x_B \to 0$, the chemical potential of B is also well approximated using $\mu_B^\emptyset + RT \ln x_B$. Consider a dilute solution of solute B in solvent A to be concrete. As in Section 9.5.2 equilibrium for B gives

$$\frac{f_B/P_0}{\gamma_B x_B} = exp\left[-\frac{\left(\mu_B^\emptyset(gas, T, P_0) - \mu_B^\emptyset(liq, T, P)\right)}{RT} \right] \equiv \kappa$$

again with κ independent of the solution composition. Now, the molecular environment of a solute B molecule in this limit consists of mostly solvent A molecules, i.e., the other type of molecules. In this limit, the partial pressure of B is experimentally found to be linear with respect to its mole fraction, that is

$$P_B = x_B K_B.$$

This observation is known as Henry's law, and the constant of proportionality K_B is known as Henry's law constant. Solutions for which this is valid are called ideal dilute solutions. This is similar to Raoult's Law, but the partial pressure is proportional to some (temperature-dependent) quantity K_B instead of the vapour pressure P_B^*. Thus, if we set $\gamma_B \to 1$ in the limit of $x_B \to 0$, we have

$$\kappa = \frac{f_B/P_0}{\gamma_B x_B} = \frac{P_B/P_0}{x_B} = \frac{K_B}{P_0}$$

Therefore, in general, for non-ideal solutions in the low concentration limit, we have

$$f_B = \gamma_B x_B K_B,$$

while for ideal dilute solutions that obey Henry's law, we simply recover $P_B = x_B K_B$. With this the definition of activity is $a_B = \gamma_B x_B \equiv \frac{f_B}{K_B}$. Henry's law constant, K_B depends upon both the solute and solvent identities. With this, a second convention to describe the chemical potential in a non-ideal solution is

$$\mu_B(liq, T, P, x_B) = \mu_B^{\emptyset}(liq, T, P) + RT \ln \gamma_B x_B$$

where $\gamma_B \to 1$ as $x_B \to 0$, with the liquid standard chemical potential

$$\mu_B^{\emptyset}(liq, T, P) \equiv \mu_B^{\emptyset}(gas, T, P_0) + RT \ln \left(\frac{f_B/P_0}{\gamma_B x_B} \right).$$
$$= \mu_B^{\emptyset}(gas, T, P_0) + RT \ln \frac{K_B}{P_0}.$$

This convention is typically used when B is the solute in the solution. Noted that with this convention for the standard state, the chemical potential of B in the solution goes to negative infinity as $x_B \to 0$ since its partial pressure goes to zero. This might be puzzling, but simply reflects the increasing difficulty of removing the last traces of any solute in a solution as its concentration decreases.

The definition of $\mu_B^{\emptyset}(liq, T, P)$ in this convention is *formally* equal to the chemical potential of B in an imaginary ideal solution for which the vapour pressure of pure B is K_B. This reference state for which $a_A = 1$ is, of course, a fictitious state; pure liquid B vapour pressure is P_B^* not K_B. A more pragmatic way to think of this definition is that it is simply the value of the standard chemical potential that is convenient to use in describing the solute chemical potential of a dilute non-ideal solution. Ultimately, the process of deciding what to use as standard state is about selecting a conveniently quantifiable reference state. We will again encounter Henry's Law and the limit of ideal dilute solutions in Chapter 11 when we examine phase diagrams.

§   *Example 9.7 How much CO_2 is there in a can of soda water?*

At 298 *K, the vapour pressure of water is $P^*_{H_2O} = 3173 \ Nm^{-2}$, and Henry's Law constant K_{CO_2} for aqueous carbon dioxide is equal to $1.61 \times 10^8 \ Nm^{-2}$. The total pressure in a can of carbonated drink is 3.5 bar. Calculate the mole fraction x_{CO_2} and mass of carbon dioxide in the drink assuming it is an ideal solution of volume 330 cm^3.*

The partial pressures of water and carbon dioxide are

$$P_{H_2O} = x_{H_2O} P^*_{H_2O} = \left(1 - x_{CO_2}\right) P^*_{H_2O}$$

$$P_{CO_2} = x_{CO_2} K_{CO_2}.$$

Then the total pressure is $P_{tot} = P_{H_2O} + P_{CO_2}$. Thus, the mole fraction of carbon dioxide is given by

$$x_{CO_2} = \frac{P_{tot} - P^*_{H_2O}}{K_{CO_2} - P^*_{H_2O}} = 2.15 \times 10^{-3}.$$

When the can is opened to atmosphere, P_{CO_2} is reduced to its ambient value of 42 Nm^{-2}; then x_{CO_2} drops to 2.61×10^{-7}. The mass of CO_2 in an unopened can is approximately 1.7 g. Almost all of this dissolved carbon dioxide comes out of solution as the fizz in your drink. This same mechanism causes decompression sickness in divers who ascent too rapidly for the dissolved nitrogen in their blood to gradually equilibrate to the lower air pressure at the surface, leading to the formation of nitrogen bubbles trapped in their tissues. §

9.5.4 Standard state concentration units

It is frequently practical to use concentration units such as molarity and molality instead of mole fractions. The molarity of a solute in a solution equal to

$$m_B = \frac{n_B}{V}$$

where n_B is the number of moles of B dissolved in a volume V of the solution. Similarly, the molality of a solute is $M_B = \frac{n_B}{\mathcal{M}_A}$ where $\mathcal{M}_A$ is the mass in kilogram of the solvent A. Using these concentration units, the activity of B is defined as $a_B = \gamma_B \frac{m_B}{m_0}$ or $a_B = \gamma_B \frac{M_B}{M_0}$ depending upon whether molarity or the molality is used as the concentration unit. Note that both the activity and the activity coefficient are dimensionless as before and their numerical values depend upon the unit chosen to specify the concentration in. The value of the reference concentration, m_0 or M_0, is chosen for relevance to the application. In most chemical systems m_0 is set at 1 mol dm^{-3}. However, in biochemical systems, the hydrogen ion concentrations are typically close to $10^{-7} \text{mol dm}^{-3}$. Thus, in biochemistry, the chemical potentials of $H^+(aq)$ and $OH^-(aq)$ are referenced to the standard molarity of m_0 equal to $10^{-7} \text{mol dm}^{-3}$, while keeping the standard molarity of one molar for other ions. These definitions of activity using molarity and molality are typically used for electrolyte solutions.

A case to note. Water is a most important solvent. Its standard state is taken to be pure liquid water, that is the activity of pure water is equal to unity with the "solvent" standard state discussed in Section 9.5.1. Then, taking the density of liquid water to be 996.95 g dm^{-3} at 298.15 K and 1 bar, $m_0 = 55.34 \text{ mol dm}^{-3}$ for liquid water. In terms of molality we get $M_0 = 55.51 \text{ mol kg}^{-1}$.

Another case to note. For any solid material, the standard state is usually taken to be the pure solid at 298.15 K and 1 bar. Since the composition of a pure solid is constant at any fixed T and P, its activity is equal to one.

9.5.5 Measuring standard chemical potentials and activities

With this convention, using molarity as the concentration unit, the chemical potential of a component B in a solution with molarity equal to m_B is then given by

$$\mu_B(liq, T, P, m_B) = \mu_B^\emptyset(liq, T, P, m_0) + RT \ln a_B$$

$$= \mu_B^\emptyset(liq, T, P, m_0) + RT \ln \left(\gamma_B \frac{m_B}{m_0}\right)$$

$$= \mu_B^\emptyset(liq, T, P, m_0) + RT \ln \left(\frac{m_B}{m_0}\right) + RT \ln \gamma_B.$$

How is γ_B in a solution measured? When the molarity m_B approaches zero, the solution behaves ideally so that the activity coefficient γ_B is equal to unity. Thus, we have

$$\mu_B^\emptyset(liq, T, P, m_0) = \lim_{m_B \to 0} \left[\mu_B(liq, T, P, m_B) - RT \ln \left(\frac{m_B}{m_0}\right)\right].$$

From Section 9.4 we know how to determine the chemical potential, even for a non-ideal gas phase. Hence, allowing the non-ideal solution to get to equilibrium with its gas phase, we can determine the chemical potential of B for a solution of any concentration m_B. In this way, by measuring the chemical potential in the limit when m_B goes to zero, we will be able to obtain the value of $\mu_B^\emptyset(liq, T, P, m_0)$. Having established the value of this standard chemical potential, we can then determine how the activity γ_B of a component in a solution depends upon its concentration. This works for any substance B for which gas-liquid equilibrium conditions are readily accessible. For example: water.

Consider water vapour in equilibrium with an aqueous solution of any salt. The equilibrium of water molecules in the gas and solution means that

$$\mu_{H_2O}(liq, T, P, m_{salt}) = \mu_{H_2O}^*(gas, T, P)$$

because the gas phase is pure water vapour. From Section 9.4 we know how to determine the chemical potential of water vapour from the independently measurable partial pressure of water. Hence, we can determine the chemical potential of water in an aqueous salt solution of any salt concentration. But this is given by

$$\mu_{H_2O}(liq, T, P, m_{salt})$$
$$= \mu^{\emptyset}_{H_2O}(liq, T, P, m_0) + RT \ln\left(\frac{m_{salt}}{m_0}\right) + RT \ln \gamma_{H_2O}.$$

At any fixed temperature and pressure, we can thus measure the chemical potential of water for any salt concentration in the solution. Then taking the limit of zero salt concentration for which the activity coefficient of water goes to one, we can determine the standard chemical potential of water at T and P:

$$\mu^{\emptyset}_{H_2O}(liq, T, P, m_0) = \lim_{m_{salt}\to 0}\left[\mu_B(liq, T, P, m_{salt}) - RT \ln\left(\frac{m_{salt}}{m_0}\right)\right].$$

Then, using this standard chemical potential and the measured values of $\mu_{H_2O}(liq, T, P, m_{salt})$, we can determine the activity of water in a salt solution of any concentration. This works for any aqueous electrolyte solution. The method also works for a gas-liquid equilibrium such as ammonia and water where both components are present in both phases. We demonstrate how to experimentally determine the value of the standard chemical potential of ammonia in aqueous solution $\mu^{\emptyset}_{NH_3(aq)}$.

§   *Example 9.8 The activity of aqueous ammonia*

Ammonia dissolves in water to form an aqueous solution $NH_3(aq)$. Thus, neglecting the (small extent of) dissociation of ammonia in water or the formation of NH_4OH, we can describe the phase equilibrium between gaseous ammonia and aqueous ammonia as

$$NH_3(gas) \rightleftharpoons NH_3(aq).$$

This equilibrium can be quantified by measuring the concentration of aqueous ammonia $M_{NH_3(aq)}$ as a function of the partial pressure *of ammonia $P_{NH_3(gas)}$. In the limit of low pressure, the ratio of the molality to the pressure is found experimentally to be*

$$\lim_{P_{NH_3(gas)} \to 0} \left(\frac{M_{NH_3(aq)}}{M_0} \right) \left(\frac{P_0}{P_{NH_3(gas)}} \right) = 0.0176$$

at a temperature of 298.15 K. Given that the Gibbs free energy of formation of ammonia gas at 298.15 K and 1 bar, directly measured by calorimetry, is $\Delta_f G^0 = -16.45$ kJ mol^{-1}, calculate the values of the standard chemical potential $\mu_{NH_3(aq)}^{\varnothing}$.

We can write the chemical potential of ammonia gas as

$$\mu_{NH_3(gas)}(T, P) = \mu_{NH_3(gas)}^{*}(T, P_0) + RT \ln a_{NH_3(gas)}$$

$$= \mu_{NH_3(gas)}^{*}(T, P_0) + RT \ln \frac{f_{NH_3(gas)}}{P_0}$$

where the activity is the ratio of its fugacity to the standard pressure. Similarly, the chemical potential of ammonia dissolved in water is

$$\mu_{NH_3(aq)}(T, P) = \mu_{NH_3(aq)}^{\varnothing}(T, P) + RT \ln a_{NH_3(aq)}$$

$$= \mu_{NH_3(aq)}^{\varnothing}(T, P, M_0) + RT \ln \left(\gamma_{NH_3(aq)} \frac{M_{NH_3(aq)}}{M_0} \right)$$

with the standard molality M_0 equal to 1 mol kg^{-1}.

When ammonia gas is in equilibrium with an aqueous solution of ammonia, the chemical potentials of the two phases are the same. Thus,

$$\mu_{NH_3(gas)}(T, P) = \mu_{NH_3(aq)}(T, P).$$

Using the expressions for $\mu_{NH_3(gas)}$ and $\mu_{NH_3(aq)}$ above gives us

$$exp \left(-\frac{\mu_{NH_3(aq)}^{\varnothing} - \mu_{NH_3(gas)}^{*}}{RT} \right) = \frac{a_{NH_3(aq)}}{a_{NH_3(gas)}}$$

$$= \left(\gamma_{NH_3(aq)} \frac{M_{NH_3(aq)}}{M_0} \right) \left(\frac{P_0}{f_{NH_3(gas)}} \right)$$

In the limit as the pressure of ammonia goes to zero, its fugacity is equal to its pressure. At the same time, the molality of aqueous ammonia also goes to zero because the solution is dilute. Thus, we expect the activity coefficient of aqueous ammonia to be equal to unity. Therefore, in the limit of zero pressure, we obtain

$$exp\left(-\frac{\mu^{\varnothing}_{NH_3(aq)} - \mu^{*}_{NH_3(gas)}}{RT}\right) = \left(\gamma_{NH_3(aq)}\frac{M_{NH_3(aq)}}{M_0}\right)\left(\frac{P_0}{f_{NH_3(gas)}}\right)$$

$$= \left(\frac{M_{NH_3(aq)}}{M_0}\right)\left(\frac{P_0}{P_{NH_3(gas)}}\right).$$

From the experimental data relating the concentration and the partial pressure, we have

$$exp\left(-\frac{\mu^{\varnothing}_{NH_3(aq)} - \mu^{*}_{NH_3(gas)}}{RT}\right) = 0.0176.$$

If you have read Section 11.3 ahead of time you will realize that this quantity is really just the equilibrium constant for the dissolution of ammonia gas in water. Thus,

$$\mu^{\varnothing}_{NH_3(aq)} - \mu^{*}_{NH_3(gas)} = -10.009 \text{ kJ mol}^{-1}.$$

Given the value of $\mu^{*}_{NH_3(gas)} = \Delta_f G^0 = -16.45$ kJ mol^{-1}, we obtain the value of -26.46 kJ mol^{-1} for $\mu^{\varnothing}_{NH_3(aq)}$. Knowing this standard chemical potential, the concentration of an aqueous ammonia solution under any ammonia gas pressure can be calculated.   §

Concentration units such as molality or molarity are practical, even though they have less fundamental significance than the mole fraction which is directly related to counting the number of microstates in the Boltzmann formula for entropy. In summary, for non-ideal solutions, we have the following general form for the chemical potential of any component A in a solution:

$$\mu_A(liq, T, P, \{c_i\}) = \mu^{\varnothing}_A(liq, T, P) + RT \ln a_A$$

$$= \mu_A^\varnothing(liq, T, P) + RT \ln\left(\gamma_A \frac{c_A}{c_0}\right)$$

$$= \mu_A^\varnothing(liq, T, P) + RT \ln\left(\frac{c_A}{c_0}\right) + RT \ln \gamma_A$$

with different choices for the standard state to define the reference chemical potential $\mu_A^\varnothing$ at unit activity depending upon the system; c_A is the concentration in any generic concentration unit, and c_0 is a standard state concentration selected for convenience; the argument $\{c_i\}$ on the left-hand side indicates that in general, the chemical potential for any one component depends upon the concentrations of all the components present. The non-ideal dependence upon temperature, pressure and concentration is accounted for by the logarithmic term in the activity coefficient.

9.6 Regular solutions: intermolecular interactions

In the previous section we have seen a general form of the equation for the chemical potential of any component in a non-ideal solution:

$$\mu_A(liq, T, P, c_A) = \mu_A^\varnothing(liq, T, P) + RT \ln a_A$$

where the activity generally depends upon temperature, pressure and composition. This dependence arises because the intermolecular interactions in a non-ideal mixture is not the same as in the pure substances. In this section we discuss a simple quantitative model that includes the effect of intermolecular interactions on the activities of the components in a non-ideal solution.

For ideal mixing at constant temperature and pressure, the internal energy and volume of mixing are zero:

$$\Delta_{mix}U = 0$$
$$\Delta_{mix}V = 0.$$

This means that the enthalpy of mixing of an ideal solution is equal to zero

$$\Delta_{mix}H = \Delta_{mix}U + P\Delta_{mix}V = 0.$$

A simple model for a binary non-ideal solution is to take the enthalpy of mixing, per mole, to be equal to

$$\Delta_{mix}H = x_A x_B \epsilon RT$$

in terms of an interaction parameter ϵ. Here we are describing the composition with mole fractions of the components A and B and the factor of RT is inserted in order to make the interaction parameter ϵ a dimensionless quantity, that is, simply a number. Note that $\Delta_{mix}H$ goes to zero for either pure A or pure B, as it should. When ϵ is positive, mixing is accompanied by an increase of the system enthalpy; heat is absorbed, mixing is endothermic. On the other hand, when ϵ is negative, mixing is exothermic.

What is the molecular picture for such a model? In any mixture of A and B, there are interactions between pairs of A molecules, pairs of B molecules, and pairs of an A molecule interacting with a B molecule. The product of the mole fractions $x_A x_B$ is a simple measure of the number of interacting AB pairs. Thus, the interaction parameter ϵ quantifies the intermolecular interaction between A and B molecules. A negative ϵ means that the intermolecular interaction between different molecules decreases the enthalpy relative to the interactions in the pure components; the interactions between unlike molecules is more attractive than the interaction between like molecules. Vice versa, a positive ϵ holds when the interactions between unlike molecules is more repulsive compared to the interaction between like molecules. With this simple model of $\Delta_{mix}H$, the Gibbs free energy of mixing can be written as

$$\Delta_{mix}G = \Delta_{mix}H - T\Delta_{mix}S.$$

If you further approximate the entropy of mixing to be the same as in an ideal solution, then

$$\Delta_{mix}G = x_A x_B \epsilon RT + RT(x_A \ln x_A + x_B \ln x_B).$$

Setting the entropy of mixing to be the same as in the ideal solution is an approximation. With this approximation, we are taking the molecular arrangement to be completely random despite the intermolecular interactions. In statistical thermodynamics this is called the mean-field approximation. We next put this approximate expression for the Gibbs free energy of mixing together with the expressions we obtained in Section 9.5.2 for the chemical potentials in a non-ideal solution using the "solvent" standard state:

$$\mu_A(liq, T, P, c_A) = \mu_A^{\emptyset}(liq, T, P) + RT \ln a_A$$
$$\mu_B(liq, T, P, c_B) = \mu_B^{\emptyset}(liq, T, P) + RT \ln a_B.$$

From these expressions for the chemical potentials of the components, the Gibbs free energy of mixing in terms of the activities of the components is

$$\Delta_{mix} G = (x_A \mu_A + x_B \mu_B) - (x_A \mu_A^{\emptyset} + x_B \mu_B^{\emptyset})$$
$$= RT(x_A \ln a_A + x_B \ln a_B)$$

because $\mu_A^{\emptyset}$ and $\mu_B^{\emptyset}$ are the chemical potentials of the pure A and B. Hence, equating our two different expressions for $\Delta_{mix} G$, we obtain

$$\epsilon x_A x_B + (x_A \ln x_A + x_B \ln x_B) = x_A \ln \gamma_A x_A + x_B \ln \gamma_B x_B$$

where we have written the activities in the form $a = \gamma x$ for each component. You can verify that this equation is correct if you set

$$\ln \gamma_A = \epsilon x_B^2$$
$$\ln \gamma_B = \epsilon x_A^2.$$

Therefore, we have here a simple model for how the activity coefficients depend upon the mole fractions in a non-ideal solution. The composition dependence of the activities can be expressed as

$$a_A = \gamma_A x_A = x_A \exp(\epsilon x_B^2)$$
$$a_B = \gamma_B x_B = x_B \exp(\epsilon x_A^2).$$

Solutions for which these model activities are valid are known as regular solutions. Additionally, using the "solvent" convention for the definition of the activity in a solution, we have seen in the previous section that the activity of a component A in a non-ideal solution is really just the ratio of its partial pressure to its pure liquid vapour pressure, taking the vapour to be an ideal gas. That is,

$$a_A = \gamma_A x_A = \frac{P_A}{P_A^*}.$$

Thus, we have explicit expressions for the partial pressures of a non-ideal binary solution:

$$P_A = x_A P_A^* \, exp(\epsilon x_B^2)$$
$$P_B = x_B P_B^* \, exp(\epsilon x_A^2).$$

You note that these expressions for the partial pressures tends toward Raoult's law for the corresponding pure liquid when the respective mole fraction tends to unity. We will use regular solutions as examples when we discuss phase diagrams for binary mixtures.

§   Example 9.9 A regular solution

To a good approximation, chloroform and acetone form a regular solution. At 300 K *the partial pressure of chloroform is* 20 mmHg *when its mole fraction in solution is* $x_{Chl} = 0.12$. *At the same temperature, it has a partial pressure of* 220 mmHg *when its mole fraction in solution is increased to* 0.80.

Calculate the interaction parameter ϵ *for the solution, the vapour pressure of chloroform and the change in its chemical potential when its mole fraction changes from* 0.12 *to* 0.80.

Since this is a regular solution, we have the partial pressures of chloroform (subscript *Chl*) and acetone (*Ace*) as follows:

$$P_{Chl} = x_{Chl} P_{Chl}^* \, exp(\epsilon x_{Ace}^2)$$
$$P_{Ace} = x_{Ace} P_{Ace}^* \, exp(\epsilon x_{Chl}^2).$$

Taking the ratio of the chloroform partial pressures at the given mole fractions, we have

$$\frac{P_{Chl}(x_{Chl} = 0.12)}{P_{Chl}(x_{Chl} = 0.80)} = \frac{20}{220} = \frac{0.12 \, exp(0.7744\epsilon)}{0.80 \, exp(0.04\epsilon)}$$

giving a value for the interaction parameter of $\epsilon = -0.6819$. Since ϵ is negative, the intermolecular interaction between chloroform and acetone is more attractive than in the pure liquids. Using this value of ϵ we can calculate the vapour pressure of pure chloroform at $300\,K$ using the data for $x_{Chl} = 0.12$

$$20 = 0.12 P_C^* \, exp(-0.7744 \times 0.6819)$$

so that the vapour pressure of chloroform is $P_{Chl}^* = 282.6$ mmHg. The activity of a component in a regular solution is the ratio of its partial pressure to its vapour pressure. Hence the activities of chloroform at mole fraction 0.12 and 0.80 are 0.0708 and 0.7785, respectively. From this, the change in the chemical potential of chloroform when its mole fraction increases from 0.12 to 0.80 is

$$RT \ln \frac{0.7835}{0.0712} = 5940 \, J \, mol^{-1}. \qquad\qquad \S$$

§   **Example 9.10 Henry's Law constant**

The vapour pressure of pure acetone is $340\,mmHg$ while the vapour pressure of pure chloroform is $280.8\,mmHg$. Thus, pure acetone is more volatile than pure chloroform.

Calculate the mole fraction of chloroform in the gas phase when the liquid composition is $x_{Chl} = 0.80$. Is acetone enriched in the gas phase relative to liquid phase? Given that chloroform-acetone is a regular solution, calculate the Henry's law constant for acetone when x_A tends to zero. Comment on this.

When $x_{Chl} = 0.80$, the partial pressures of chloroform and acetone are

$$P_{Chl} = 220 \text{ mmHg (from Example 9.7)}$$

$$P_{Ace} = 0.2 \times 340 \, exp(-0.6819 \times 0.64)$$

$$= 44.0 \text{ mmHg.}$$

Hence, the total gas phase pressure is 264 mmHg and its chloroform mole fraction is 0.83. Hence, the mole fraction of acetone in the gas phase is lower than in the liquid phase, even though pure acetone is more volatile than pure chloroform. This is unexpected if you are thinking only in terms of ideal solutions.

The Henry's law constant for acetone is given by the tangent to its partial pressure curve when x_{Ace} equals to zero. This can be calculated from

$$\left(\frac{\partial P_{Ace}}{\partial x_{Ace}}\right)_T = \frac{\partial}{\partial x_{Ace}}[x_{Ace}P_{Ace}^* \exp(\epsilon(1 - x_{Ace})^2)]_T$$

$$= P_{Ace}^* \exp(\epsilon(1 - x_{Ace})^2)[1 - 2\epsilon x_{Ace}(1 - x_{Ace})]$$

$$= 171.9 \text{ mmHg for } x_{Ace} \text{ equal to zero.}$$

Hence, Henry's law constant for acetone in a solution with chloroform is equal to 171.9 mmHg, which is less than the vapour pressure of chloroform. At a mole fraction of x_{Chl} equal to 0.8, the chloroform-acetone solution is an ideal dilute solution of acetone where the *effective* vapour pressure of acetone is less than the vapour pressure of chloroform. Thus, the equilibrium gas phase for a solution with $x_{Chl} = 0.8$ is not enriched with acetone relative to the liquid phase even though acetone is more volatile than chloroform. §

9.7 Electrolytes

9.7.1 μ for electrolyte solutions

In Section 9.5 we have discussed how to determine the chemical potential of a substance in a solution by setting the solution in equilibrium with a gas phase containing the same substance. In the case of electrolytes, there are two problems with this approach.

First, under typical conditions of temperature and pressure, electrolytes are not present in the gas phase in equilibrium with a solution. For example, at typical temperatures and pressures, you do not find sodium and chloride ions in the gas phase that is at equilibrium with an aqueous solution of sodium chloride. However, the chemical potential of the electrolyte can be directly measured through electrochemistry. We will consider this in Section 13.2.

Second, under normal conditions, electrolyte solutions are electrically neutral. In the solution, the total charge from the cations ions is equal and opposite to the total charge from the anions. We cannot change the number of cations in a neutral solution while keeping the number of anions constant, and vice versa. This makes it impossible to measure the chemical potential of any individual species of electrolyte. We discuss this point here.

Consider n moles of an ionic compound $A_{v_+}B_{v_-}$ dissolved in n_{solv} moles of a solvent to give cation A^{Z+} and anion B^{Z-} of ionic charges Z_+ and Z_-, respectively. The number of moles of the cation and the anion are $n_{diss}v_+$ and $n_{diss}v_-$, respectively for n_{diss} moles of $A_{v_+}B_{v_-}$ that dissociates. Then at constant temperature and pressure, the change in the Gibbs free energy of the solution with n, n_{solv} and n_{diss} is

$$dG = \left(\mu_{A_{v_+}B_{v_-}}\right)d(n - n_{diss}) + (v_+\mu_{A^{Z+}})dn_{diss} + (v_-\mu_{B^{Z-}})dn_{diss}$$

$$+\mu_{solv}dn_{solv}$$

where $\mu_{A_{v_+}B_{v_-}}$ is the chemical potential of the undissociated compound, $\mu_{A^{Z+}}$ and $\mu_{B^{Z-}}$ are the chemical potentials of the ions, and μ_{solv} is the chemical potential of the solvent. If we hold n, the total number of moles of the compound, and n_{solv} constant, then the dissociation equilibrium is attained when

$$\left(\frac{\partial G}{\partial n_{diss}}\right)_{T,P,n,n_{solv}} = 0.$$

Hence, equilibrium for the reaction

$$A_{v_+}B_{v_-} \rightleftharpoons v_+A^{Z+} + v_-B^{Z-}$$

is reached when

$$\mu_{A_{\nu_+}B_{\nu_-}}(T,P,\{m\}) = \nu_+\mu_{A^{z+}}(T,P,\{m\}) + \nu_-\mu_{B^{z-}}(T,P,\{m\})$$

with $\{m\}$ indicating the molarities of the ions; we have chosen to use units of molarity but any other concentration unit can be used.

Following the discussion in Section 9.5.5, the chemical potential of ionic species is given by

$$\mu_{A^{z+}}(T,P,m) = \mu^{\emptyset}_{A^{z+}}(T,P,m_0) + RT \ln a_{A^{z+}}$$

$$\mu_{B^{z-}}(T,P,m) = \mu^{\emptyset}_{B^{z-}}(T,P,m_0) + RT \ln a_{B^{z-}}.$$

The standard chemical potentials $\mu^{\emptyset}_{A^{z+}}$ and $\mu^{\emptyset}_{B^{z-}}$ are defined for states in which each species has unit activity. We have seen in Section 8.3 for a multicomponent system that the chemical potential of any component is defined as the derivative of the Gibbs free energy with respect to the number of moles of that component, keeping the number of moles of all other components constant. That is

$$\mu_1 = \left(\frac{\partial G}{\partial n_1}\right)_{T,P,n_2}$$

for component 1, for example, in a two-component system. With electrolyte solutions being neutral, it is not possible to experimentally measure the chemical potential $\mu_{A^{z+}}$ because $n_{A^{z+}}$ cannot be varied independently of $n_{B^{z-}}$. Since $\mu_{A^{z+}}$ cannot be measured, similarly it is also not possible to measure the activity of an individual species of electrolyte $a_{A^{z+}}$. This is why writing the chemical potentials of the ionic species in terms of *measurable* quantities requires some discussion beyond what we have covered in Section 9.5.5.

We start by noting that we can experimentally prepare solutions with any concentration of the electrolyte by varying the ratio n/n_{solv} so that the chemical potential of $A_{\nu_+}B_{\nu_-}$ can be measured experimentally. Using the standard form of the chemical potentials of the individual ionic species in the equilibrium condition above, we have

$$\mu_{A_{\nu_+} B_{\nu_-}} = \left(\nu_+ \mu^{\emptyset}_{A^{Z+}} + \nu_- \mu^{\emptyset}_{B^{Z-}}\right) + \nu_+ RT \ln a_{B^{Z-}} + \nu_- RT \ln a_{A^{Z+}}$$

$$= \left(\nu_+ \mu^{\emptyset}_{A^{Z+}} + \nu_- \mu^{\emptyset}_{B^{Z-}}\right) + RT \ln \left[\left(\frac{m_{A^{Z+}}}{m_0}\right)^{\nu_+} \left(\frac{m_{B^{Z-}}}{m_0}\right)^{\nu_-}\right]$$

$$+ RT \ln \left(\gamma^{\nu_+}_{A^{Z+}} \gamma^{\nu_-}_{B^{Z-}}\right)$$

$$= \nu_+ \left(\mu^{id}_{A^{Z+}} + RT \ln \gamma_{A^{Z+}}\right) + \nu_- \left(\mu^{id}_{B^{Z-}} + RT \ln \gamma_{B^{Z-}}\right),$$

where we have then defined the ideal chemical potential of each of the species A^{Z+} and B^{Z-} as

$$\mu^{id}_{A^{Z+}} = \mu^{\emptyset}_{A^{Z+}}(T, P, m_0) + RT \ln \left(\frac{m_{A^{Z+}}}{m_0}\right)$$

$$\mu^{id}_{B^{Z-}} = \mu^{\emptyset}_{B^{Z-}}(T, P, m_0) + RT \ln \left(\frac{m_{B^{Z-}}}{m_0}\right).$$

We now further define the mean ionic molarity $m_\pm$ as the geometric mean of the molarities of the cation and anion. That is,

$$m_\pm = \left(m^{\nu_+}_{A^{Z+}} m^{\nu_-}_{B^{Z-}}\right)^{1/(\nu_+ + \nu_-)}.$$

And we similarly define the mean activity coefficient as

$$\gamma_\pm = \left(\gamma^{\nu_+}_{A^{Z+}} \gamma^{\nu_-}_{B^{Z-}}\right)^{1/(\nu_+ + \nu_-)},$$

which is the geometric mean of the activities $\gamma_{A^{Z+}}$ and $\gamma_{B^{Z-}}$. With these definitions of $m_\pm$ and $\gamma_\pm$, we have

$$\mu_{A^{Z+}} = \mu^{id}_{A^{Z+}} + RT \ln \gamma_{A^{Z+}}$$

$$= \mu^{\emptyset}_{A^{Z+}} + RT \ln \left(\frac{m_\pm}{m_0}\right) + RT \ln \gamma_\pm$$

$$\mu_{B^{Z-}} = \mu^{id}_{B^{Z-}} + RT \ln \gamma_{B^{Z-}}$$

$$= \mu^{\emptyset}_{B^{Z-}} + RT \ln \left(\frac{m_\pm}{m_0}\right) + RT \ln \gamma_\pm.$$

That is, the chemical potentials of each of the ions present in the solution can be expressed in terms of the mean molarity and mean activity.

To see how you measure $\mu_{A_{\nu_+}B_{\nu_-}}$ we consider the reaction of solid silver chloride with hydrogen to form solid silver and aqueous hydrogen and chloride ions

$$AgCl(s) + \frac{1}{2}H_2(gas) \rightleftharpoons Ag(s) + H^+(aq) + Cl^-(aq).$$

This is an electrochemical reaction which we will encounter again in Chapter 13. For each mole of the reaction, the Gibbs free energies of the products and reactant are, respectively,

$$\mu_{prod} = \mu_{Ag(s)} + \mu_{H^+(aq)} + \mu_{Cl^-(aq)}$$
$$\mu_{react} = \mu_{AgCl(s)} + \frac{1}{2}\mu_{H_2(g)}.$$

Thus, the change in the Gibbs free energy per mole of this reaction is given by

$$\Delta\mu = \mu_{Ag(s)} + \mu_{H^+(aq)} + \mu_{Cl^-(aq)} - \mu_{AgCl(s)} - \frac{1}{2}\mu_{H_2(g)}$$
$$= \mu^{\emptyset}_{Ag(s)} + \mu^{\emptyset}_{H^+(aq)} + \mu^{\emptyset}_{Cl^-(aq)} - \mu^{\emptyset}_{AgCl(s)} - \frac{1}{2}\mu^{\emptyset}_{H_2(g)}$$
$$+RT\ln\left(\frac{a_{Ag(s)}a_{H^+(aq)}a_{Cl^-(aq)}}{a_{AgCl(s)}a^{\frac{1}{2}}_{H_2(g)}}\right).$$

With $\mu^{\emptyset}_{Ag(s)}$ and $\mu^{\emptyset}_{AgCl(s)}$ being the chemical potentials of pure solid silver and solid silver chloride at standard temperature and pressure, respectively, their activities $a_{AgCl(s)}$ and $a_{Ag(s)}$ are equal to unity. We can independently measure the activity of hydrogen gas from its fugacity as discussed in Section 9.4. At standard temperature and pressure, hydrogen gas is close to ideal so that its activity is approximately one. Hence, at standard temperature and pressure, we have

$$\Delta\mu = \Delta\mu^{\emptyset} + 2RT \ln\left(\frac{m_{\pm}}{m_0}\right) + 2RT \ln \gamma_{\pm}$$

where

$$\Delta\mu^{\emptyset} = \mu^{\emptyset}_{Ag(s)} + \mu^{\emptyset}_{H^+(aq)} + \mu^{\emptyset}_{Cl^-(aq)} - \mu^{\emptyset}_{AgCl(s)} - \frac{1}{2}\mu^{\emptyset}_{H_2(g)}.$$

For $H^+(aq)$ and $Cl^-(aq)$, the mean molarity and mean activity coefficient are, as defined above,

$$m_{\pm} = \left(m_{H^+(aq)}m_{Cl^-(aq)}\right)^{1/2}$$

$$\gamma_{\pm} = \left(\gamma_{H^+(aq)}\gamma_{Cl^-(aq)}\right)^{1/2}.$$

The molarities $m_{H^+(aq)}$ and $m_{Cl^-(aq)}$ can be measured directly, by chemical titrations, for example. Then, the mean molarity can be determined from these concentrations. However, the mean activity coefficient is a little trickier to determine.

As it turns out, you can directly measure the chemical potential *difference* $\Delta\mu$ through electrochemistry as we will describe in Chapter 12. If you measure $\Delta\mu$ for a range of molarities, you can then determine the value of $\Delta\mu$ in the limit as the concentrations $m_{H^+(aq)}$ and $m_{Cl^-(aq)}$ tend to zero, as we have discussed in Section 9.5.5. In this limit, the solution is ideal so that the activity coefficients $\gamma_{H^+(aq)}$ and $\gamma_{Cl^-(aq)}$ tend towards unity. Thus, we determine the value of $\Delta\mu^{\emptyset}$ through

$$\Delta\mu^{\emptyset} = \lim_{m_{\pm}\to 0}\left\{\Delta\mu - 2RT \ln\left(\frac{m_{\pm}}{m_0}\right)\right\}.$$

Having determined the value of $\Delta\mu^{\emptyset}$ at any temperature and pressure, we can then determine the mean activity coefficient through the electrochemically measured values of $\Delta\mu$. If you examine the form of the defining equation for the difference of the chemical potentials,

$$\Delta\mu = \Delta\mu^{\emptyset} + 2RT \ln \left(\frac{m_{\pm}}{m_0}\right) + 2RT \ln \gamma_{\pm}$$

you can see why it is advantageous to define the mean activity coefficient. With $\Delta\mu$, $\Delta\mu^{\emptyset}$ and the molarities directly measurable, it means that the mean activity coefficient $\gamma_{\pm}$ is a well-defined and measurable physical quantity. You also note that it is the product of powers of the activity coefficients of the individual electrolytes. For the system we are considering, we have $\gamma_{\pm}$ equal to $\left(\gamma_{H^+(aq)}\gamma_{Cl^-(aq)}\right)^{1/2}$. While the quantities $\gamma_{H^+(aq)}$ and $\gamma_{H^+(aq)}$ are not measurable physical quantities for a neutral electrolytic system, their product is. This is the reason for defining the mean activity coefficient.

From electrochemical data for the silver chloride-hydrogen gas reaction, $\Delta\mu^{\emptyset}$ is equal to -21.41 kJ mol^{-1}. We have

$$\Delta\mu^{\emptyset} = \mu^{\emptyset}_{Ag(s)} + \mu^{\emptyset}_{H^+(aq)} + \mu^{\emptyset}_{Cl^-(aq)} - \mu^{\emptyset}_{AgCl(s)} - \frac{1}{2}\mu^{\emptyset}_{H_2(g)}.$$

The standard chemical potentials of solid silver and hydrogen gas are zero since these are the stable forms of these elements at standard state. Thus,

$$\Delta\mu^{\emptyset} = \mu^{\emptyset}_{H^+(aq)} + \mu^{\emptyset}_{Cl^-(aq)} - \mu^{\emptyset}_{AgCl(s)} = -21.41 \text{ kJ mol}^{-1}.$$

We also can obtain the standard Gibbs free energy of solid silver chloride from calorimetry; see Example 7.1. This is equal to -109.81 kJ mol^{-1}. Hence,

$$\mu^{\emptyset}_{H^+(aq)} + \mu^{\emptyset}_{Cl^-(aq)} = -21.41 - 109.81 = -131.22 \text{ kJ mol}^{-1}.$$

The sum of the chemical potentials of the electrolytes is measurable, just as the product of their activity coefficients is. The next step is to set an arbitrary reference value for the aqueous ions. By convention, the standard chemical potential of the aqueous hydrogen ion

$$\mu^{\emptyset}_{H^+(aq)} = 0$$

is taken as the reference value for other ions. With this reference state, we end up with

$$\mu^{\varnothing}_{Cl^-(aq)} = -131.22 \text{ kJ mol}^{-1}.$$

As a second example, let us extend this to the dissociation of pure water. In any aqueous solution, water molecules dissociate to hydrogen and hydroxide ions

$$H_2O(liq) \rightleftharpoons H^+(aq) + OH^-(aq).$$

The chemical potential of the products and reactant are, respectively,

$$\mu_{prod} = \mu_{H^+(aq)} + \mu_{OH^-(aq)}$$
$$\mu_{react} = \mu_{H_2O(liq)}.$$

Thus, the change in the Gibbs free energy per mole of this reaction is given by

$$\Delta\mu = \mu_{H^+(aq)} + \mu_{OH^-(aq)} - \mu_{H_2O(liq)}$$

$$= \mu^{\varnothing}_{H^+(aq)} + \mu^{\varnothing}_{OH^-(aq)} - \mu^{\varnothing}_{H_2O(liq)} + RT\ln\left(\frac{a_{H^+}a_{OH^-}}{a_{H_2O}}\right)$$

$$= \Delta\mu^{\varnothing} + 2RT\ln\left(\frac{m_{\pm}}{m_0}\right) + 2RT\ln\gamma_{\pm}$$

$$-RT\ln\left(\frac{m_{H_2O}}{55.34}\right) - RT\ln\gamma_{H_2O}$$

where

$$m_{\pm} = (m_{H^+}m_{OH^-})^{1/2}$$
$$\gamma_{\pm} = (\gamma_{H^+}\gamma_{OH^-})^{1/2}.$$

We have used the mean molarity and activity coefficient for the ions $H^+(aq)$ and $OH^-(aq)$, and the standard molarity m_0 equal to $55.34 \text{ mol dm}^{-3}$ for liquid water. For pure water, the activity coefficient of water γ_{H_2O} is just equal to unity, and at standard temperature of 298.15 K and pressure of 1 bar, m_{H_2O} is simply equal to 55.34 mol dm^{-3}, so the last two terms are equal to zero. We will see in Section 12.8 that we can directly measure $\Delta\mu$ through electrochemistry. To obtain the value of $\Delta\mu^\emptyset$, we consider the value of $\Delta\mu$ in the limit when the mean molarity $m_\pm$ goes to zero. In this limit the ions behave ideally so that $\gamma_\pm$ is equal to unity. Therefore, the standard change in chemical potential $\Delta\mu^\emptyset$ can be obtained from

$$\Delta\mu^\emptyset = \lim_{m_\pm \to 0} \left\{ \Delta\mu - 2RT \ln\left(\frac{m_\pm}{m_0}\right) \right\}.$$

All the quantities in the curly brackets are measurable. Using this approach $\Delta\mu^\emptyset$ can be determined for the dissociation of water Experimentally, $\Delta\mu^\emptyset$ is found to be equal to 79.89 kJ mol^{-1}. But this is defined in terms of the chemical potentials of the individual electrolytes through

$$\Delta\mu^\emptyset = \mu^\emptyset_{H^+(aq)} + \mu^\emptyset_{OH^-(aq)} - \mu^\emptyset_{H_2O(liq)}.$$

For standard temperature and pressure the standard chemical potential $\mu^\emptyset_{H_2O(liq)}$ is just the standard Gibbs free energy of formation of liquid water, which is equal to -237.13 kJ mol^{-1} from calorimetry. With the reference value of $\mu^\emptyset_{H^+(aq)}$ equal to zero, then we have

$$\mu^\emptyset_{OH^-(aq)} = \Delta\mu^\emptyset + \mu^\emptyset_{H_2O(liq)}$$
$$= 79.89 - 237.13 = -157.24 \text{ kJ mol}^{-1}.$$

Using this procedure, and selecting the appropriate electrochemical reactions, the standard chemical potentials of all aqueous ions can be determined relative to the arbitrary reference value of $\mu^\emptyset_{H^+(aq)} = 0$. Of course, in addition to the standard temperature and pressure of 298.15 K and 1 bar, the standard electrolyte concentration m_0 needs to be specified. With the values of these standard chemical potential established, and with the molarities of the ions directly measurable, the value of the mean activity coefficient for any electrolyte can then be determined from the defining equations:

$$\mu_{A^{z+}} = \mu_{A^{z+}}^{\emptyset} + RT \ln \left(\frac{m_{\pm}}{m_0}\right) + RT \ln \gamma_{\pm}$$

$$\mu_{B^{z-}} = \mu_{B^{z-}}^{\emptyset} + RT \ln \left(\frac{m_{\pm}}{m_0}\right) + RT \ln \gamma_{\pm}.$$

It is worth repeating that we go through the trouble of defining the mean molarities and mean activities because while the chemical potential of $A_{\nu_+} B_{\nu_-}$ is experimentally measurable, the quantities $\mu_{A^{z+}}$ and $\mu_{B^{z-}}$ are not. Ultimately, this is because you cannot keep $n_{A^{z+}}$ constant while varying $n_{B^{z-}}$, and vice versa, in a neutral electrolyte solution.

9.7.2 The Debye-Hückel model

Now that we have discussed the meaning of the mean activity coefficient, we can discuss the simplest model that gives an approximation for activity coefficients in terms of the concentration of the electrolytes in a solution. Comparing the ideal gas model to the van der Waals gas, you can appreciate that non-ideal behaviour arises from the interactions between molecules. Similarly, the technical obstacle in understanding electrolyte solutions arises from having to account for the interactions between ions in solution. This is addressed, in an instructive approach, by the Debye-Hückel model. We first describe qualitatively the interaction potential used in the model, and then provide the expression for the activity coefficient.

Neutral molecules in solution interact, many rather strongly but all through rather short-range interactions. In contrast, ionic interactions are long ranged because the Coulomb interaction decreases slowly with distance compared to intermolecular interactions between electrically neutral molecules. The interaction potential $\phi(r)$ in the Debye-Hückel model is taken to be a shielded Coulomb potential which depends upon distance r as

$$\phi(r) \propto \frac{exp(-\kappa r)}{r}$$

where the r^{-1} distance dependence of the Coulomb potential is diminished by an exponential factor $exp(-\kappa r)$, which approximately describes the dielectric shielding due to other molecules and ions surrounding any particular ion in the solution. The quantity κ^{-1} has units of length, and is called the Debye screening length. It quantifies how rapidly the electrostatic potential of an ion decreases with distance as a result of the dielectric shielding in the solution. It is determined by the ionic strength I of the solution and the dielectric permittivity ε of the solvent:

$$\kappa^2 \equiv \left(\frac{e^2}{\varepsilon k_B T}\right)(2N_A \rho I)$$

where N_A is the Avogadro number. The ionic strength is defined as

$$I \equiv \frac{1}{2}(M_+ z_+^2 + M_- z_-^2) = \frac{1000}{2\rho}(m_+ z_+^2 + m_- z_-^2)$$

with the concentration of the cations (subscript $+$) and anions (subscript $-$) in molality M or molarity m units; ρ is the density of the pure solvent in units of kg m^{-3} and e is the electronic charge. If there are several ionic species i in the solution, then

$$I \equiv \frac{1}{2}\sum_i M_i z_i^2$$

where z_i is the charge on ion i. The unit of ionic strength is mol kg^{-1}, that is, it has the same units as molality. As the ionic strength of a solution increases, the electrostatic potential of an ion decreases more rapidly with distance from the ion; qualitatively, more ions are providing the dielectric shielding.

For distances less than a "hard-sphere" parameter a, the interionic interaction is taken to be infinitely repulsive. This parameter a plays a role rather similar to the finite volume parameter b in the factor $(V - b)$ in the van der Waals equation of state. It prevents the ionic particles from approaching closer to each other than $r = a$. With these parameters the activity coefficient of species i in the Debye-Hückel model is given by

$$\ln \gamma_i = -A_\gamma z_i^2 \frac{I^{1/2}}{\left(1 + aB_\gamma I^{1/2}\right)}$$

where the constants A_γ and B_γ are

$$A_\gamma \equiv \frac{1}{8\pi} \left(\frac{e^2}{\varepsilon k_B T}\right)^{3/2} (2N_A\rho)^{1/2}$$

$$B_\gamma \equiv \frac{\kappa}{I^{1/2}} = \left(\frac{e^2}{\varepsilon k_B T}\right)^{1/2} (2N_A\rho)^{1/2}.$$

Note that A_γ and B_γ are properties of the solvent and not the properties of any specific electrolyte in the solution. At very low electrolyte concentrations, the product κa is much smaller than unity, so the Debye-Hückel model reduces to the limiting model

$$\ln \gamma_i = -A_\gamma z_i^2 I^{1/2}.$$

To connect these results to the mean activity coefficients in a solution with cations of charge z_+ and anions of charge z_-, we note that

$$|\nu_+ z_+| = |\nu_- z_-|.$$

We thus have the following result for the mean activity coefficient, which is the quantity that is measurable:

$$\ln \gamma_\pm = \ln\left(\gamma_+^{\nu_+} \gamma_-^{\nu_-}\right)^{1/(\nu_+ + \nu_-)} = -A_\gamma |z_+ z_-| \frac{I^{1/2}}{\left(1 + aB_\gamma I^{1/2}\right)}.$$

In the limit of zero concentration, we have the limiting Debye-Hückel model

$$\ln \gamma_\pm = -A_\gamma |z_+ z_-| I^{1/2}.$$

The rule-of-thumb is that the limiting Debye-Hückel model is accurate for molalities lower that 0.01.

§   *Example 9.11 Debye-Hückel model for water*

Water is a most important solvent.   How rapidly does the electrostatic field of dissolved ions decay with distance? Liquid water   at   298.15 K   has   permittivity   equal   to 6.954×10^{-10} C^2 N^{-1} m^{-2} and a density equal to 996.95 g dm^{-3}.

Calculate the parameters A_γ and B_γ in the Debye-Hückel model using the density of water. Calculate the Debye screening length for water. Then calculate the activities of the hydrogen and hydroxide ions in pure water at 298.15 K taking the excluded radii to be 9 Å and 3 Å, respectively.

$$k_B = 1.381 \times 10^{-23} \text{ J K}^{-1}$$
$$N_A = 6.022 \times 10^{23} \text{ mol}^{-1}$$
$$e = 1.602 \times 10^{-19} \text{ C.}$$

The Debye-Hückel parameters for water are

$$A_\gamma = 1.170 \text{ kg}^{\frac{1}{2}} \text{ mol}^{-\frac{1}{2}}$$
$$B_\gamma = 3.281 \times 10^9 \text{ m}^{-1} \text{ kg}^{\frac{1}{2}} \text{ mol}^{-\frac{1}{2}}.$$

The screening parameter, which depends upon the ionic strength of the solution, is given by

$$\kappa^2 = B_\gamma^2 I = \left(\frac{e^2}{\varepsilon k_B T} \right) (2 N_A \rho I).$$

Taking the concentration of hydrogen ions in water to be 10^{-7} mol dm^{-3}, the molality of hydrogen and hydroxide ions are both equal to

$$\frac{10^{-7}}{0.99695} = 1.00306 \times 10^{-7} \text{ mol kg}^{-1}.$$

Hence, the ionic strength of water is

$$I \equiv \frac{1}{2}(M_+z_+^2 + M_-z_-^2)$$

$$= 1.00306 \times 10^{-7} \text{ mol kg}^{-1}.$$

Therefore, κ^2 for water is $1.0795 \times 10^{12} \text{ m}^{-2}$. This gives a Debye screening length of

$$\kappa^{-1} = 0.9625 \times 10^{-6} \text{ m}.$$

The Debye screening length is about a micrometer. This gives a idea of how effectively the electrostatic potential of an ion decreases with distance in pure water.

With the values of these parameters, the activity coefficients for $H^+(aq)$ and $OH^-(aq)$ are

$$\gamma_{H^+} = 0.99963$$
$$\gamma_{OH^-} = 0.99963.$$

Thus, using the mole fractions in place of activities in writing the chemical potentials of $H^+(aq)$ and $OH^-(aq)$ gives quite good approximations for pure water at room temperature.   §

§   **Example 9.12 Activity of $Na^+(aq)$ and $Cl^-(aq)$**

Neglecting the comparatively small dissociation of water, calculate the ionic strength of a solution with 0.01 mole of NaCl dissolved per kg of water. Assume that sodium chloride is fully dissociated.

Take the excluded radius to be 4 Å for Na^+ and 3 Å for Cl^-. Calculate γ_{Na^+}, γ_{Cl^-} and the mean activity coefficient $\gamma_\pm$.

The molalities are

$$M_{Na^+} = M_{Cl^-} = 0.01 \text{ mol kg}^{-1}.$$

Thus, the ionic strength of the solution is

$$I \equiv \frac{1}{2}\left(M_{Na^+}z_{Na^+}^2 + M_{Cl^-}z_{Cl^-}^2\right)$$
$$= 0.01 \text{ mol kg}^{-1}.$$

Using the Debye-Hückel model, we get

$$\ln \gamma_{Na^+} = -1.170\left[\frac{0.01^{1/2}}{1 + (\times 10^{-10} \times 3.281 \times 10^9 \times 0.01^{1/2})}\right].$$

This gives the activity coefficient

$$\gamma_{Na^+} = 0.902.$$

Similarly, the activity coefficient for the chloride ion is

$$\gamma_{Cl^-} = 0.899.$$

Hence, the mean activity coefficient is

$$\gamma_\pm = 0.900.$$

Figure 9.5 gives a visual sense of how activity coefficients depend upon the molality of the solution within the Debye-Hückel model.

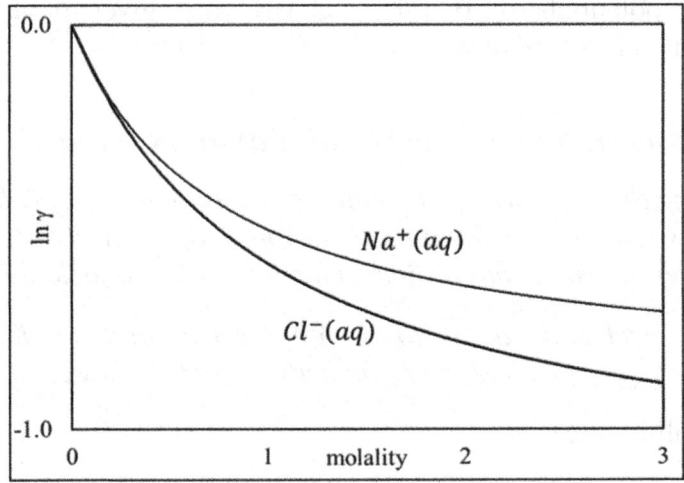

Figure 9.5 ln γ for Na⁺(aq) and Cl⁻(aq) in the Debye-Hückel model.

§ *Example 9.13 Ionic strength in a mixed salt solution*

In an aqueous solution with 0.001 mol kg^{-1} *of sodium carbonate and* 0.002 mol kg^{-1} *of sodium chloride, both ionic compounds dissociate completely. Calculate the ionic strength of the solution and the activity of each of the ions present.*

The ionic strength of the solution is

$$I \equiv \frac{1}{2} \sum_i M_i z_i^2$$

$$= \frac{1}{2}[2(0.001)(1^2) + (0.001)(-2)^2$$

$$+(0.002)(1^1) + (0.002)(-1)^2]$$

$$= 0.005 \text{ mol kg}^{-1}.$$

This gives activity coefficients of 0.927, 0.743 and 0.926 for Na^+, CO_3^{2-} and Cl^-, respectively, in this solution. Then the activities are

$$a_{Na^+} = \frac{\gamma_{Na^+} M_{Na^+}}{M_0} = 0.927 \times 0.004 = 3.708 \times 10^{-3}$$

$$a_{CO_3^{2-}} = \frac{\gamma_{CO_3^{2-}} M_{CO_3^{2-}}}{M_0} = 0.743 \times 0.001 = 7.43 \times 10^{-4}$$

$$a_{Cl^-} = \frac{\gamma_{Cl^-} M_{Cl^-}}{M_0} = 0.926 \times 0.002 = 1.852 \times 10^{-3}.$$

We will need such calculations in our discussion of equilibrium in electrolyte systems in Chapter 12. For example, accurate calculations of the solubility of salts require us to use the activities rather than the mole fractions of ions. The interactions between ions in solution cannot be neglected in many systems, much like many gas phase systems require non-ideal gas equations of state for better than the most elementary estimates of thermodynamic properties. §

9.8 Rubber bands

We have thus far discussed systems with only volume expansion work. It is good learning experience to consider a model system in which non-PV work is involved. We start with qualitative observations of rubber bands and see where Thermodynamics lead us.

Consider a rubber band of length L under tension F so that the differential work done on the rubber band is $dW = FdL$. Applying the First and Second Laws we get:

$$dU = dQ + dW$$
$$= dQ + FdL$$
$$= TdS + FdL.$$

Although rubber bands can undergo volume changes and thus can perform volume expansion work, we do not include a $-PdV$ term in line 2 because the compressibility is typically very small so that the PdV term would be much smaller than FdL. From line 3 we see that FLT-data on rubber bands is analogous to PVT-data for fluids.

We start with two empirical observations. First, to isothermally stretch a rubber band to a longer length requires a larger tension. Thus, isothermal stretching of a rubber band is characterised by a $\left(\frac{\partial L}{\partial F}\right)_T$ which is positive. Second, the change in length of a rubber band with temperature while under constant tension is characterised by $\left(\frac{\partial L}{\partial T}\right)_F$. This is, interestingly, a negative quantity. That is, a rubber band under constant tension shortens when it is warmed up, an old observation that is known as the Gough-Joule effect. If we take a rubber band to be a homogenous substance, then the Gibbs Phase Rule tells us that we have two independent intensive variables so that the tension is a function of only L and T:

$$F = F(L, T).$$

Thus, simply mathematically, we have from Euler's chain rule for reciprocal derivatives (see Chapter 2):

$$\left(\frac{\partial L}{\partial F}\right)_T \left(\frac{\partial T}{\partial L}\right)_F \left(\frac{\partial F}{\partial T}\right)_L = -1.$$

This tells us that $\left(\frac{\partial F}{\partial T}\right)_L$ is a positive quantity; the tension of a rubber band held at constant length increases with its temperature.

At this point, we bring in the thermodynamics we have learned. In the same way for systems doing only PV-work, we define the Helmholtz free energy as

$$A = U - TS$$

so that for rubber bands, we have:

$$dA = -SdT + FdL$$

as a fundamental thermodynamic equation, where

$$-S = \left(\frac{\partial A}{\partial T}\right)_L$$

$$F = \left(\frac{\partial A}{\partial L}\right)_T$$

giving the Maxwell relation:

$$\left(\frac{\partial F}{\partial T}\right)_L = -\left(\frac{\partial S}{\partial L}\right)_T.$$

Therefore, from the two empirical observations above, Thermodynamics allows us to deduce that $\left(\frac{\partial S}{\partial L}\right)_T$ is a negative quantity. This has physical content: the entropy of a rubber band decreases as it is isothermally stretched. This is consistent with our microscopic understanding of a rubber band as an interconnected network of polymer chains in which each monomer unit in a chain is oriented randomly relative to its connected neighbouring units. Stretching a rubber band "straightens" out each of the polymer chains reducing this random relative orientation of the monomer units.

Compared to a straighter chain, there are many more configurations W of chains in which each monomer unit is randomly oriented relative to its connected neighbouring units. Hence, Boltzmann's entropy formula tells us that the entropy of a rubber band decreases when it is stretched.

Similarly, the Gibbs free energy is defined as:

$$G = U - FL - TS,$$

noting the negative sign of the FL term since work done on the system is FdL compare to $-PdV$ for PV-work. This gives us

$$dG = -SdT - LdF,$$

$$-S = \left(\frac{\partial G}{\partial T}\right)_F$$

$$-L = \left(\frac{\partial G}{\partial F}\right)_T$$

giving the Maxwell relation:

$$\left(\frac{\partial L}{\partial T}\right)_F = \left(\frac{\partial S}{\partial F}\right)_T.$$

Going further, if we consider the internal energy of the rubber band as a function of its length and temperature, we have:

$$dU = \left(\frac{\partial U}{\partial T}\right)_L dT + \left(\frac{\partial U}{\partial L}\right)_T dL.$$

The first derivative is the constant length heat capacity C_L which is analogous to $C_V = \left(\frac{\partial U}{\partial T}\right)_V$ and $\left(\frac{\partial U}{\partial L}\right)_T$ is analogous to $\pi_T = \left(\frac{\partial U}{\partial V}\right)_T$, both of which we have already examined for gases in some detail. Both C_L and π_T can depend upon L and T. In an ideal gas, $\left(\frac{\partial U}{\partial V}\right)_T = 0$ reflecting the zero intermolecular interactions. In a rather similar way, a rubber band for which $\left(\frac{\partial U}{\partial L}\right)_T = 0$ is frequently referred to as an ideal rubber band. Since $dU = TdS + FdL$, we have for an ideal rubber band

$$\left(\frac{\partial U}{\partial L}\right)_T = T\left(\frac{\partial S}{\partial L}\right)_T + F = 0.$$

Therefore, the internal energy of an ideal rubber band depends only upon its temperature,

$$U = U(T),$$

thus giving $dU = C_L dT$. This means that for an ideal rubber band, the internal energy change in an isothermal process is equal to zero. This is analogous to the internal energy of an ideal gas. Hence, if you isothermally stretch an ideal rubber band, we have

$$dU = dQ + dW = 0.$$

The work done on the ideal rubber band in isothermal stretching is exactly equal to the heat lost to the surroundings. This is what happens when you slowly stretch a rubber band so that its temperature remains constant throughout the process. Going back to the microscopic picture above, in this isothermal stretching process the entropy of the rubber band decreases, thus losing heat to the surroundings. This amount of heat lost to the surroundings is exactly equal to the work done on the rubber band by the stretching force F. Conversely, when a stretched rubber band is allowed to isothermally shorten, then it does work FdL at the expense of increasing the entropy of the orientation of its polymer chains. This is why a rubber band is sometimes considered an entropy engine – it can be used as a device to absorb heat from its surroundings, and that heat then used to perform useful work by shortening the rubber band against an external force. You can certainly construct a heat engine using rubber bands operating between two different temperatures.

How might we deduce an equation of state, relating F, L and T for an ideal rubber band? We can use $dU = TdS + FdL$, written as the fundamental equation for entropy:

$$dS = \frac{1}{T}dU - \frac{F}{T}dL.$$

This gives the Maxwell relation:

$$\left[\frac{\partial}{\partial L}\left(\frac{1}{T}\right)\right]_U = -\left[\frac{\partial}{\partial U}\left(\frac{F}{T}\right)\right]_L.$$

For an ideal rubber band, the internal energy depends only upon temperature so that the left-hand side is zero; when U is held constant, T remains constant. Hence, the right-hand side is:

$$\left[\frac{\partial}{\partial U}\left(\frac{F}{T}\right)\right]_L = \frac{dT}{dU}\left[\frac{\partial}{\partial T}\left(\frac{F}{T}\right)\right]_L = \frac{1}{C_L}\left[\frac{\partial}{\partial T}\left(\frac{F}{T}\right)\right]_L = 0.$$

Thus, $\frac{F}{T}$ is only a function of L, not of T. Hence, any equation of state for the ideal rubber band must be of the form:

$$F = \alpha T g(L)$$

with g only a function of the length and α a (positive) constant. One simple equation of state is thus:

$$F = \alpha T L.$$

With this ideal rubber band equation of state, we can calculate C_L. We have the Maxwell relation obtained earlier:

$$\left(\frac{\partial F}{\partial T}\right)_L = -\left(\frac{\partial S}{\partial L}\right)_T.$$

Taking the derivative with respect to the temperature gives:

$$\left(\frac{\partial^2 F}{\partial T^2}\right)_L = -\left[\frac{\partial}{\partial T}\left(\frac{\partial S}{\partial L}\right)_T\right]_L$$

$$= - \left[\frac{\partial}{\partial L} \left(\frac{\partial S}{\partial T} \right)_L \right]_T$$

$$= - \left[\frac{\partial}{\partial L} \left(\frac{C_L}{T} \right) \right]_T.$$

Thus, if the equation of state is $F = \alpha T L$, then we have

$$0 = \left(\frac{\partial C_L}{\partial L} \right)_T$$

so that the constant length heat capacity is independent of length. It can depend upon temperature, $C_L = C_L(T)$.

The entropy due to the relative orientations of the monomers in the rubber band decreases when it is stretched. Thus, if you stretch a rubber band so that the heat generated by the stretch does not have time to dissipate into the surroundings, or if the rubber band is appropriately insulated, you would expect its temperature to go up. A demonstration of this is commonly described online whereby you are instructed to place a rubber band on your (thermally sensitive) lips and then rapidly stretch it. For an adiabatic process:

$$dS = \frac{1}{T} dU - \frac{F}{T} dL = 0.$$

Thus, using $dU = C_L dT$ and the equation of state $F = \alpha T L$,

$$\frac{C_L}{T} dT = \frac{F}{T} dL = \alpha L dL,$$

$$\int_{T_1}^{T_2} \frac{C_L}{T} dT = \int_{L_1}^{L_2} \alpha L dL,$$

for an adiabatic stretch from L_1 to L_2, accompanied by a temperature change from T_1 to T_2. If we further take the constant length heat capacity C_L to be independent of temperature, then we have

$$C_L \ln \left(\frac{T_2}{T_1} \right) = \frac{1}{2} \alpha (L_2^2 - L_1^2).$$

Thus, when a rubber band is stretched adiabatically, the temperature goes up. This equation gives the relationship between T and L for an ideal rubber band. You can compare it to the corresponding equation for the ideal gas adiabat: $T_2 V_2^{\frac{R}{C_V}} = T_1 V_1^{\frac{R}{C_V}}$.

As it turns out, the equation of state $F = \alpha T L$ does not fit the behaviour of real rubber bands very well. A better fit is given by

$$F = \alpha T \left(\frac{L}{L_0} - \frac{L_0^2}{L^2} \right),$$

where L_0 is the length of the rubber band when the tension is zero. From this equation of state, we get

$$\left(\frac{\partial F}{\partial T} \right)_L = \alpha \left(\frac{L}{L_0} - \frac{L_0^2}{L^2} \right) > 0$$

for $L > L_0$, i.e, tension greater than zero. Using the Maxwell relation

$$\left(\frac{\partial S}{\partial L} \right)_T = - \left(\frac{\partial F}{\partial T} \right)_L$$

we see that the entropy of the rubber band decreases with length, consistent with the microscopic picture of the decrease in the orientational entropy of the monomers when a rubber band is stretched. Taking the second derivative of F with respect to T keeping L constant, and applying the Maxwell relation, would similarly convince you that the heat capacity C_L is again independent of the length. More complicated equations of state where the tension depends upon higher powers of the temperature, for example $F = \alpha T^2 g(L)$ are useful rubber band models to fit experimental data. For these models, $\frac{F}{T}$ depends upon the temperature. Hence, these would be non-ideal rubber bands in the sense that the internal energy depends upon both the temperature and the length.

9.9 Paramagnets

When a material containing atoms/molecules with unpaired electrons is placed inside a magnetic field $\mathcal{H}$, a magnetic moment M is induced in the material. The induced magnetic moment is due to the spin of the unpaired electrons, a fraction of which align to the external field $\mathcal{H}$ generating an induced field within the material. Such a material is known as a paramagnet. The induced moment is generally small and depends linearly on the external field. The differential work done on the system is given by $dW = \mu_0 \mathcal{H} dM$, where μ_0 is the permeability of free space. The external field could, for example be generated by an electrical current flowing through a solenoid. The work dW corresponding to the increase in magnetic moment dM is thus the additional energy required to power the solenoid to align the electron spins inside the material. We have previously encountered the paramagnetic system in our discussion of thermodynamic phenomena at temperatures close to absolute zero. In particular, we have seen the important role of adiabatic demagnetization, where the gradual reduction of the applied magnetic field $\mathcal{H}$ in a thermally isolated system is used to lower its temperature. In this section, we focus on this phenomenon of cooling through adiabatic demagnetization. We would thus like to calculate $\left(\frac{\partial T}{\partial \mathcal{H}}\right)_S$, that is, how the temperature of a paramagnet changes with the applied field in a reversible adiabatic process.

Since the work done on the system is $\mu_0 \mathcal{H} dM$, the fundamental equation for the internal energy of the system consisting of the magnetic field and the paramagnet is:

$$dU = TdS + \mu_0 \mathcal{H} dM.$$

Thus, we define the Helmholtz free energy as

$$A = U - TS$$

so that for paramagnets, we have:

$$dA = -S dT + \mu_0 \mathcal{H} dM,$$

$$-S = \left(\frac{\partial A}{\partial T}\right)_M$$

$$\mu_0 \mathcal{H} = \left(\frac{\partial A}{\partial M}\right)_T$$

giving the Maxwell relation:

$$\mu_0 \left(\frac{\partial \mathcal{H}}{\partial T}\right)_M = -\left(\frac{\partial S}{\partial M}\right)_T.$$

Additionally, anticipating a need for the Maxwell relation from the Gibbs free energy, we have

$$G = U - \mu_0 \mathcal{H} M - TS$$

giving us

$$dG = -S dT - \mu_0 M d\mathcal{H},$$

$$-S = \left(\frac{\partial G}{\partial T}\right)_{\mathcal{H}}$$

$$-\mu_0 M = \left(\frac{\partial G}{\partial \mathcal{H}}\right)_T$$

giving the Maxwell relation:

$$\mu_0 \left(\frac{\partial M}{\partial T}\right)_{\mathcal{H}} = \left(\frac{\partial S}{\partial \mathcal{H}}\right)_T.$$

We will also make use of the empirical equation of state which was discovered very early on by Pierre Curie:

$$\mathcal{H} = \frac{1}{\mathbf{C}} TM$$

where $\mathbf{C}$ is known as the Curie constant and depends upon the properties of the paramagnetic material. Thus, the induced magnetic moment is proportional to the applied field and inversely proportional to the temperature. This equation of state, known as Curie's law is a good approximation for small magnetic moment and high enough temperatures.

In order to calculate $\left(\frac{\partial T}{\partial \mathcal{H}}\right)_S$, we make use of the Euler chain rule for derivatives:

$$\left(\frac{\partial T}{\partial \mathcal{H}}\right)_S \left(\frac{\partial \mathcal{H}}{\partial S}\right)_T \left(\frac{\partial S}{\partial T}\right)_{\mathcal{H}} = -1.$$

With the Curie equation, we have

$$\left(\frac{\partial S}{\partial \mathcal{H}}\right)_T = \mu_0 \left(\frac{\partial M}{\partial T}\right)_{\mathcal{H}} = -\mu_0 \frac{M}{T}$$

giving us the reciprocal of the second derivative in the Euler chain rule. From the Second Law, the third derivative is simply equal to the constant field heat capacity $C_{\mathcal{H}}$ divided by the temperature, that is:

$$\left(\frac{\partial S}{\partial T}\right)_{\mathcal{H}} = \frac{1}{T} C_{\mathcal{H}}.$$

Therefore, we have:

$$\left(\frac{\partial T}{\partial \mathcal{H}}\right)_S = -\left(\frac{\partial S}{\partial \mathcal{H}}\right)_T \bigg/ \left(\frac{\partial S}{\partial T}\right)_{\mathcal{H}}$$

$$= \mu_0 \frac{M}{C_{\mathcal{H}}}.$$

Because all the quantities on the right-hand side are positive, we see that when the magnetic field is reduced adiabatically, the paramagnet cools down. This gives the thermodynamic basis for cooling by adiabatic demagnetization. This phenomenon was sketched in Figures 6.5 and 6.6 by the horizontal dashed lines. With the thermodynamic framework in Chapter 7, we are now able to understand why that cooling occurs.

You should notice that the thermodynamic equations describing the paramagnet have exactly the same structure as those for the rubber band in the previous section. Simply map F to $\mu_0 \mathcal{H}$ and L to M. Similarly, these systems have the same structure as systems with only PV-work, with the correspondence:

$$-P \leftrightarrow F \leftrightarrow \mu_0 \mathcal{H}$$
$$V \leftrightarrow L \leftrightarrow M.$$

Thus, making a correspondence to the adiabatic demagnetization of a paramagnet, if you reversibly and adiabatically reduce the applied force F on a rubber band, the rubber band cools down. We have seen early on in Section 3.1.4 that an adiabatic expansion of an ideal gas lowers its temperature. Now, you can show simply by using this correspondence that for a rubber band:

$$\left(\frac{\partial T}{\partial F}\right)_S = -\left(\frac{\partial S}{\partial F}\right)_T \bigg/ \left(\frac{\partial S}{\partial T}\right)_F$$

$$= \frac{L}{C_F}.$$

Both the length L and the heat capacity C_F are positive quantities, the latter from the general consideration that an equilibrium state must be stable to fluctuations in temperature.

To be complete, we should briefly mention the matter of accurately describing the heat capacity of the paramagnet. In a solid state paramagnet the magnetic spins are coupled to the lattice vibrations so that there are contributions to the heat capacity from the thermal excitations of the lattice vibrations and the magnetic moment. A quantum mechanically accurate description of these thermal lattice excitations gives a heat capacity for the paramagnet which goes to zero as the temperature approaches zero, as prescribed by the Third Law. This is the case in the right-hand panel of Figure 6.6; quantum mechanics is needed to understand this.

10

The Van der Waals Fluid

The ideal gas is deficient as a physical model because no matter how low the temperature and/or how high the pressure, it does not condense to form a liquid phase. This is because the intermolecular interactions are taken to be zero. In the previous chapter, we have introduced the chemical potentials of non-ideal gases and solutions which include the influence of intermolecular interactions. Using these chemical potentials, we discussed how to calculate, for example, the fugacity of non-ideal gases using the van der Waals equation of state. Here, we expand upon that by providing a more detailed discussion of the van der Waals fluid because this is an important elementary model for understanding the phase transformation from gas to liquid. In Section 7.10, we looked at the gas-liquid phase transformation using one equation to describe each of the two phases. Historically, the van der Waals equation of state was the first to describe this phase transformation using the same equation of state for both the gas and liquid phases. It is not accurate quantitatively as a description of specific substances, especially in their liquid states, but it illustrates the essential role of intermolecular interactions. In this chapter, we further examine this equation of state, starting with its description of phase change.

10.1 The cubic equation of state

The van der Waals equation gives the pressure as a cubic function of the molar volume, with the general shape of the isotherms being determined by whether the temperature is above or below a critical temperature T_c. We illustrate this with the three isotherms in Figure 10.1, one above, one below and one at the critical temperature T_c. Below the critical temperature, each isotherm has a maximum and a minimum. At and above the critical temperature, the isotherms are monotonically decreasing in pressure as the volume increases; there is only one value of the molar volume for each value of the pressure. The critical temperature isotherm has an inflection point as illustrated in Figure 10.1. The pressure at which this occurs is known as the critical pressure P_c. At any temperature below the critical temperature T_c, the equation of state gives three possible solutions for the molar volume for each pressure. For $0.85T_c$, these are indicated by the points B, D and F. With the general framework discussed in Chapters 7 and 8 we can make sense of these three possibilities for the molar volume.

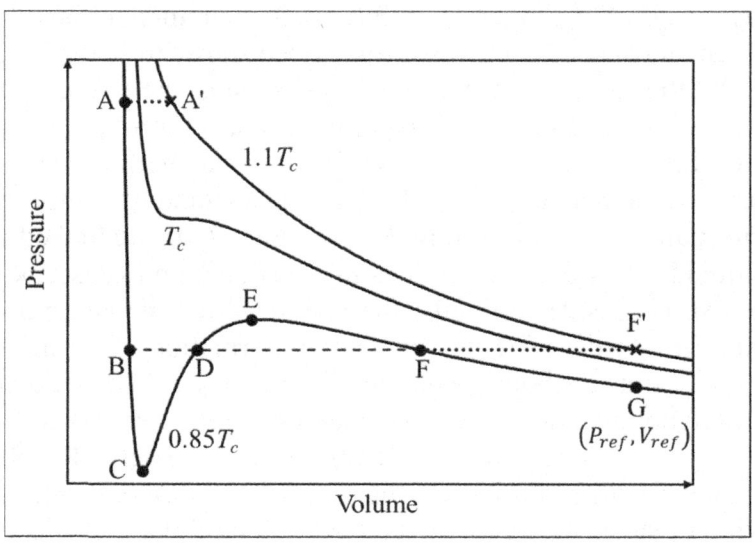

Figure 10.1: *Isotherms of the van der Waals equation of state at $0.85T_c$, at the critical temperature T_c and at $1.1T_c$. Below the critical temperature, each value of pressure corresponds to three possible values of the volume.*

10.2 Vapour-liquid equilibrium

From the fundamental equation for the Gibbs free energy, we have $\left(\frac{\partial \mu}{\partial P}\right)_T = v$. Thus, integrating the molar volume with respect to pressure gives us the chemical potential at T and P relative to the chemical potential at a reference state with pressure P_{ref}:

$$\mu(T,P) = \mu(T,P_{ref}) + \int_{P_{ref}}^{P} v dP'.$$

Using the van der Waals equation of state and taking the reference state to be any arbitrary state at (P_{ref}, V_{ref}), we have the following expression for the chemical potential of a van der Waals fluid:

$$\mu(T,P) = \mu(T,P_{ref}) + \left[v \left(\frac{RT}{v-b} - \frac{a}{v^2} \right) \right] \Big|^{v}_{v_{ref}}$$

$$-RT \ln \left(\frac{v-b}{v_{ref}-b} \right) - \left(\frac{a}{v} - \frac{a}{v_{ref}} \right),$$

where the pressure dependence of v is implicit on the right-hand side. Using this expression, the graph of $\mu(T,P) - \mu(T,P_{ref})$ is illustrated in Figure 10.2 for the temperature $0.85T_c$. We have arbitrarily picked as the reference state (P_{ref}, V_{ref}) the point G in Figure 10.1. The states A to G in Figure 10.2 are the same states A to G in Figure 10.1. As the system moves along the isotherm from state A to state C in Figure 10.1, the chemical potential decreases along the corresponding curve ABC in Figure 10.2. Similarly, as the system moves from state E to state G along the isotherm, the chemical potential moves along the curve EFG in Figure 10.2. Because the chemical potential curves ABC and EFG intersect at the point B, which is also F, the states B and F have the same chemical potential. Thus, applying the equilibrium condition for systems at fixed T and P, the state at point B on the isotherm is in equilibrium with the state at point F. Thus, the horizontal dashed line in Figure 10.1 connects the two states B and F which are in equilibrium with each other.

It might not be clear that the choice of the pressure at which to draw the horizontal dashed line is crucial. If this dashed line were drawn at a pressure below that at B and F, then it would connect points slightly below B and F on the isotherm; these latter points are not states that are in equilibrium with each other. This is similarly the case if the dashed line had been drawn above BF. Thus, the horizontal dashed line in Figure 10.1 is at a pressure that is selected so that the two points B and F which are connected by it have the same chemical potential.

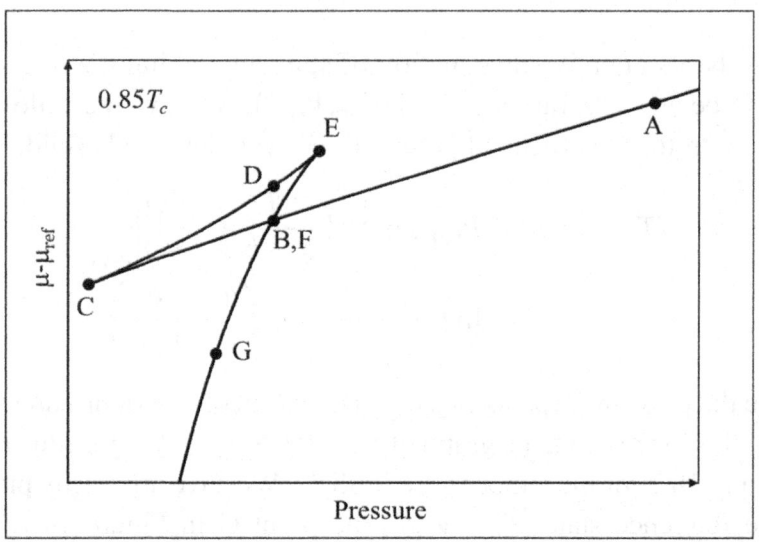

Figure 10.2: *The relative chemical potential of points on the van der Waals isotherms at $0.85T_c$. The states from A to G are the same states labelled correspondingly in Figure 9.1. States B and F have equal chemical potential; they are in equilibrium with each other.*

The chemical potential curves ABC and EFG in Figure 10.2 illustrate for the van der Waals fluid exactly what we sketched in general in Figure 7.2: reducing the pressure of a system results in the phase change from liquid to gas when the chemical potential of the gas phase drops below that of the liquid phase. For pressures greater than the pressure at the states B and F, the liquid phase is more stable than the gas phase. Conversely, for pressure lower than at states B and F, the gas phase is more stable than the liquid phase. Hence, the branches ABC and EFG of the van der Waals isotherm describe the liquid and gas phases, respectively, of the van der Waals fluid. For pressure values along the isotherm section B to C, the liquid phase chemical potential is higher than the gas phase chemical potential, so that the gas phase is more stable than the liquid phase. As the pressure decreases from B to C the volume increases. Thus, the liquid phase along BC is mechanically stable, even though it is less stable than the gas phase. The states from B to C along the isotherm are referred to as metastable. If the pressure on a liquid is carefully reduced below its boiling point without providing any nucleation centres for boiling to occur, we get superheated liquid metastable states; these are the liquid states along BC. On the other hand, from C to D, the volume given by the van der Waals equation of state increases as the pressure increases. Thus, the states along C to D are not stable mechanically. These states have never been produced in the laboratory. Similarly, the gas phase states along EF are metastable, and the states along ED are not stable. To represent the metastable liquid and gas states in Figures 7.1 and 7.2, we have drawn the curves for the Gibbs free energy for each phase to extend just slightly into the region where that phase is less stable than the other phase.

It is useful to discuss the Maxwell equal area construction. We have seen that the value of the pressure at which to draw the horizontal dashed line in Figure 10.1 is crucial. Only at the correct pressure would the liquid state B and the gas state F have equal chemical potentials and be in phase equilibrium with each other. In the equal area construction, the choice of the pressure of the horizontal line BF is made by selecting it such that the two areas BCDB and FEDF enclosed between the isotherm and the dashed line in Figure 10.1 are equal. How does this connect with the approach we have taken above?

Consider the expression for the chemical potential:

$$\mu(T,P) = \mu(T,P_{ref}) + \int_{P_{ref}}^{P} v\,dP'.$$

If we integrate from state B to state F, this is:

$$\mu_F = \mu_B + \int_{state\ B}^{state\ F} v_{vdW}\, dP,$$

where integral is from state B to state F of the molar volume of the van der Waals equation of state. We rewrite this by doing an integration by parts and noting that states B and F are at the same pressure P,

$$\mu_F = \mu_B + (vP)\big|_{(P,v_B)}^{(P,v_F)} - \int_{v_B}^{v_F} P_{vdW}\, dv'$$

$$= \mu_B + P(v_F - v_B) - \int_{v_B}^{v_F} P_{vdW}\, dv'$$

$$= \mu_B + \int_{v_B}^{v_F} (P - P_{vdW})\, dv'.$$

In the first line, we have changed the integration variable from P to v using integration by parts. In the third line, we have simply combined the contributions from the constant pressure P and the pressure given by the van der Waals equation of state P_{vdW}. Thus, the integrand is simply the difference between the pressure along the van der Waals isotherm and the constant pressure P of the horizontal line. This gives us the important point that in order for μ_B to be equal to μ_F, the two areas BCDB (giving a positive integral) and FEDF (giving a negative integral) enclosed by the isotherm and the horizontal dashed line must be equal and thus cancel in the integration.

Both this equal-area argument and our approach above depend upon the integration of pressure with respect to volume of the portion of the isotherm from C to E. This integration is at best a mathematical procedure based on the assumption that all points along the van der Waals isotherm is a thermodynamically meaningful state. In particular, a portion of any isotherm where the volume increases with pressure does not represent any set of equilibrium states. The rigorous way would be to find a path from state B to state F which integrates through equilibrium states of the van der Waals fluid in order to demonstrate that the chemical potentials at B and F are equal. One possible path is to use isobaric paths to connect the isotherm that states B and F are on to another isotherm that is above the critical temperature, and integrate along that. For example, all points along the path BAA'F'F in Figure 10.1 are well-defined stable equilibrium states of the van der Waals equation of state. This is a little tedious but would be completely rigorous thermodynamically. The same result, that is, $\mu_B = \mu_F$ would be obtained.

Even though it does not give a quantitatively accurate description of specific substances, especially for the liquid phase, the van der Waals equation of state provides a conceptually accurate account of how intermolecular interactions lead to phase equilibrium between liquid and gas. To complete the description of the van der Waals fluid, we sketch five isotherms in the PV phase diagram in Figure 10.3 for temperatures from $0.85T_c$ to $1.05T_c$.

Each isotherm below the critical temperature has a maximum and a minimum which converge at the inflection point when the temperature is equal to the critical temperature. As the temperature increases to T_c, the liquid and gas molar volumes become closer in value, until at T_c $v_l = v_g$ at the critical point (P_c, T_c). Because, the molar volume is single-valued with respective to pressure for temperatures at and above the critical temperature, the chemical potential of the van der Waals gas increases monotonically with pressure. Hence, above the critical temperature no phase change occurs from a gas phase to a liquid phase unlike the case of isotherms below the critical temperature. For temperatures above the critical temperature, the van der Waals equation of state does not distinguish between a liquid phase and a gas phase. The critical point (P_c, T_c) described qualitatively by the van der Waals equation of state is the critical point we have encountered in Figures 8.2 and 8.3 which illustrate typical one-component phase diagrams.

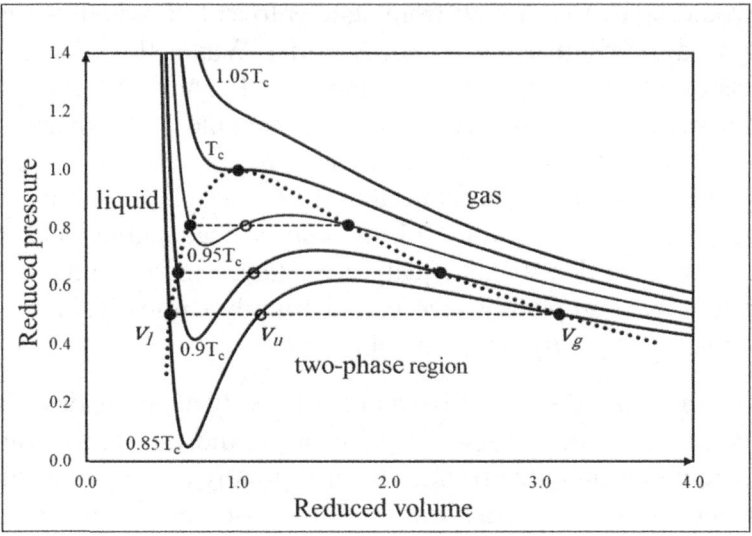

Figure 10.3: *For pressures and temperatures below the dotted envelope, the gas and liquid phases of the van der Waals fluid coexist.*

10.3 The critical point and the law of corresponding states

Below the critical temperature, for each value of the pressure, the cubic equation gives three possible solutions for the molar volume: v_l, v_u and v_g for liquid, thermodynamically unstable and gas phases, respectively. v_l and v_g are indicated by the filled circles while the unstable v_u are indicated by the empty circles in Figure 10.3 for the three isotherms below the critical temperature. The states B, D and F in Figures 10.1 and 10.2 correspond to v_l, v_u and v_g, respectively. The envelope indicated by the dotted line, which joins the points v_l and v_g for isotherms below the critical temperature, is referred to as the vapour-liquid coexistence curve. It delineates the region on the phase diagram where liquid and gas can co-exist. The region of temperature and pressure below the coexistence curve is the two-phase part of the van der Waals phase diagram. In the region of the phase diagram above the coexistence region and to the left of the critical isotherm, only the liquid phase is stable. In the region of the phase diagram to the right of the coexistence envelope or the critical isotherm only the gas phase is stable. These three regions are labelled in the figure.

The above qualitative features of the van der Waals phase diagram are common to all the so-called cubic equations of state, including the quantitatively more accurate ones such as the modified Benedict-Webb-Rubin equations. Thus, understanding the van der Waals equation of state is of some fundamental importance in dealing with processes involving gas-liquid equilibrium. For any van der Waals fluid, the critical temperature and pressure can be found by setting equal to zero the first and second derivatives of pressure with respective to volume:

$$\frac{dP}{dv} = -\frac{RT}{(v-b)^2} + \frac{2a}{v^3} = 0$$
$$\frac{d^2P}{dv^2} = \frac{2RT}{(v-b)^3} + \frac{6a}{v^4} = 0.$$

These define the inflection point of the critical isotherm, and thus solving these equations gives the critical point:

$$T_c = \frac{8a}{27Rb}$$

$$P_c = \frac{a}{27b^2}$$

$$V_c = 3b.$$

With these critical point parameters, we define dimensionless reduced temperature, pressure and volume by:

$$T_r = \frac{T}{T_c}$$

$$P_r = \frac{P}{P_c}$$

$$V_r = \frac{V}{V_c}.$$

These reduced variables are convenient to express equations of state in, and the numerical values used to label the axes in Figure 10.3 are these reduced variables. In terms of these variables, the van der Waals equation of state is

$$\left(P_r + \frac{3}{V_r^2}\right)(3V_r - 1) = 8T_r,$$

for any van der Waals substance, regardless of the values of the interaction parameters a and b. All van der Waals gases and liquids are described by this same equation. This property is known as the law of corresponding states, namely that all van der Waals gases, regardless of their differing intermolecular interactions, can be described by one universal equation. Indeed, if the compressibility of any van der Waals gas is calculated at the critical point, we have the common value for all gases:

$$Z_c = \frac{P_c V_c}{RT_c} = \frac{3}{8} = 0.375.$$

Real gases have critical compressibilities approximately equal to 0.3, which is not too different from the van der Waals universal estimate.

To the extent that real gases are only approximately described by the van der Waals equation of state, the law of corresponding states is only approximate for real gases (it is even more inaccurate for liquids). Nevertheless, it provides a conceptually insightful way to approximately capture the properties of all gases within one simple quantitative description. If the pressure, temperature and volume of gases are scaled by the values at their respective critical points, all gases are approximately the same thermodynamically.

10.4 Using the fundamental equations

We wrap up our discussion of the van der Waals equation of state by looking at its fundamental thermodynamic variables. We have seen that complete thermodynamic information is available if any one of the thermodynamic potentials is known in terms of its natural variables. Here, we consider how to proceed if we only know the PVT equation of state. Say, we have fitted PVT data for a gas and found that it is well-described by the van der Waals equation of state. What can we learn about its other thermodynamic properties?

Taking the internal energy as a function of temperature and volume, we have

$$
\begin{aligned}
dU &= \left(\frac{\partial U}{\partial T}\right)_V dT + \left(\frac{\partial U}{\partial V}\right)_T dV \\
&= C_V dT + \left(\frac{\partial U}{\partial V}\right)_T dV.
\end{aligned}
$$

In Section 7.7.2 and Example 7.9 we have already seen that

$$
\left(\frac{\partial U}{\partial V}\right)_T = T\left(\frac{\partial P}{\partial T}\right)_V - P = \frac{an^2}{V^2}.
$$

Thus, we have the differential of the internal energy:

$$
dU = C_V dT + \frac{an^2}{V^2} dV
$$

where the second term depends only upon the volume. In principle, the heat capacity can depend upon both the temperature and volume, i.e., $C_V = C_V(T, V)$. We examine this by taking its derivative with respect to the volume,

$$
\begin{aligned}
\left(\frac{\partial C_V}{\partial V}\right)_T &= \left[\frac{\partial}{\partial V}\left(T\left(\frac{\partial S}{\partial T}\right)_V\right)\right]_T \\
&= \left[T\frac{\partial}{\partial T}\left(\frac{\partial S}{\partial V}\right)_T\right]_V \\
&= \left[T\frac{\partial}{\partial T}\left(\frac{\partial P}{\partial T}\right)_V\right]_V \\
&= T\left(\frac{\partial^2 P}{\partial T^2}\right)_V \\
&= 0
\end{aligned}
$$

for the van der Waals equation of state. In the first line, we have used the definition of the heat capacity in terms of the temperature derivative of the entropy from our discussion of the Clausius inequality. In the second line, we have exchanged the order of the derivative, and in fourth line, used the Maxwell relation from the Helmholtz free energy. Since the pressure is linear in temperature in the van der Waals equation of state, we have the final result.

Hence, C_V for the van der Waals gas is not dependent upon its volume, just like for an ideal gas. Now, at large molar volume, the van der Waals gas behaves like an ideal gas. Therefore, the heat capacity of the van der Waals gas must be the same as that of an ideal gas in the limit of large volume. However, since C_V is independent of volume as we have just seen, it means that the van der Waals gas has the same constant volume heat capacity as the ideal gas, for any volume. Therefore, we have found that the constant volume heat capacity of the van der Waals gas is a constant, independent of temperature and volume. As a reminder of our discussions in Section 6.3, it is for this reason that the van der Waals gas, just like the ideal gas, is not a good model at low temperatures when quantum mechanics requires that $C_V \to 0$ as temperature goes to zero.

Since C_V is constant, we are thus able to easily integrate dU from a reference state at (T_{ref}, V_{ref}),

$$U(T,V) - U(T_{ref}, V_{ref}) = \int_{T_{ref}}^{T} C_V dT' + \int_{V_{ref}}^{V} \frac{an^2}{V'^2} dV'$$

$$= C_V(T - T_{ref}) - an^2 \left(\frac{1}{V} - \frac{1}{V_{ref}} \right).$$

It is useful to write the first term on the right in terms of the change the internal energy of the ideal gas when temperatures changes from T_{ref} to T:

$$\Delta U_{vdW}(T,V) = U_{ideal}(T) - U_{ideal}(T_{ref}) - an^2 \left(\frac{1}{V} - \frac{1}{V_{ref}} \right).$$

§   *Example 10.1 Reversible isothermal compression of a van der Waals gas*

A van der Waals fluid is compressed reversibly and isothermally from V_i to V_f. Calculate the change in internal energy, and the work and heat exchange needed. Compare the work done to that for an ideal gas for the same change in volume in a similar reversible isothermal compression.

The internal energy change is

$$\Delta U = C_V\left(T_f - T_i\right) - an^2\left(\frac{1}{V_f} - \frac{1}{V_i}\right)$$

$$= -an^2\left(\frac{1}{V_f} - \frac{1}{V_i}\right),$$

since the temperature stays constant. At the same time, the work done on the gas in a reversible compression is

$$\Delta W = -\int_{V_i}^{V_f} PdV = -\int_{V_i}^{V_f}\left(\frac{nRT}{V' - nb} - \frac{an^2}{V'^2}\right)dV'$$

$$= -nRT\ln\left(\frac{V_f - nb}{V_i - nb}\right) - an^2\left(\frac{1}{V_f} - \frac{1}{V_i}\right).$$

And thus, the heat transferred to the gas is

$$\Delta Q = \Delta U - \Delta W$$

$$= nRT\ln\left(\frac{V_f - nb}{V_i - nb}\right).$$

Examine the two terms in ΔW. For compression, V_f is smaller than the initial volume V_i. Hence, the first term is positive while the second term is negative. Compare the first term in ΔW to the work done in compressing an ideal gas,

$$\Delta W_{ideal} = -nRT \ln\left(\frac{V_f}{V_i}\right).$$

The repulsive intermolecular interaction due to the finite volume of the molecules gives an additional positive contribution to the compression work needed in comparison to compressing an ideal gas because

$$\Delta W - \Delta W_{ideal} = -nRT \ln\left(\frac{1 - \dfrac{nb}{V_f}}{1 - \dfrac{nb}{V_i}}\right)$$

is positive. Conversely, the attractive intermolecular interaction gives a negative contribution. The signs of these contributions to the work needed for compressing a gas are as expected. §

§   *Example 10.2 Irreversible adiabatic expansion of a van der Waals gas*

We expand a van der Waals gas adiabatically and irreversibly into a vacuum from volume V_f to V_i. Discuss how its temperature changes.

Work and heat are both zero, and thus ΔU is zero. Hence,

$$\Delta U = C_V(T_f - T_i) - an^2\left(\frac{1}{V_f} - \frac{1}{V_i}\right) = 0.$$

Therefore, the temperature change is equal to:

$$T_f - T_i = \frac{an^2}{C_V}\left(\frac{1}{V_f} - \frac{1}{V_i}\right).$$

A van der Waals gas cools down in an irreversible, adiabatic expansion. The decrease in temperature is proportional to the strength of the attractive intermolecular interaction. This is as expected since the energy for the increase in the intermolecular potential energy as a result of the expansion can come only from the kinetic energy of the thermal motion of the molecules. §

The expressions above for ΔU, ΔW and ΔQ are valid for single phase systems of the van der gas or the van der Waals liquid; but not for a two-phase system. The quantitative description of the gas phase can be reasonably accurate. However, the van der Waals equation of state is not a good approximation for the liquid phase.

Notice that the approach used in Examples 10.1 and 10.2 is the same as what we have used in Examples 3.1 and 3.3 for the ideal gas. The only difference is that we have to deal with a slight bit more algebra with the van der Waals equation of state. We next consider the reversible adiabatic expansion of a van der Waals gas from volume V_i to V_f. Since the process is adiabatic, we have $dU = dW$, so that the process is described by equating the general expressions for dU and dW:

$$C_V dT + \frac{an^2}{V^2} dV = -P dV$$

and with the pressure given by the van der Waals equation,

$$C_V dT = -\left(P + \frac{an^2}{V^2}\right) dV = -\frac{nRT}{(V - nb)} dV.$$

This is almost the same equation that holds for the reversible adiabatic expansion of an ideal gas, except that instead of $(V - nb)$ the ideal gas has V in the denominator on the right-hand side. Thus, collecting variables and integrating this from the initial state (T_i, V_i) to the final state (T_f, V_f) we have:

$$T_i(V_i - nb)^{\frac{R}{C_V}} = T_i(V_i - nb)^{\frac{R}{C_V}}.$$

This is the equation for states along an adiabat of a van der Waals gas in terms of temperature and volume. Combining this with the van der Waals equation of state, we obtain states along the adiabat in terms of pressure and volume:

$$\left(P_i + \frac{an^2}{V_i^2}\right)(V_i - nb)^{\frac{C_V + R}{C_V}} = \left(P_f + \frac{an^2}{V_f^2}\right)(V_f - nb)^{\frac{C_V + R}{C_V}}.$$

Similarly, an expression in terms of P and T can be obtained to relate states along the adiabat. The form of the equations of these adiabats correspond to those of the ideal gas which we found in Example 3.4.

We finally consider the entropy for the van der Waals fluid. Writing the entropy as a function of T and V, we have

$$
\begin{aligned}
dS &= \left(\frac{\partial S}{\partial T}\right)_V dT + \left(\frac{\partial S}{\partial V}\right)_T dV \\
&= \frac{C_V}{T} dT + \left(\frac{\partial P}{\partial T}\right)_V dV \\
&= \frac{C_V}{T} dT + \frac{nR}{V - nb} dV
\end{aligned}
$$

where we have again used $T\left(\frac{\partial S}{\partial T}\right)_V = C_V$ and the Maxwell relation for the Helmholtz free energy in the second line. Integrating from a reference state with temperature T_{ref} and volume V_{ref}, we get:

$$
\begin{aligned}
S(T,V) - S\left(T_{ref}, V_{ref}\right) &= C_V \ln \frac{T}{T_{ref}} + nR \ln \frac{(V - nb)}{\left(V_{ref} - nb\right)} \\
&= C_V \ln \frac{T}{T_{ref}} + nR \ln \frac{V}{V_{ref}} + nR \ln \left(\frac{1 - \dfrac{nb}{V}}{1 - \dfrac{nb}{V_{ref}}}\right).
\end{aligned}
$$

Notice from the first line that if $S(T,V) = S\left(T_{ref}, V_{ref}\right)$, then we recover the equation of the adiabat; any reversible adiabatic process is isentropic.

If we do the same calculation for the ideal gas equation of state, we obtain

$$dS = \left(\frac{\partial S}{\partial T}\right)_V dT + \left(\frac{\partial S}{\partial V}\right)_T dV$$

$$= \frac{C_V}{T} dT + \frac{nR}{V} dV,$$

so that

$$S_{ideal}(T,V) - S_{ideal}(T_{ref}, V_{ref}) = C_V \ln \frac{T}{T_{ref}} + nR \ln \frac{V}{V_{ref}}.$$

Let us choose the temperature T_{ref} to be sufficiently high and the volume V_{ref} to be sufficiently large so that the van der Waals gas behaves as an ideal gas at the reference state. With this choice of reference state, the entropy of the van der Waals gas in terms of the entropy of the ideal gas is:

$$\Delta S_{vdW}(T,V) = \Delta S_{ideal}(T,V) + nR \ln \left(\frac{1 - \frac{nb}{V}}{1 - \frac{nb}{V_{ref}}}\right).$$

With these expressions for the internal energy and the entropy, we can write the Helmholtz and Gibbs free energies as follows:

Helmholtz free energy $\Delta A = \Delta U - \Delta(TS)$

$$A(T,V) - A(T_{ref}, V_{ref})$$

$$= C_V(T - T_{ref}) - an^2\left(\frac{1}{V} - \frac{1}{V_{ref}}\right)$$

$$- nRT \ln \frac{(V - nb)}{(V_{ref} - nb)} - C_V \ln \frac{T}{T_{ref}}$$

Gibbs free energy $\Delta G = \Delta U - \Delta(TS) + \Delta(PV)$

$$G(T,V) - G\left(T_{ref}, V_{ref}\right)$$

$$= C_V\left(T - T_{ref}\right) - an^2\left(\frac{1}{V} - \frac{1}{V_{ref}}\right)$$

$$-nRT \ln\frac{(V - nb)}{\left(V_{ref} - nb\right)} - C_V \ln\frac{T}{T_{ref}}$$

$$+V\left(\frac{nRT}{V - nb} - \frac{an^2}{V^2}\right) - V_{ref}\left(\frac{nRT}{V_{ref} - nb} - \frac{an^2}{V_{ref}^2}\right).$$

For T_{ref} set equal to T, this reduces to the chemical potential which we used in our discussion in Section 10.2 of vapour-liquid phase equilibrium in the van der Waals fluid. This wraps up the important learning point in this chapter. That is, the framework discussed in Chapter 7 can be straightforwardly applied to obtain thermodynamic properties of any substance from its PVT-equation of state.

Applications of Phase Equilibrium

11.1 Colligative properties of dilute solutions

11.1.1 Lowering of the freezing point

Ice freezes out from an aqueous salt solution at a lower temperature than from pure water. The aqueous salt solution also has a higher boiling point than pure water. And, the osmotic pressure of seawater is higher than that of pure water. It is as if water molecules find it more difficult to escape from an aqueous salt solution than from pure water. These properties can be collectively understood as being due merely to how many solute particles are present in the solution, irrespective of their chemical identity. We examine this by considering the chemical potential of solutions.

Freezing point depression: we consider a solution consisting of a solvent and a dissolved solute in equilibrium with pure solid solvent. An example is an aqueous solution of common salt in equilibrium with ice. Let use denote the solvent by A and the solute by B. We see from Chapter 7 that this equilibrium is quantitatively described by the condition:

$$\mu_A^*(solid, T, P) = \mu_A(soln, T, P, x_A)$$

where $\mu_A^*(solid, T, P)$ is the chemical potential of the pure solid solvent. The right-hand side is the chemical potential of the solvent in the aqueous solution with a mole fraction of x_A. In general, we write the chemical potential of A in the solution as

$$\mu_A(soln, T, P, x_A) = \mu_A^{\emptyset}(liq, T, P) + RT \ln a_A$$

where the convention is to define $\mu_A^{\emptyset}(soln)$, the standard state chemical potential of A in the solution, as the chemical potential of the pure liquid $\mu_A^*(liq, T, P)$; as in Section 9.5.2. The activity of A is then equal to unity for the pure solvent. Thus, equilibrium between the solution and the solid solvent gives us:

$$\frac{\mu_A^*(liq, T, P) - \mu_A^*(solid, T, P)}{RT} = -\ln a_A$$

$$\frac{\Delta_{fus}\mu_A^*(T, P)}{RT} = -\ln a_A$$

where $\Delta_{fus}\mu_A^*(T, P)$ is the change in the chemical potential of A when it melts at temperature T and pressure P. Thus, equilibrium relates the three intensive variables T, P and a_A to each other, giving two independent intensive variables. This is consistent with the Gibbs Phase rule. For a two-component (solute and solvent) system c is equal to 2 and with two phases (solution and solid solvent) p is also equal to 2, so that the number of degrees of freedom is

$$f = c + p - 2 = 2.$$

With the two independent variables in the equilibrium condition, how does the equilibrium temperature vary with composition of the aqueous solution if we hold pressure constant? A quantitative approach to this question would be to take the temperature derivative of the equilibrium condition:

$$\left[\frac{\partial}{\partial T}\left(\frac{\Delta_{fus}\mu_A^*}{RT}\right)\right]_P = -\left[\frac{\partial}{\partial T}(\ln a_A)\right]_P.$$

Since the chemical potential is simply the molar Gibbs free energy, from looking at the term on the left-hand side of the equation, you will immediately realize why we introduced the Gibbs-Helmholtz equation in Section 7.12:

$$\left[\frac{\partial}{\partial T}\left(\frac{\Delta G}{RT}\right)\right]_P = -\frac{\Delta H}{RT^2}.$$

Therefore, the temperature dependence of the equilibrium condition is:

$$\frac{\Delta_{fus}h_A}{RT^2} = \left[\frac{\partial}{\partial T}(\ln a_A)\right]_P$$

where $\Delta_{fus}h_A$ is the molar enthalpy of fusion of the pure solvent at (T, P). By referencing $\Delta_{fus}h_A$ to the <u>pure</u> solvent we are taking the "solvent" standard state discussed in Section 9.5.2. Integrating both sides of this equation:

$$\int_{T_{fus}^*}^{T} \frac{\Delta_{fus}h_A}{RT'^2} dT' = \int_1^{a_A} d\ln a_A'$$

where the lower limit of the integrals is the thermodynamic state corresponding to pure solvent so at $a_A = x_A = 1$ and, therefore T_{fus}^* is the melting point of the pure solvent at pressure P. Upon integration we have

$$\Delta_{fus}h_A\left(\frac{1}{T_{fus}^*} - \frac{1}{T}\right) = R\ln a_A$$

assuming that the molar enthalpy of fusion of A is independent of temperature. If we have an ideal solution, then

$$\Delta_{fus}h_A\left(\frac{1}{T_{fus}^*} - \frac{1}{T}\right) = R\ln x_A$$
$$= R\ln(1 - x_B)$$

because $x_A = 1 - x_B$. Then using the Taylor series expansion for $\ln(1 - x)$ for small x we get

$$\Delta_{fus}h_A\left(\frac{1}{T_{fus}^*} - \frac{1}{T}\right) = -R\left(x_B + \frac{1}{2}x_B^2 + \frac{1}{3}x_B^3 + \cdots\right)$$

for dilute solutions. Therefore, we have the approximation:

$$\Delta_{fus}h_A\left(\frac{1}{T} - \frac{1}{T_{fus}^*}\right) \cong Rx_B.$$

Both $\Delta_{fus}h_A$ and x_B are positive quantities, so that we have $T_{fus}^* > T$; ice requires a lower temperature to freeze out of an aqueous solution of salt than the freezing point of pure water. From this, the lowering of the freezing point $\Delta T = \left(T_{fus}^* - T \right)$ is given by:

$$\Delta T \cong x_{solute} \frac{RT_{fus}^{*2}}{\Delta_{fus}h_{H_2O}},$$

where we have approximated $\Delta T \ll T_{fus}^*$. This is frequently written in terms of a composition variable such as the molality of the solute M_{solute}. For a dilute solution,

$$x_{solute} = \frac{n_{solute}}{n_{solute} + n_{solvent}} \cong \frac{n_{solute}}{n_{solvent}}$$

$$= \frac{n_{solute}}{m_{solvent}} \times \frac{m_{solvent}}{n_{solvent}}$$

$$= M_{solute} \times \frac{M_{solvent}}{1000}.$$

where $m_{solvent}$ is the mass in kg of the solvent present in the solution, $M_{solvent}$ is the molar mass of the solvent, and the factor of 1000 converts between gram and kilogram. Hence, freezing point constants, (also known as the cryoscopic constant) are reported as K_f, where

$$\Delta T = M_{solute} \times \frac{M_{solvent}}{1000} \frac{RT_{fus}^{*2}}{\Delta_{fus}h_{H_2O}}$$

$$= M_{solute} K_f.$$

You can see that the freezing point constant K_f is dependent only upon solvent properties T_{fus}^* and $\Delta_{fus}h_{H_2O}$; freezing point constants are properties of the solvent, not the solute, at least for an ideal solution. The decrease in the freezing point temperature is the product of K_f and the number of moles of the solute present per unit mass of the solution. It is not dependent upon the chemical identity of the solute.

§ *Example 11.1 Cryoscopic constant of water*

Calculate the cryoscopic constant of water from its enthalpy and freezing point.

The enthalpy of fusion of water is 6.008 kJ mol^{-1} at the normal (pressure of 1 bar) freezing point of water of 273.15 K. The molar mass of water $\mathcal{M}_{H_2O}$ is 18.015. Hence, the cryoscopic constant for water is

$$K_f = \frac{\mathcal{M}_{H_2O}}{1000} \frac{R{T^*_{fus}}^2}{\Delta_{fus}h_{H_2O}} = 1.86 \text{ K kg mol}^{-1}. \qquad \S$$

Note that the mole fraction x_B or the molality M_B of solute counts the number of solute particles present in the solution. For example, one mole of sucrose gives one mole of sugar molecules in solution, while one mole of common salt gives two moles of ions, Na^+ and Cl^-, in solution. Thus, the freezing point depression due to one mole of common salt is twice as large as that due to one mole of sucrose. In order to melt ice that poses a skidding hazard for driving in winter in places with cold climates, common salt is sprinkled on the roads so that the ice melts even when the temperature is below the freezing point of pure water. For this purpose, it is more effective to use, for example, one mole of $CaCl_2$ compared to one mole of $NaCl$ because the former produces three moles of ions in solution compared to two for the latter.

§ *Example 11.2 Freezing point of salt solutions*

You have an aqueous solution of one mole of sodium chloride dissolved in one kilogram of water, and a second aqueous solution with one mole of calcium chloride dissolved in one kilogram of water. Assuming complete dissociation of both salts in water, calculate the freezing points of these solutions.

The molality of the ions in the $NaCl$ solution is equal to 2 mol kg^{-1} where we have counted both the sodium and chloride ions. Using the cryoscopic constant of water calculated in Example 11.1, the freezing point depression for this solution is

$$\Delta T = M_{solute} K_f = 3.72 \text{ K}.$$

Thus, the freezing point for a one molal solution of sodium chloride is $-3.72°C$. Similarly, the freezing point for a solution with one mole of calcium chloride dissolved in one kg of water is $-5.58°C$.

§   *Example 11.3 Freezing point of seawater*

Approximate seawater as a salt solution with 35 g of sodium chloride per kg of water. At what temperature does the sea start to freeze?

The cryoscopic constant of water is $1.86 \text{ K kg mol}^{-1}$. The molality of the solute in the solution is

$$M_{solute} = \frac{35}{58.44} \text{ mol kg}^{-1}.$$

Hence, for seawater, the freezing point lowering relative to pure water is $2 \times \frac{35}{58.44} \times 1.86 \text{ K}$. Thus, the sea starts to freeze when then temperature falls below approximately $-2.23°C$. Thus, if a glacier in contact with seawater starts melting, the temperature of the sea is above $-2.23 °C$.

11.1.2 Elevation of the boiling point

Boiling point elevation: we now consider equilibrium between a solution and its pure solvent vapour. You should note that the approach is exactly parallel to that in the treatment of freezing point depression. Similar to the discussion of freezing point depression, we consider a non-volatile solute, so that the gas phase consists only of solvent molecules, just as the solid phase in freezing point depression consists of only the pure solvent. Hence, the equilibrium condition is as follows,

$$\mu_A^*(gas, T, P) = \mu_A(soln, T, P, x_A),$$

relating the three independent intensive variables T, P and x_A. Using the same standard state for the chemical potential of a solution as above,

$$\mu_A(soln, T, P, x_A) = \mu_A^{\emptyset}(liq, T, P) + RT \ln a_A$$

we now obtain:

$$\frac{\mu_A^*(gas, T, P) - \mu_A^*(liq, T, P)}{RT} = \ln a_A,$$

$$\frac{\Delta_{vap}\mu_A^*(T, P)}{RT} = \ln a_A.$$

$\Delta_{vap}\mu_A^*(T, P)$ is the change in the chemical potential of solvent A upon vaporization at temperature T and pressure P. This is the same as the corresponding equation for freezing point depression except for the sign of the logarithmic term. Now, we take the derivative with respect to temperature in order to cast this equation in a recognizable form to apply the Gibbs-Helmholtz equation. This gives

$$\left[\frac{\partial}{\partial T}\left(\frac{\Delta_{vap}\mu_A^*}{RT}\right)\right]_P = -\frac{\Delta_{vap}h_A}{RT^2} = \left[\frac{\partial}{\partial T}(\ln a_A)\right]_P.$$

Assuming that the enthalpy of vaporization $\Delta_{vap}h_A$ is constant with temperature, we integrate from the pure solvent with boiling point T_{vap}^* to a solution of activity a_A with boiling point T, to get

$$\frac{\Delta_{vap}h_A}{R}\left(\frac{1}{T_{vap}^*} - \frac{1}{T}\right) = -\int_1^{a_A} d\ln a_A'.$$

If we have an ideal solution, then

$$\Delta_{vap}h_A\left(\frac{1}{T_{vap}^*} - \frac{1}{T}\right) = -R\ln x_A$$

$$\cong Rx_B.$$

Vaporization enthalpy is positive, mole fraction x_B is positive. Hence, $T_{vap}^* < T$. For example, an aqueous salt solution boils at a higher temperature than pure water. Writing this in terms of the boiling point elevation $\Delta T = \left(T - T_{vap}^*\right)$, we have:

$$\Delta T \cong x_B \frac{RT_{vap}^{*}{}^{2}}{\Delta_{vap}h_A}$$

taking $\Delta T \ll T_{vap}^*$. The boiling point elevation constant K_b (sometimes known as the ebullioscopic constant) is frequently defined with units of molality similarly to the freezing point constant. This gives the boiling point elevation of:

$$\Delta T = M_{solute} \times \frac{\mathcal{M}_{solvent}}{1000} \frac{RT_{vap}^{*}{}^{2}}{\Delta_{vap}h_{H_2O}}$$

$$= M_{solute}K_b,$$

where M_{solute} is the molality of the solute in the solution.

§   ***Example 11.4 Ebullioscopic constant of water***

The enthalpy of vaporization of water is 40.656 kJ mol⁻¹ at the normal boiling point of water at 373.15 K. Calculate the ebullioscopic constant of water.

$$K_b = \frac{\mathcal{M}_{solvent}}{1000} \frac{RT_{vap}^{*}{}^{2}}{\Delta_{vap}h_{H_2O}} = 0.513 \text{ K kg mol}^{-1}. \qquad §$$

We pause here to discuss the *form* of the equations describing freezing point depression and boiling point elevation of dilute solutions of solute mole fraction x_{solute}:

$$\Delta h \left(\frac{1}{T^*} - \frac{1}{T}\right) = \pm R \ln x_{solute}.$$

To arrive at these equations, we have applied the condition that at thermodynamic equilibrium the chemical potential of the solvent is the same in both of the phases in each system. This comes ultimately from the Second Law which tells us that the entropy of the universe does not change when a system is at equilibrium. Let us see examine how the entropies of the surroundings and the system change when ice melts into the pure liquid phase compared to melting into a solution.

When ice melts into the pure solvent A at temperature T^*_{fus} the entropy of the surroundings decreases by $\Delta_{fus}h_A/T^*_{fus}$ because an amount of heat $\Delta_{fus}h_A$ is absorbed from the surroundings at temperature T^*_{fus}. At the same time, the entropy of the system increases because there are more ways of arranging the solvent molecules in the liquid phase than in the solid phase. These entropy changes balance out so that $\Delta S_{sys} + \Delta S_{surr} = \Delta S_{universe} = 0$.

On the other hand, when ice melts into a solution of composition x_{solute}, there is an additional positive contribution to ΔS_{sys} due to the entropy of mixing, which is zero for the pure solvent. From the Gibbs formula for the entropy of mixing in Section 9.2, this additional increase per mole in ΔS_{sys} is

$$\Delta S_{mix} = -R[x_{solute} \ln x_{solute} + (1 - x_{solute}) \ln(1 - x_{solute})],$$

assuming an ideal solution. For dilute solutions, $x_{solute} \to 0$ and

$$\Delta S_{mix} \cong -R \ln(1 - x_{solute}) \cong R x_{solute}.$$

To balance this larger increase in the system entropy, the melting temperature T has to be smaller than T^*_{fus} so that the entropy decrease of the surroundings is larger by $\Delta h \left(\dfrac{1}{T} - \dfrac{1}{T^*_{fus}} \right)$ compared to melting into the pure solvent at T^*_{fus}. Thus, for melting into a solution of mole fraction x_{solute}, the equilibrium temperature T must be such that

$$\Delta h \left(\frac{1}{T} - \frac{1}{T^*_{fus}} \right) = R x_{solute},$$

which is what we have obtained above in a more formal way. A similar argument can be made for boiling point elevation, briefly sketched as follows. When boiling a solution, there is a smaller increase in the system entropy compared to boiling the pure solvent. This is due to a mixing entropy contribution for the solution. At the same time, we get a smaller decrease in the entropy of the surroundings if boiling occurs at a higher temperature than T^*_{vap}. If you reread Section 9.2, you can see that both boiling point elevation and freezing point depression of a dilute solution are really direct consequences of the entropy of mixing; this is an important point to understand.

11.1.3 The osmotic pressure

Osmotic pressure of solutions: we have considered the pure solvent in the solid phase in equilibrium with the solution. We have also considered the pure solvent in the gas phase in equilibrium with the solution. Now we consider the pure liquid phase solvent in equilibrium with the solution at temperature T. This equilibrium can be maintained by separating the aqueous salt solution from water using a rigid semi-permeable membrane through which only water molecules can pass, and with an additional pressure Π applied to the solution. This is illustrated in Figure 11.1.

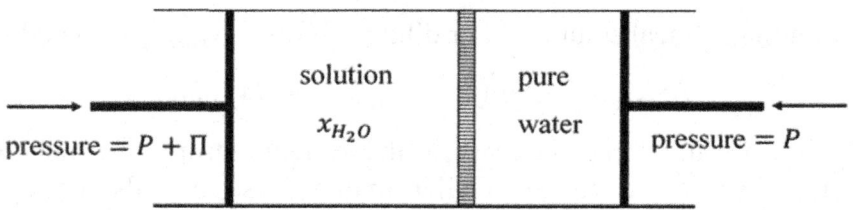

Figure 11.1: *Equilibrium between an aqueous solution of osmotic pressure Π in the left compartment and pure water in the right compartment.*

The osmotic pressure of the solution is the additional pressure Π that needs to be applied to it so that the solution is in equilibrium with the pure water, at which point there is no net flow of water through the semi-permeable membrane. If the applied pressure is lower than Π, water molecules move from the pure water compartment to the solution compartment. Conversely, if the applied pressure is higher than Π, water molecules are driven from the solution into the pure water compartment. In the process of reverse osmosis the additional pressure Π is applied to sea water in the left compartment in order to extract pure water into the right compartment through a semi-permeable membrane.

In order for there to be equilibrium between the two phases, we have:

$$\mu_{H_2O}^*(liq, T, P) = \mu_{H_2O}\left(soln, T, P + \Pi, x_{H_2O}\right)$$
$$= \mu_{H_2O}^{\emptyset}(liq, T, P + \Pi) + RT \ln a_{H_2O},$$

with the standard chemical potential $\mu_{H_2O}^{\emptyset}$ being the chemical potential of pure water $\mu_{H_2O}^*$. If the solution is ideal, we simply replace the activity a_{H_2O} with the mole fraction x_{H_2O}. Collecting the terms for the chemical potential of the pure liquid, we obtain:

$$-RT \ln a_{H_2O} = \mu_{H_2O}^*(liq, T, P + \Pi) - \mu_{H_2O}^*(liq, T, P)$$
$$= \Delta_\Pi \mu_{H_2O}^*(liq, T)$$

with $\Delta_\Pi \mu_{H_2O}^*$ is the change in the chemical potential of the pure liquid solvent when its pressure is increased from P to $P + \Pi$, keeping the temperature constant. We know from Chapter 7 that $\left(\frac{\partial G}{\partial P}\right)_T = V$ from the fundamental equation for the Gibbs free energy. We then have:

$$\Delta_\Pi \mu_{H_2O}^*(liq, T) = \int_P^{P+\Pi} v \, dP' \cong \Pi v$$

with v being the molar volume of the pure solvent in the liquid phase, which we approximate to be independent of the pressure. Hence, for equilibrium with the activity of water equal to a_{H_2O}, we get:

$$\Pi v = -RT \ln a_{H_2O}.$$

For an ideal solution, we replace the activity with the mole fraction:

$$\Pi v \cong -RT \ln x_{H_2O} \cong x_{solute} RT$$

where we have used

$$\ln x_{H_2O} = \ln(1 - x_{solute}) \cong -x_{solute}$$

with x_{solute} being the mole fraction of solute particles in solution. This is related to the mole fraction of the salt by $x_{solute} = \mathcal{N} x_{salt}$ where $\mathcal{N}$ is the number of moles of ions per mole of the salt. For sodium chloride, $\mathcal{N} = 2$, since there is one mole of Na^+ and one mole of Cl^- in solution per mole of dissolved $NaCl$. Hence, for a dilute solution

$$\Pi v = \mathcal{N} x_{salt} RT.$$

This equation gives the osmotic pressure of a salt solution at temperature T and mole fraction x_{salt}, if the salt solution can be approximated as an ideal solution.

For solutions that are sufficiently dilute to be ideal, it is clear that freezing point depression, boiling point elevation and the osmotic pressure are proportional to the number of solute particles present per unit volume of the solution. This suggests a comparison to ideal gases where the pressure is proportional to the number of gas particles per unit volume. Considering the osmotic pressure, there is an especially simple comparison with the ideal gas equation of state.

Let the number of moles of solution be n so that $n = n_{H_2O} + n_{salt}$, the sum of the number of moles of water and salt. The mole fraction of salt is equal to

$$x_{salt} = \frac{n_{salt}}{n_{H_2O} + n_{salt}} = \frac{n_{salt}}{n},$$

while molar volume of the solution is approximately

$$v \cong \frac{V_{soln}}{n}.$$

The osmotic pressure equation is then given by:

$$\Pi V_{soln} \cong \mathscr{N} n x_{salt} RT = \mathscr{N} n_{salt} RT = n_{solute} RT,$$

which has exactly the form of the ideal gas equation of state $PV = nRT$; the solute particles behave analogously to an ideal gas in the solution. Since it is convenient to use moles per unit volume for concentrations of liquid solutions, this equation is frequently written as:

$$\Pi = \frac{n_{solute}}{V_{soln}} RT = CRT,$$

where C is the number of moles of solute per unit volume of the solution. This is the van't Hoff equation for the osmotic pressure of the salt solution. It is approximate.

§   *Example 11.5 Osmotic pressure of seawater*

Seawater has about 35g of salt per kilogram of water. Approximately modelling seawater as a solution of sodium chloride with this concentration, let's calculate the osmotic pressure of seawater at a temperature of 298.15 K.

With a molar mass of 58.44 for $NaCl$, the salt concentration is:

$$C \cong \frac{35}{58.44} \text{ mol dm}^{-3}.$$

Hence, the osmotic pressure of seawater is approximately equal to:

$$\Pi = 2 \left(\frac{35}{58.44}\right) \times 1000 \times 8.314 \times 298.15 = 2.97 \times 10^6 \text{ N m}^{-2},$$

which is about 29 atmospheres pressure! This is the pressure needed in reverse osmosis to push water molecules from seawater through a semi-permeable membrane to recover fresh water.   §

11.2 Solution-pure solid equilibrium

Although substances in their liquid phases are frequently miscible with each other, many pairs of substances, such as sodium chloride and water, do not form solid solutions (or alloys) with each other. For example, sodium chloride crystals in equilibrium with a saturated solution of aqueous sodium chloride contain essentially no water molecules. The phase diagram of such a pair of substances is simple and illustrative of how we read phase diagrams in general. In this section, we examine such a phase diagram, the general structure of which is illustrated in Figure 11.2.

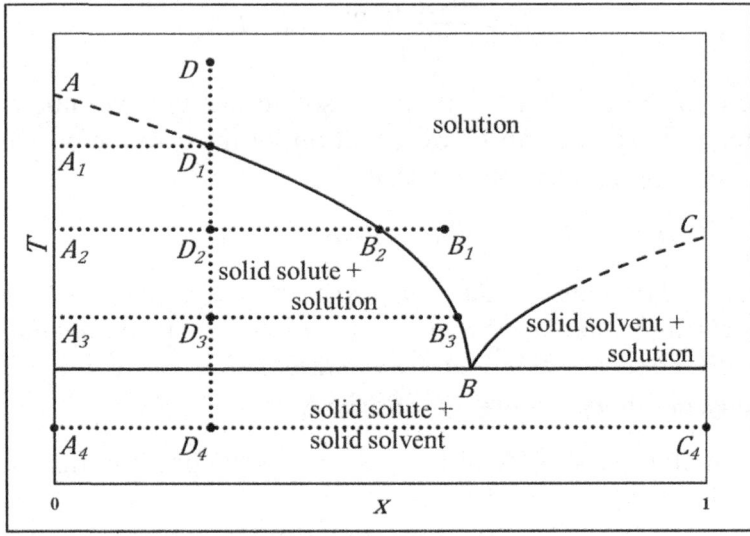

Figure 11.2: *Temperature-composition phase diagram of a binary system in which the crystals of the two components are not miscible.*

This is a temperature-composition phase diagram. It shows the phase equilibrium between the two components for a specific pressure. The vertical axis is the temperature and the horizontal axis is the composition of the two-component system in terms of the mole fraction of one of the components, arbitrarily labelled as the solvent; $x = 1$ is pure solvent, $x = 0$ is pure solute. The equilibrium curves AB and BC describe the composition dependence of the temperature at which the solute or solvent, respectively, freezes out of the liquid solution. Thus, in the region of the phase diagram above these curves the equilibrium state of the system is a liquid solution consisting of the solute and the solvent. The point A, at a temperature $T_A = T^*_{solute}$, indicates the freezing point of the pure solute at pressure P. The point C, at temperature $T_C = T^*_{solvent}$, is the freezing point of the pure solvent at pressure P.

When we discussed freezing point lowering in Section 11.1, we obtained approximate equations for the phase equilibrium lines AB and BC in the composition regions where the solution composition is close to the pure solute or the pure solvent, respectively. These regions are indicated by the dashed portions of AB and BC. We have seen that when $x_{solvent}$ is close to unity, the freezing temperature T of the pure solvent depends upon the composition in the following way:

$$\Delta T = T^*_{solvent} - T \cong -\frac{RT^*_{solvent}{}^2}{\Delta_{fus}H_{solvent}} \ln x_{solvent}$$

where $\Delta_{fus}H_{solvent}$ is the enthalpy of fusion of the solid solvent into the solution. This equation gives the dashed portion of the curve BC.

To remind ourselves,

$$x_{solvent} = 1 - \sum_i x_i$$

where $\sum_i x_i = x_{solute}$ is the sum of the mole fractions of all species i that are present in the solution due to the dissolution of the solute. For example, for an aqueous solution of sodium chloride, we have

$$x_{solute} = x_{Na^+} + x_{Cl^-}$$
$$x_{solvent} = x_{H_2O}.$$

We neglect the much smaller mole fractions x_{H^+} and x_{OH^-} contributions from the dissociation of water molecules in the solution. For pure water when $x_{H_2O} = 1$, the enthalpy change $\Delta_{fus}H_{solvent}$ is simply the enthalpy of fusion of ice. For a salt solution, $\Delta_{fus}H_{solvent}$ is the enthalpy of dissolution of ice into the salt solution.

Similarly, at the other end of the composition range when $x_{solvent}$ is close to zero, the freezing point T of the solute out of the solution depends upon the solution composition in the following way:

$$\Delta T = T^*_{solute} - T \cong -\frac{R T^{*}_{solute}{}^{2}}{\Delta_{fus} H_{solute}} \ln x_{solute}$$

where $\Delta_{fus} H_{solute}$ is the enthalpy of fusion of the solid solute into the solution. For an aqueous solution of sodium chloride, x_{solute} is the sum of x_{Na^+} and x_{Cl^-}, as noted above. This equation gives the dashed portion of the equilibrium curve AB.

For solution compositions that are not in these dilute regions, the ideal solution approximation generally does not hold. A quantitative description of the portion of the curves AB and BC drawn with solid lines needs a similar treatment as in Section 11.1, but with the chemical potentials in terms of the activities $a_{solute} = \gamma_{solute} x_{solute}$ and $a_{solvent} = \gamma_{solvent} x_{solvent}$ which we have defined in Chapter 9:

$$\mu_{solvent} = \mu^{\emptyset}_{solvent}(T, P) + RT \ln a_{solvent}$$

$$\mu_{solute} = \mu^{\emptyset}_{solute}(T, P) + RT \ln a_{solute}$$

with $\gamma_{solvent} \to 1$ and $\gamma_{solute} \to 1$ as $x_{solvent} \to 1$. This is sufficient for a qualitative understanding of phase diagrams. We will not pursue any quantitative treatment, which may be somewhat involved even for simple reasonable approximations for the dependence of the activity coefficients upon temperature, pressure and composition. For electrolytes, the Debye-Hückel model in Section 9.7.2 can be used to approximate the activities of the ions in solution.

Consider a system at the state indicated by the point D, for which the overall mole fraction of the system is x_D and its temperature is T_D. When its temperature is decreased, the system traverses the vertical dotted line through D. When the system is cooled from D to any temperature above D_1, the equilibrium state of the system is a liquid solution because T_{D_1} is the freezing point of the solution with composition x_D. At D_1 pure solute, denoted by the state A_1, begins to freeze out of the solution. Thus, the concentration of the solute in the solution decreases. When the temperature is lowered further to T_{D_2}, the state of the system consists of the pure solute (state A_2) in equilibrium with a solution of composition x_{B_2} (state B_2). States A_2 and B_2 are joined by the horizontal line $A_2 D_2 B_2$ at temperature T_{D_2}. In general, such lines on phase diagrams that connect states in phase equilibrium with each other are called tie-lines. Thus, as a solution of concentration x_D is cooled from temperature T_{D_1} to T_{D_2} to T_{D_3}, pure solute freezes out while the solute concentration in the solution decreases from state D_1 to B_2 to B_3 along the curve AB. These states are in equilibrium with pure solute states A_1, A_2 and A_3 to which they are connected by tie-lines at temperatures T_{D_1}, T_{D_2} and T_{D_3}, respectively.

In the previous section, we have discussed equilibrium curves such as AB in terms of the lowering of the freezing point of the solute as a function of the composition of the solution. It is also commonly thought of as the solubility of the solute as a function of the temperature. Thus, AB simply relates the saturation concentration of the solute as a function of the temperature. For example, if we start from state B_1 and increase the concentration of the solute at constant temperature (T_{D_2}), the solution becomes saturated at the state B_2. If we further add solid solute to the solution, no more solute will dissolve. The equilibrium state then consists of the excess solid solute together with the saturated solution, i.e. states A_2 and B_2, as we have above. What happens along equilibrium curve BC is the same, just switching the roles of the solute and the solvent.

When a solution is cooled below the curve BC, pure solvent freezes out, and the solution gets concentrated in the solute. For any mole fraction along either AB or BC, when the temperature is lowered to T_B or below, both solute and solvent freeze out of solution. For temperatures below T_B, the equilibrium state consists of two separate solid phases: pure solute crystals at state A_4 and pure solvent crystals at state C_4, for example. No liquid phase exists at equilibrium below T_B. Such a state B is known as the eutectic point of the system.

At the eutectic point, freezing of both pure solids occur in the proportions such that the composition of the solution does not change due to the solids freezing out. To summarise the above discussion, we have labelled the regions of the phase diagram in Figure 11.2. Above the equilibrium curves AB and BC we have "*solution*". The region bounded by AB and the horizontal line at the eutectic temperature T_B is "*solid solute + solution*". Similarly, the region bounded by BC and the horizontal line at T_B is "*solid solvent + solution*". For temperatures below T_B, we have "*solid solute + solid solvent*".

11.3 From tie-line to the lever rule

At any temperature and composition in the two-phase regions of the phase diagram, how much of each phase is there at equilibrium? Consider a system whose state is indicated by the point D_2 in the phase diagram in Figure 11.2. This is a system which is at a temperature T_{D_2} and has an overall composition of x_{D_2}. The tie-line (dotted horizontal line) through the state D_2 tells us that at equilibrium, the system consists of two phases: a solution of composition x_{B_2} and the pure solid solute with composition x_{A_2}. The mole fraction x_{A_2} is zero in this case, but we will retain the symbol so that our result will be a general one for the equilibrium between two phases of any composition. Let the total number of moles in the system be N_{tot}. This phase separates into N_{soln} moles of solution and N_{solid} moles of solid:

$$N_{tot} = N_{soln} + N_{solid}.$$

The total number of moles of solvent in the system is:

$$N_{solvent} = x_{D_2} N_{tot},$$

and this is distributed between the solution and the solid, so that:

$$x_{D_2} N_{tot} = x_{B_2} N_{soln} + x_{A_2} N_{solid},$$

using the equilibrium compositions of the solution and the solid phases from the tie-line. We then substitute into this expression the total number of moles in terms of the number of moles of solution and solid, getting:

$$x_{D_2}(N_{soln} + N_{solid}) = x_{B_2}N_{soln} + x_{A_2}N_{solid}.$$

Upon rearranging the terms, this gives us:

$$(x_{D_2} - x_{A_2})N_{solid} = (x_{B_2} - x_{D_2})N_{soln}.$$

You might have encountered this relationship. It is known as the lever-rule, the terminology being visually appealing if you think of the mole fraction differences $(x_{D_2} - x_{A_2})$ and $(x_{B_2} - x_{D_2})$ as the arms and the number of moles N_{solid} and $N_{solution}$ as the loads of a balanced lever. It is simply an accounting of the number of moles of each phase present, the only thermodynamic input being the overall composition of the system and the compositions of the equilibrium phases, read off the phase diagram. It is applicable in general, and we will encounter the lever rule again when we discuss other types of phase diagrams. For example, if we have a system in a state D_4 in Figure 11.2, with overall composition given by x_{D_4}, then the number of moles of the solid solvent phase and the solid solute phase are related by:

$$(x_{D_4} - x_{A_4})N_{\substack{solid \\ solute}} = (x_{C_4} - x_{D_4})N_{\substack{solid \\ solvent}}.$$

11.4 Binary gas-liquid equilibrium

When a gas mixture is in equilibrium with a solution, then generally:

$$\mu_A(soln, T, P, x_A) = \mu_A(gas, T, P, y_A).$$

Following from this equilibrium condition, if the gas mixture is ideal and the solution is ideal, then we find, from the discussion in

Section 9.3, that Dalton's law of partial pressure applies to the gas phase and Raoult's law applies to the solution, so that we have:

$$P_A = y_A P = x_A P_A^*$$

where y_A and x_A are mole fractions of the gas and liquid phases, respectively; P_A is the partial pressure of A; P_A^* is the vapour pressure of pure liquid A. We use these equations to understand the phase diagram for gas-liquid equilibrium in two-component systems. Specifically, we have:

$$P_A = x_A P_A^*$$
$$P_B = x_B P_B^* = (1 - x_A)P_B^*.$$

Therefore, the total pressure as a function of x_A is

$$P(x_A) = P_A + P_B = x_A(P_A^* - P_B^*) + P_B^*.$$

We see from this that the total pressure is a straight-line, depending linearly upon the solution mole fraction x_A. Thus, the total pressure of the system is a function of the composition and the temperature, the latter through the vapour pressures of the pure liquids A and B. Similarly, the partial pressures P_A and P_B are also linearly dependent upon the composition. This is illustrated in the pressure-composition diagram in Figure 11.3 for the temperature T. As sketched, at that temperature the pure component A has a higher vapour pressure than the pure component B, and thus A is more volatile.

How about the dependence of the total pressure P upon y_A, the mole fraction of A in the gas phase? In the above, we have used Raoult's law in examining the component partial pressures of the solution. Let us bring Dalton's law into the discussion:

$$P_A = x_A P_A^* = y_A P$$

$$P_B = (1 - x_A)P_B^* = (1 - y_A)P.$$

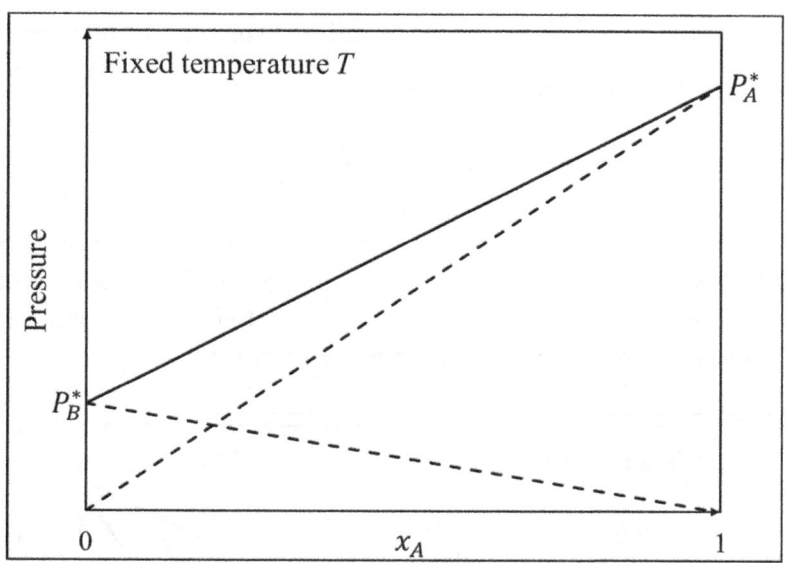

Figure 11.3: *Vapour pressure at temperature T of an ideal binary solution as a function of the mole fraction of component A. The solid line is the total pressure, the dashed lines are the partial pressures of A and B. P_A^* and P_B^* are the vapour pressures of the pure components.*

We already have an expression for the total pressure P in terms of the mole fraction x_A, so substituting that into the first equation here, we have:

$$x_A P_A^* = y_A[x_A(P_A^* - P_B^*) + P_B^*]$$

which immediately gives

$$y_A = \frac{x_A P_A^*}{P_B^* + (P_A^* - P_B^*)x_A},$$

and allows us to write the equilibrium mole fraction of the solution x_A in terms of the equilibrium mole fraction of the gas phase y_A:

$$x_A = \frac{y_A P_B^*}{P_A^* + (P_B^* - P_A^*)y_A}.$$

But we have $P = \frac{x_A P_A^*}{y_A}$, so that the total pressure can now be written in terms of the gas-phase mole fraction y_A:

$$P(y_A) = \frac{P_A^* P_B^*}{P_A^* + (P_B^* - P_A^*)y_A}.$$

Unlike the total pressure in terms of the solution mole fraction $P(x_A)$, the dependence of the total pressure upon the gas phase mole fraction $P(y_A)$ is not a straight-line. We sketch these two expressions for the total pressure together in one diagram in Figure 11.4.

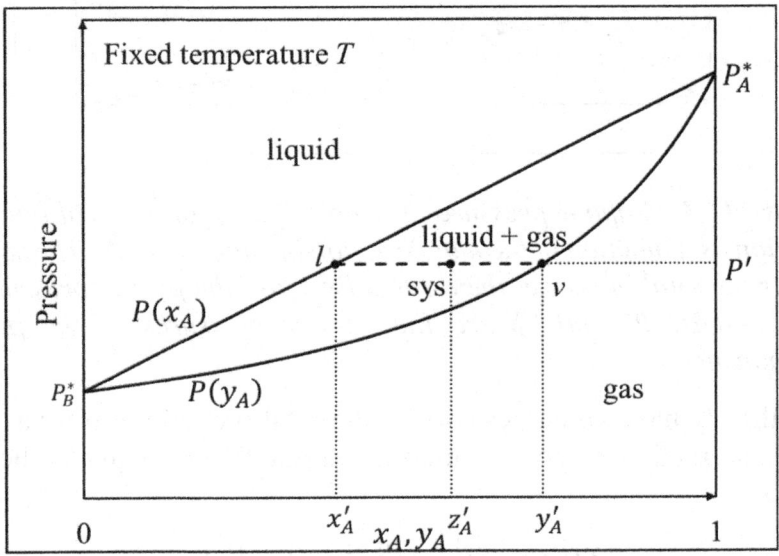

Figure 11.4: *P-xy diagram for an ideal-solution in equilibrium with its ideal gas mixture. The tie-line for pressure P′ is drawn as the dashed horizontal line.*

The horizontal axis here serves to indicate both x_A and y_A, the pressure dependence upon the liquid and gas phase mole fractions are given by the line $P(x_A)$ and the curve $P(y_A)$, respectively. This type of phase diagram showing the dependence of the total pressure upon both the liquid and gas compositions is called a *P-xy* phase diagram.

How do we read this pressure-composition phase diagram? At pressures above the liquid line, the equilibrium state of the system is a liquid solution of the components A and B. This is a single phase, two-component state, which we have labelled "liquid". At pressures below the gas-phase curve, the equilibrium state is a gas phase of the two components A and B, which we have labelled "gas". The equilibrium state in the region, labelled "liquid + gas", bounded by the liquid line and the gas-phase curve consists of two phases each with two components A and B. Applying the Gibbs Phase Rule to this region, we have

$$
\begin{aligned}
f &= c + p - 2 \\
&= 2 + 2 - 2 \\
&= 2.
\end{aligned}
$$

Hence, of the four variables describing the system, namely P, T, x_A and y_A, only two are independent. For a fixed temperature as we have sketched in Figures 11.3 and 11.4, if we specify the pressure, then both x_A and y_A are determined; the compositions of both liquid and gas phase are determined once temperature and pressure are specified. This is illustrated in Figure 11.4 using the idea of tie-lines from Section 11.3.

At the point l on the liquid curve the pressure is P' and the mole fraction of A in the solution is x'_A. At the point v on the gas-phase curve at the same pressure P', the mole fraction of A in the gas phase is y'_A. The horizontal dashed line connecting points l and v is the tie-line at temperature T and pressure P'; at this temperature and pressure a solution of mole fraction x'_A is in equilibrium with a gas of mole fraction y'_A. Similarly, at any other pressure between P^*_A and P^*_B, a tie-line connects the respective liquid and gas states that are at phase equilibrium with each other. Because A is more volatile than B, we intuitively expect x'_A to be less than y'_A. Thus, at any temperature and pressure in the two-phase region of the phase diagram, we have a liquid in equilibrium with a gas that is enriched in the more volatile component.

The compositions of the phases in equilibrium can be directly read off the phase diagram. How about the amounts of each of the phases present? We use the lever rule that was discussed in Section 11.3 above. For a system with the overall composition given by the point *sys* with overall system mole fraction of A equal to z'_A, the lever rule gives us:

$$(z'_A - x'_A)N_{liq} = (y'_A - z'_A)N_{gas},$$

where N_{liq} and N_{gas} are the number of moles in the liquid and gas phases, respectively.

11.5 Binary gas-liquid equilibrium: *T-xy* diagram

In the previous section, we have seen how the compositions of the liquid and gas depend upon pressure for a fixed temperature. How do these compositions depend upon the temperature at a fixed pressure? The Clapeyron equation discussed in Sections 8.10 and 8.11 describes how the boiling temperature of a pure liquid depends upon the pressure of the gas it is in equilibrium with. Approximating the gas phase as an ideal gas, neglecting the molar volume of the liquid compared to the molar volume of the gas, and taking the enthalpy of vaporization $\Delta_{vap}H$ to be constant, we have the approximate but explicit Clausius-Clapeyron equation:

$$\ln\left(\frac{P^*(T)}{P^*(T_0)}\right) = -\frac{\Delta_{vap}H}{R}\left(\frac{1}{T} - \frac{1}{T_0}\right).$$

$P^*(T)$ and $P^*(T_0)$ are the vapour pressures at temperatures T and T_0. Applying this to each of the two components in binary gas-liquid equilibrium, we have the following (approximate) vapour pressures for the pure component liquids:

$$P_A^*(T) = P_A^*(T_0)e^{-\frac{\Delta_{vap}H_A}{R}\left(\frac{1}{T} - \frac{1}{T_0}\right)}$$

$$P_B^*(T) = P_B^*(T_0)e^{-\frac{\Delta_{vap}H_B}{R}\left(\frac{1}{T} - \frac{1}{T_0}\right)}.$$

Hence, applying Raoult's law, the total pressure of a two-component system is:

$$P(T) = P_A + P_B = x_A P_A^* + (1 - x_A)P_B^*$$

$$= x_A \left[P_A^*(T_0)e^{-\frac{\Delta_{vap}H_A}{R}\left(\frac{1}{T} - \frac{1}{T_0}\right)} - P_B^*(T_0)e^{-\frac{\Delta_{vap}H_B}{R}\left(\frac{1}{T} - \frac{1}{T_0}\right)} \right]$$

$$+ P_B^*(T_0)e^{-\frac{\Delta_{vap}H_B}{R}\left(\frac{1}{T} - \frac{1}{T_0}\right)}.$$

For total pressure P, this equation gives x_A as a function of T:

$$x_A = \frac{P - P_B^*(T_0)e^{-\frac{\Delta_{vap}H_B}{R}\left(\frac{1}{T} - \frac{1}{T_0}\right)}}{P_A^*(T_0)e^{-\frac{\Delta_{vap}H_A}{R}\left(\frac{1}{T} - \frac{1}{T_0}\right)} - P_B^*(T_0)e^{-\frac{\Delta_{vap}H_B}{R}\left(\frac{1}{T} - \frac{1}{T_0}\right)}},$$

which we have written explicitly to show that the temperature dependence of the liquid composition is generally not a linear one even for an ideal solution in equilibrium with an ideal gas mixture. Similarly, expressing y_A in terms of x_A and the vapour pressures of the two components, we can also see that the equilibrium temperature dependence of the gas composition is generally not a linear one. These two temperature-composition dependences of the liquid and gas phase are sketched in Figure 11.5, in a T-xy phase diagram. The curves are labelled $T(x_A)$ and $T(y_A)$ for the dependence of the equilibrium temperature upon the mole fractions of the liquid and gas phase, respectively. At temperatures above the gas curve, the equilibrium state is a single gas phase. Below the liquid curve, the equilibrium state is a single liquid phase. For temperatures in the region enclosed by the curves, the equilibrium state consists of a two-component liquid coexisting with a two-component gas. Just as in Figure 11.4, we draw horizontal tie-lines to connect a liquid state to the gas state with which it is in equilibrium. The tie-lines are now constant temperature, rather than constant pressure, lines.

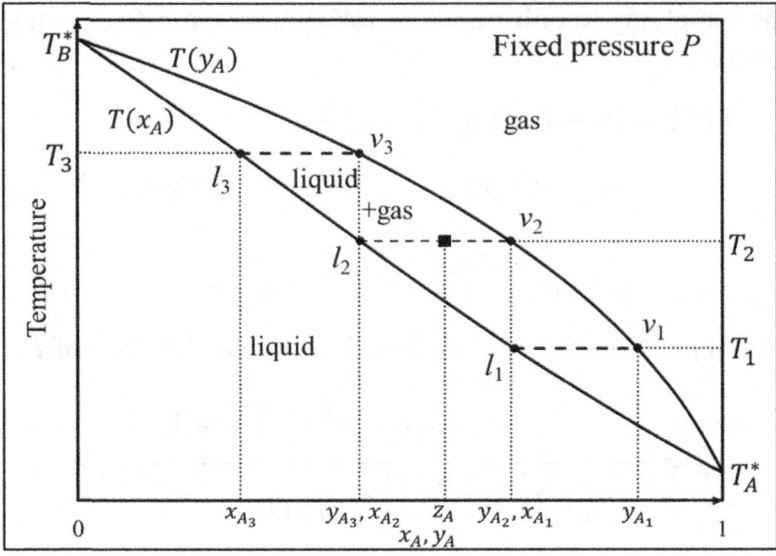

Figure 11.5: *T-xy diagram for an ideal-solution in equilibrium with its ideal gas mixture. Three tie-lines are drawn as dashed lines. Dotted lines indicate temperature and composition for each point.*

We have drawn three tie-lines at temperatures T_1, T_2 and T_3. For example, the tie-line at T_2 connects the states l_2 and v_2; liquid of composition x_{A_2} is in equilibrium with gas of composition y_{A_2}. Similarly to what we have pointed out in the P-xy diagram, notice that the gas phase is enriched with the more volatile component A. The lever rule applies in the same way as above. For the system with composition specified by the overall mole fraction z_A, indicated by a square in Figure 11.5, the lever rule gives:

$$\left(z_A - x_{A_2}\right)N_l = \left(y_{A_2} - z_A\right)N_g,$$

where N_l and N_g are the number of moles of the liquid and gas phases at equilibrium in a two-phase system.

11.6 Distillation

Distillation is a method frequently used to separate the components in a liquid mixture by taking advantage of the different volatilities of the components. We use the T-xy diagram in Figure 11.5 to see how this works. We start with a system that has the overall mole fraction equal to z_A and temperature T_2, which places it in the two-phase region, as illustrated in Figure 11.5. This system phase separates into a liquid phase and a gas phase at equilibrium. Relative to the overall composition, the former is richer in the less volatile component B, while the latter is richer in the more volatile component A.

Taking just the gas phase from the equilibrium mixture at T_2 and lowering its temperature to T_1, we can condense a liquid of mole fraction $x_{A_1} = y_{A_2}$; this is indicated by following the vertical dotted line connecting v_2 to l_1. The key point is that at T_1 and P, liquid of this composition will be in equilibrium with a gas phase of mole fraction $y_{A_1} > x_{A_1}$ with further enrichment in the gas phase of component A. Hence, by repeating this condensation and evaporation at progressively lower temperatures, we obtain a gas phase that is progressively enriched in the more volatile component.

Going back to the liquid phase we get from equilibrium at temperature T_2, i.e. in the state l_2 with mole fraction x_{A_2}, this liquid phase is enriched in component B relative to the starting overall composition z_A. Following the dotted vertical line and raising the temperature of this liquid to T_3, we then have equilibrium between a gas phase of mole fraction $y_{A_3} = x_{A_2}$, and a liquid of mole fraction $x_{A_3} < y_{A_3}$. Thus, repeated vaporization and condensation can be used to produce a liquid that is enriched in the less volatile component B. This enrichment of A in the gas phase at increasingly lower temperatures, and of B in the liquid phase at increasingly higher temperatures is the basis of fractional distillation.

A schematic illustration of a fractional distillation column is shown in Figure 11.6. A fractional distillation column consists essentially of physical structures such as a column packed with material (inert beads) to facilitate phase equilibrium between a gas phase that rises up the column and a liquid phase that trickles down the column. The temperature decreases up the column so that the top of the column is at a lower temperature than the bottom.

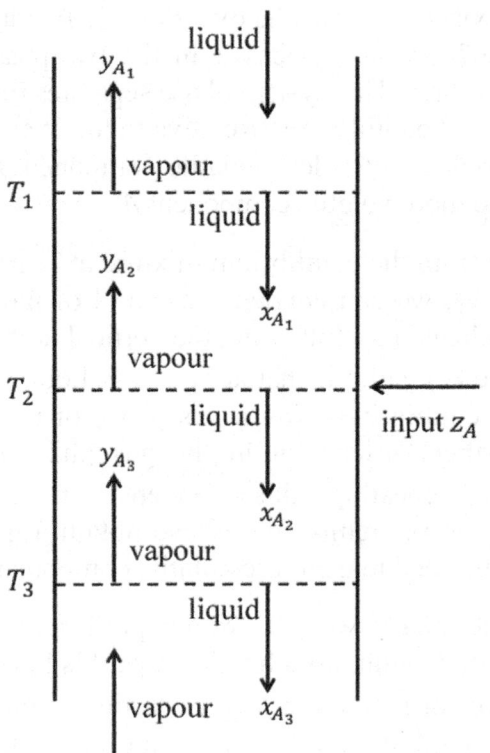

Figure 11.6: *Schematic diagram of a distillation column. Three theoretical plates represented by the dashed line are drawn, at temperatures T_1, T_2 and T_3.*

The effectiveness of such a column is rated in terms of the number of theoretical plates needed to achieve a specific enrichment of the gas with A and the liquid with B. The former is extracted off the top of the column and the latter from the bottom of the column. In the illustration we have three theoretical plates which produces a gas with mole fraction y_{A_1} at the top and a liquid with mole fraction x_{A_3} at the bottom of this section of the column. These theoretical plates, at temperatures T_1, T_2 and T_2 correspond to the tie-lines drawn in Figure 11.5.

You can imagine using a fractional distillation column equipped with as many theoretical plates as needed in order to extract a gas that is enriched in A from the top, and a liquid that is enriched in B at the bottom. In a longer column with more theoretical plates, the extracted gas becomes more enriched with A, and similarly, the extracted liquid is more enriched with B. This works generally for systems with phase diagrams that are qualitatively the same as that in Figure 11.5. However, merely applying fractional distillation does not work with some non-ideal solutions where it is not possible to further enrich the gas or the liquid beyond a particular concentration. We will discuss this later.

11.7 Mixing in non-ideal systems

What happens thermodynamically when molecules are mixed? In Section 9.2, we examined the mixing process by considering the change in the Gibbs free energy at fixed temperature and pressure:

$$\Delta_{mix}G = \Delta_{mix}(U + PV - TS)$$
$$= \Delta_{mix}U + P\Delta_{mix}V - T\Delta_{mix}S.$$

We saw that for ideal gas mixtures and ideal solutions, both $\Delta_{mix}U$ and $\Delta_{mix}V$ are zero. In a non-ideal mixture, the important term $\Delta_{mix}U$ is non-zero. Typically, $\Delta_{mix}V \cong 0$ even for non-ideal mixtures, so that it is the internal energy of mixing which plays the most important role in the deviation of the properties of non-ideal mixtures from the ideal case. If the molecules in the mixture are in a less energetically favourable environment compared to the pure liquids or gases, then $\Delta_{mix}U > 0$; this is referred to as a positive deviation from ideal behaviour. This means that the intermolecular interactions in the mixture are on average less attractive than in the separate pure substances.

Conversely, if the mixture has on average more attractive intermolecular interactions than the pure components, then $\Delta_{mix}U < 0$ and we have a negative deviation from ideal behaviour. In Figure 11.7, we sketch the vapour pressure as a function of the mole fraction x_A for non-ideal solutions. The top panel is for positive deviation and the bottom panel is for negative deviation. For the former, the intermolecular interactions are more repulsive than the pure liquids and hence the vapour pressure of each component in the solution is higher than expected from Raoult's law: the pressure-composition dependence is concave downwards. Conversely, for the negative deviation case, for which the pressure-composition dependence is concave upwards. Using the regular solution model discussed in Section 9.6, we can describe these cases quantitatively in a simple approximate way:

$$P_A = x_A P_A^* \exp\left(\epsilon x_B^2\right)$$

$$P_B = x_B P_B^* \exp(\epsilon x_A^2).$$

For positive deviation and more repulsive interactions in the mixture than the pure liquids, the interaction parameter ϵ in the regular solution model is positive. Thus, the vapour pressures of the components are larger than they would be if the mixture were ideal. Conversely for the case of negative deviation.

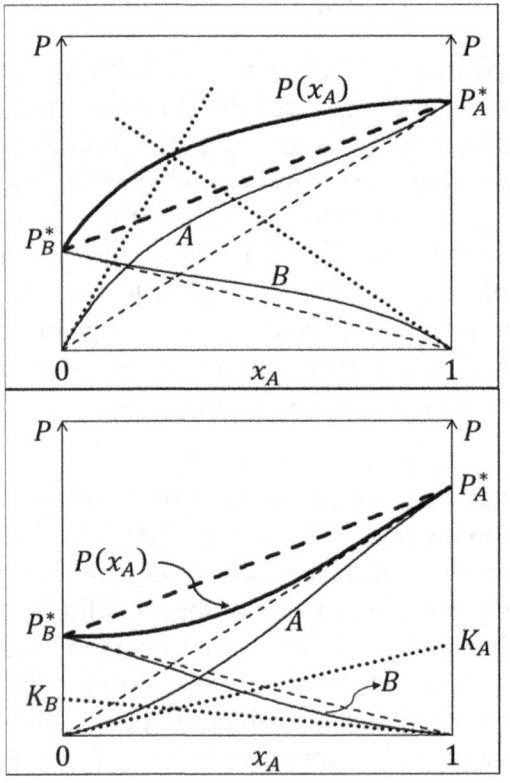

Figure 11.7: *Schematic vapour pressure curves for binary regular solutions with interaction parameters ϵ equal to 1 for positive deviation (top panel) and ϵ equal to -1 for negative deviation (bottom panel). The thick solid curves are for the total pressure $P(x_A)$ while the thin solid curves are for components A and B. The dotted lines are tangents for Henry's law. Intercepts for these are labelled only for the lower panel. The dashed lines are drawn for comparison with an ideal solution; thick dashed line for the ideal total pressure.*

These vapour pressure curves illustrate the idea of ideal dilute solutions. That is, even in a non-ideal solution, there are two composition regimes for which so-called ideal dilute solution behaviour is observed. If a solution has x_A close to unity, it consists of mostly A (solvent) molecules with relatively few B (solute) molecules. Hence, the molecular neighbourhood of any A molecule approaches that of the pure A liquid and, therefore, the vapour pressure curve $P(x_A)$ tangentially approaches the estimate from Raoult's law. Thus, even in a non-ideal solution

$$P_A(x_A) \rightarrow x_A P_A^*$$

as $x_A \rightarrow 1$. Similarly, for component B in the regime where x_A approaches zero, we observe

$$P_B(x_A) \rightarrow (1 - x_A) P_B^*$$

so that its vapour pressure curve $P_B(x_A)$ tangentially approaches Raoult's law for component B illustrated by the dashed line. Thus, the (thin solid) curve for each component in Figure 11.7 tangentially approaches Raoult's law behaviour for x_A and x_B, respectively, approaching unity.

At the other end of the concentration regime for component A, i.e. as x_A goes to zero, the average molecular neighbourhood of an A molecule consists of mostly B molecules. The vapour pressure of A is again ideal in the sense that $P_A(x_A)$ is linear with respect to x_A. This is shown by the dotted straight-lines in Figure 11.7. In this regime,

$$P_A(x_A) \rightarrow x_A K_A$$

so that the vapour pressure of A is proportional to the mole fraction of A as in Raoult's law, but with a different constant of proportionality K_A. This is Henry's law for ideal dilute solutions for the dilute component A in its solution with B. We have discussed Henry's law in Section 9.5.3. For this solution, the Henry's law constant K_A is dependent upon the interactions of an A molecule with mostly B neighbours. This is similarly the case for $P_B(x_A)$ as

x_A approaches unity. Henry's law for the dilute solution of B in solution with A is:

$$P_B(x_A) \rightarrow (1 - x_A)K_B,$$

with K_B depending upon the interactions of a B molecule with mostly A neighbours. In the lower panel of Figure 11.7, we label the Henry's law intercepts K_A and K_B. In the concentration regime where A obeys Raoult's law, B obeys Henry's law, and vice versa. Thus, the (thin solid) curve for the partial pressure of each component in Figure 11.7 tangentially approach the dotted lines for x_A and x_B, respectively, approaching zero. Drawing in the same diagram both the liquid and gas curves, $P(x_A)$ and $P(y_A)$ respectively, we have the *P-xy* diagram for non-ideal systems illustrated in Figure 11.8 for positive and negative deviations from ideal behaviour.

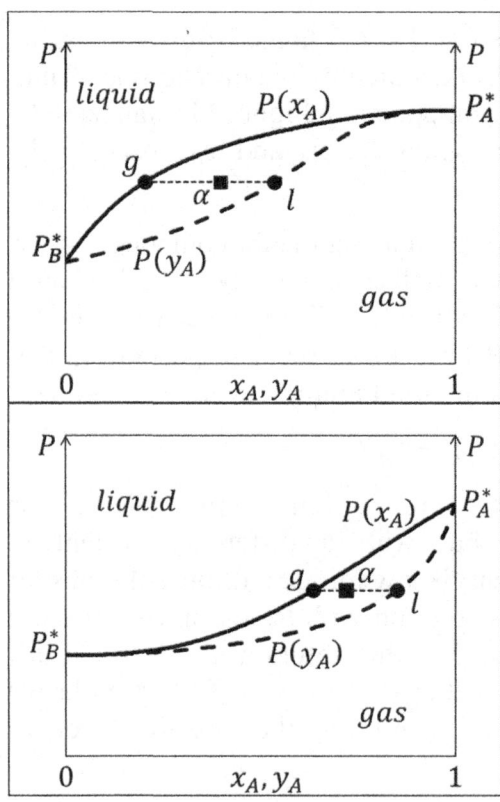

Figure 11.8: Schematic P-xy diagrams for regular solutions showing positive deviation (top panel) and negative deviation (bottom panel) from ideal solution behavior. In each panel the pressure-composition curve for both liquid and gas phases are drawn, $P(x_A)$ and $P(y_A)$, respectively. A tie-line is also drawn to illustrate the co-existing liquid and gas phases with compositions at points l and g for a system with an overall composition at any point α along the tie-line.

The horizontal axis is now the mole fractions for both liquid and gas phases, x_A and y_A. We have labelled the single-phase regions in these phase diagrams. For example, any pressure-composition point above the $P(x_A)$ curve exists completely as liquid. Similarly, any point below the $P(y_A)$ curve in these phase diagrams exists completely in the gas phase. The two-phase region in each diagram is between the liquid and gas curves. Any system with pressure and overall composition in this two-phase region, for example α, will phase separate into a liquid phase and a gas phase, the compositions of which are indicated by the points l and g connected by a horizontal tie-line at the pressure of the system. The relative amounts of the liquid and gas phases can be calculated using the lever-rule discussed in Section 11.4.

Note that the pressure-composition or P-xy diagrams in Figure 11.7 and 11.8 are at a specific value of temperature. For a specific value of pressure, the temperature-composition or T-xy diagrams are similar, with the $T(x_A)$ and $T(y_A)$ curves for positive deviation from ideal behaviour being concave upwards. For negative deviations, these are concave downwards.

Non-ideal phase behaviour can differ considerably from ideal systems. This is illustrated in Figure 11.9 where we sketch (schematically) the T-xy phase diagram for a two-component system such as the ethanol-water system. For the ethanol-water solution, there is a non-pure composition known as the azeotrope, labelled as az in the figure, for which equilibrium mole fractions of the liquid and gas are equal. That is, at the azeotropic point

$$x_{az} = y_{az}.$$

For ethanol and water at one atmosphere pressure, this occurs at a mass fraction of ethanol equal to 0.956 and a temperature of 78.5°C. Notice that if we insist on a thermodynamic state for which x is equal to y, then this additional constraint must be included in the Gibbs phase rule so that

$$f = c - p + 2 - c_{add} = 1,$$

not the usual $f = 2$ for a two-component, two-phase system. Thus, specifying a pressure of one atmosphere completely determines x_{az}, y_{az} and the temperature of the temperature of the azeotropic state.

Because $x_{az} = y_{az}$, when an azeotropic liquid is vaporised the gas phase has the same composition as the liquid. Hence, the enrichment of the vapour with the more volatile component does not occur at the azeotropic point. This means that distillation cannot further separate the components in an azeotropic mixture. If you begin with a liquid mole fraction x_A that is less than the azeotropic mole fraction x_{az}, it is only possible to enrich a distillate up to composition x_{az}. The tie-line at the azeotrope is a point, and thus fractional distillation cannot "get past" the azeotropic composition. Hence, starting with a fermentation mixture with ethanol mass fraction less than 0.956, it is not possible to obtain an ethanol-water mixture that is "stronger" in the alcohol using only fractional distillation.

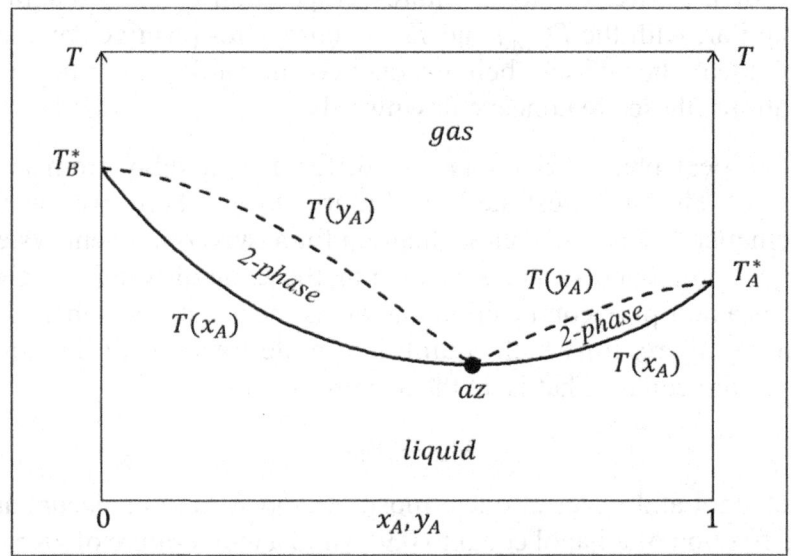

Figure 11.9: A schematic T-xy phase diagram showing a binary system with an azeotropic point denoted by az. The solid line are the liquid phase temperature-composition curves, and the dashed lines are the gas phase temperature-composition curves.

12

Chemical Equilibrium

In your chemistry lab in school, you must have precipitated calcium carbonate using the reaction between aqueous solutions of sodium carbonate and calcium chloride:

$$CaCl_2(aq) + Na_2CO_3(aq) \rightarrow CaCO_3(solid) + 2NaCl(aq).$$

It is seemingly an irreversible reaction to form solid limestone and an aqueous solution of common salt. Indeed, up until the start of the 19th century chemical reactions were thought to occur in only one direction. Berthollet, the 19th century French chemist, was rather familiar in his laboratory with the reaction:

$$CaCl_2(aq) + Na_2CO_3(solid) \rightarrow CaCO_3(solid) + 2NaCl(aq)$$

where solid sodium carbonate reacts with a solution of calcium chloride to form solid calcium carbonate. He had thought this was the natural one-way direction of the chemical reaction until he was surprised to find the reverse reaction

$$CaCO_3(solid) + 2NaCl(aq) \rightarrow CaCl_2(aq) + Na_2CO_3(solid)$$

occurring at the edges of salt lakes in regions where the ground is limestone. This observation led him to formulate the idea that chemical equilibrium results from the balance of chemical reactions occurring in opposite directions, that is:

$$CaCO_3(solid) + 2NaCl(aq) \rightleftharpoons CaCl_2(aq) + Na_2CO_3(solid).$$

By adjusting the concentrations of the reactants and products the direction of chemical reactions can be reversed, much like adjusting the external pressure on a gas allows us to compress or expand a volume of gas. Just as the notion of reversible expansion of gases is useful for analysing the thermodynamics of gases, a reversible chemical reaction is useful for understanding the thermodynamic transformations of chemically reacting systems. Today we are completely familiar with this idea. In this chapter, we discuss how to figure out the temperature, pressure and composition at which the forward and backward reactions balance to reach chemical equilibrium. As in phase equilibrium, the chemical potential plays the central role.

12.1 The general condition for chemical equilibrium

At constant temperature and pressure the Gibbs free energy is minimized at equilibrium. For phase equilibrium the discussion in Section 8.6 gave us

$$\mu_i^1 = \mu_i^2 = \cdots = \mu_i^p$$

for chemical species i present in phases 1, 2, ..., p in a system. This comes from considering the fundamental equation

$$dG = -SdT + VdP + \mu_{gas}dn_{gas} + \mu_{liq}dn_{liq}.$$

For gas-liquid equilibrium at constant T and P,

$$dG = \left(\mu_{gas} - \mu_{liq}\right)dn_{gas},$$

since $dn_{gas} = -dn_{liq}$. Hence, at the minimum of the Gibbs free energy we have $\left(\dfrac{\partial G}{\partial n_{gas}}\right)_{eq} = 0$, and thus

$$\mu_{gas} = \mu_{liq}.$$

In order to establish the phase equilibrium condition, we looked for the minimum in G with respect to the transfer of material from one phase to the other. Similarly, to establish the condition for chemical equilibrium we look at the transfer of material from the reactant state to the product state in a chemical reaction.

Consider the generic chemical reaction:

$$\nu_A A + \nu_B B \rightleftharpoons \nu_C C + \nu_D D$$

where, conventionally, A and B are referred to as reactants, C and D as products, and their respective stoichiometric coefficients are denoted by the ν's. Increasing the concentration of the reactants drives the reaction in the forward direction, from left to right. Vice versa, increasing the concentration of the products drives the reaction in the reverse direction. How is the balance between the forward and reverse reactions determined by thermodynamics?

We define a quantity ξ that measures the extent of reaction, such that when $d\xi$ moles of the reaction occur, proceeding from the reactants to the products, we have:

$$\nu_A d\xi \text{ moles of } A \text{ consumed,}$$

$$\nu_B d\xi \text{ moles of } B \text{ consumed,}$$

$$\nu_C d\xi \text{ moles of } C \text{ produced,}$$

$$\nu_D d\xi \text{ moles of } D \text{ produced.}$$

Hence, the corresponding change in the Gibbs free energy of the system is:

$$dG = (-\nu_A \mu_A - \nu_B \mu_B + \nu_C \mu_C + \nu_D \mu_D) d\xi$$

so that at equilibrium when G is minimized, we have $\frac{\partial G}{\partial \xi} = 0$. Since the quantity in the brackets is simply the molar Gibbs free energy change for the reaction, for chemical equilibrium at T and P, we have the following condition for the change in the Gibbs free energy per mole of reaction:

$$\Delta_{rxn} G = \nu_C \mu_C + \nu_D \mu_D - \nu_A \mu_A - \nu_C \mu_C = 0.$$

This equation determines the equilibrium state for all chemical reactions. How do we use it in connection with what we have learned about the chemical potentials of different systems in Chapter 9?

In general, we write the chemical potential of any of the reacting species in terms of its activity as

$$\mu_A(T, P) = \mu_A^\varnothing(T) + RT \ln a_A,$$

with $\mu_A^\varnothing(T)$ equal to the chemical potential of an appropriate standard state for A, which has unit activity. Hence the Gibbs free energy of reaction is

$$\Delta_{rxn}G(T, P) = \Delta_{rxn}G^\varnothing(T) + RT \ln Q$$

where the reaction quotient Q is the ratio of the activities of the reactants and products, each raised to the power of its respective stoichiometric coefficient:

$$Q \equiv \frac{a_C^{\nu_C} a_D^{\nu_D}}{a_A^{\nu_A} a_B^{\nu_B}},$$

and $\Delta_{rxn}G^\varnothing(T)$ given by

$$\Delta_{rxn}G^\varnothing(T) = \nu_C \mu_C^\varnothing + \nu_D \mu_D^\varnothing - \nu_A \mu_A^\varnothing - \nu_B \mu_B^\varnothing.$$

For the reaction to be at equilibrium at T and P, we have $\Delta_{rxn}G(T, P) = 0$. Therefore, at equilibrium the concentrations of the reacting species must be such that

$$0 = \Delta_{rxn}G^\varnothing(T) + RT \ln Q,$$

which gives

$$Q_{eq} = \left(\frac{a_C^{\nu_C} a_D^{\nu_D}}{a_A^{\nu_A} a_B^{\nu_B}} \right)_{eq} = exp\left(-\frac{\Delta_{rxn}G^\varnothing(T)}{RT} \right)$$

where the subscript eq emphasizes that it is the reaction quotient at equilibrium. We then define the equilibrium constant for the reaction as

$$K(T) \equiv exp\left(-\frac{\Delta_{rxn}G^\varnothing(T)}{RT} \right).$$

Thus, at equilibrium, the reaction quotient is equal to the equilibrium constant.

$$Q_{eq} = K(T).$$

To go further, we have to consider the appropriate standard states used to define the chemical potentials of the reacting species.

As a short aside on the sign convention for the stoichiometric coefficient, note that the stoichiometric coefficients ν_i we have defined above are all positive numbers. Frequently a different sign convention is used for the stoichiometric coefficients in which the equilibrium condition is written as:

$$\zeta_A \mu_A + \zeta_B \mu_B + \zeta_C \mu_C + \zeta_D \mu_D = 0$$

where the stoichiometric coefficients for the reactants A and B are negative, that is, $\zeta_A = -\nu_A$ and $\zeta_B = -\nu_B$, whence

$$K(T) = x_A^{\zeta_A} x_B^{\zeta_B} x_C^{\zeta_C} x_D^{\zeta_D} = \prod_i x_A^{\zeta_A}$$

which can be written in a more concise form; this is simply a difference in the sign convention for the stoichiometric coefficient.

12.2 Equilibrium in reacting ideal gases

For gases, the standard state chemical potentials can be written as the chemical potentials of the pure gases at fugacity equal to the standard pressure P_0 so that the activity is equal to one; see Section 9.4. Thus, the standard Gibbs free energy of reaction is

$$\Delta_{rxn} G^{\varnothing}(T) = \nu_C \mu_C^{\varnothing} + \nu_D \mu_D^{\varnothing} - \nu_A \mu_A^{\varnothing} - \nu_B \mu_B^{\varnothing}$$

$$= \nu_C \mu_C^*(T, f_C = P_0) + \nu_D \mu_D^*(T, f_D = P_0)$$

$$- \nu_A \mu_A^*(T, f_A = P_0) - \nu_B \mu_B^*(T, f_B = P_0).$$

And, the activities are defined in terms of the fugacities:

$$\frac{a_C^{v_C} a_D^{v_D}}{a_A^{v_A} a_B^{v_B}} = \frac{\left(\frac{f_C}{P_0}\right)^{v_C} \left(\frac{f_D}{P_0}\right)^{v_D}}{\left(\frac{f_A}{P_0}\right)^{v_A} \left(\frac{f_B}{P_0}\right)^{v_B}}.$$

If the gases are ideal, then the standard Gibbs free energy of reaction becomes

$$\Delta_{rxn} G^{\emptyset}(T) = v_C \mu_C^*(T, P_0) + v_D \mu_D^*(T, P_0)$$

$$-v_A \mu_A^*(T, P_0) - v_B \mu_B^*(T, P_0)$$

$$= \Delta_{rxn} G^*(T, P_0).$$

This is the difference in the chemical potentials of the pure species at temperature T and pressure P_0, each weighted by its respective stoichiometric coefficient. Note that $\Delta_{rxn} G^*$ depends only upon the temperature. It does not depend upon the pressure P of the reacting system. The dependence upon P_0 is simply a dependence upon what is arbitrarily chosen as a standard state pressure, and not a dependence upon the pressure of the reacting system. For a reacting ideal gas mixture, the reaction quotient is

$$Q = \frac{\left(\frac{f_C}{P_0}\right)^{v_C} \left(\frac{f_D}{P_0}\right)^{v_D}}{\left(\frac{f_A}{P_0}\right)^{v_A} \left(\frac{f_B}{P_0}\right)^{v_B}} = \frac{\left(\frac{P_C}{P_0}\right)^{v_C} \left(\frac{P_D}{P_0}\right)^{v_D}}{\left(\frac{P_A}{P_0}\right)^{v_A} \left(\frac{P_B}{P_0}\right)^{v_B}}.$$

Hence, equilibrium for a reacting ideal gas mixture is given by

$$Q = \left(\frac{P_C}{P_0}\right)^{v_C} \left(\frac{P_D}{P_0}\right)^{v_D} \bigg/ \left(\frac{P_A}{P_0}\right)^{v_A} \left(\frac{P_B}{P_0}\right)^{v_B} = exp\left(-\frac{\Delta_{rxn} G^*}{RT}\right) = K_P(T).$$

In the reacting ideal gas mixture, the reaction quotient is expressed in terms of the partial pressures, and thus the subscript P for the equilibrium constant. Since $\Delta_{rxn} G^*(T, P_0)$ depends only upon temperature, the equilibrium constant K_P is a function of temperature only. If the reaction mixture has lower product concentrations and higher reactant concentrations compared to the equilibrium state, then $Q < K_P(T)$, and

$$\Delta_{rxn}G(T,P) = \Delta_{rxn}G^*(T,P_0) + RT \ln Q$$

$$= RT \ln \left(\frac{Q}{K_P}\right),$$

so that $\Delta_{rxn}G$ is negative. Then, the reaction spontaneously proceeds in the forward direction to form more products in order to decrease the Gibbs free energy of the system. Conversely, if $Q > K_P(T)$, then $\Delta_{rxn}G$ is positive, and the reaction will spontaneously convert reactants to products. Equilibrium is reached only when Q is equal to the equilibrium constant, here $K_P(T)$.

Writing the partial pressures in terms of the mole fractions and the total pressure of the system, we have:

$$K_P(T) = \left(\frac{P}{P_0}\right)^{\nu_C + \nu_D - \nu_A - \nu_B} \frac{y_C^{\nu_C} y_D^{\nu_D}}{y_A^{\nu_A} y_B^{\nu_B}}$$

where the second factor on the right is in terms of the mole fractions y_i of the reacting ideal gas mixture. Thus, we obtain the dependence of the equilibrium composition upon pressure through the first factor on the right-hand side and upon temperature through the equilibrium constant on the left. This equation can be rearranged as follows:

$$\frac{y_C^{\nu_C} y_D^{\nu_D}}{y_A^{\nu_A} y_B^{\nu_B}} = K_P(T) \left(\frac{P_0}{P}\right)^{\nu_C + \nu_D - \nu_A - \nu_B} \equiv K_y(T,P)$$

giving the equilibrium constant $K_y(T,P)$ which depends upon both temperature and pressure. Clearly, the equilibrium mole fractions of the component species in a reacting ideal gas mixture depend upon both temperature and pressure. Note that both K_P and K_y are dimensionless numbers with no units.

§   **Example 12.1 Chemical equilibrium in reacting gases**

Consider the reaction of nitrogen and hydrogen to form ammonia:

$$N_2(g) + 3H_2(g) \rightleftharpoons 2NH_3(g).$$

Calculate the standard Gibbs free energy of reaction and the equilibrium constants K_P and K_y for this reaction at a temperature of 298.15 K and a pressure of 5 bar. Does simply mixing hydrogen and nitrogen gas at 298.15 K and 5 bar produce ammonia gas?

The standard Gibbs free energy of formation are

$$\Delta_f G^\varnothing(N_2(g)) = 0 \text{ kJ mol}^{-1}$$

$$\Delta_f G^\varnothing(H_2(g)) = 0 \text{ kJ mol}^{-1}$$

$$\Delta_f G^\varnothing(NH_3(g)) = -16.45 \text{ kJ mol}^{-1}.$$

The standard Gibbs free energy of reaction is

$$\Delta_{rxn} G^\varnothing = 2\Delta_f G^\varnothing(NH_3(g)) - \Delta_f G^\varnothing(N_2(g)) - 3\Delta_f G^\varnothing(H_2(g))$$

$$= -32.9 \text{ kJ mol}^{-1}.$$

The equilibrium constants at a temperature of 298.15 K and a pressure P bar are:

$$K_P(298.15 \text{ K}) = exp\left[-\frac{\Delta_{rxn} G^\varnothing}{298.15R}\right]$$

$$= 5.81 \times 10^5$$

and

$$K_y(298.15 \text{ K}, 5 \text{ bar}) = \left(\frac{P_0}{P}\right)^2 K_P(298.15 \text{ K})$$

$$= 2.32 \times 10^4,$$

which depends inversely upon the square of the system pressure. Thus, the equilibrium composition at 298.15 K is given by,

$$\left(\frac{P_{NH_3}}{P_0}\right)^2 \Big/ \left(\frac{P_{N_2}}{P_0}\right)\left(\frac{P_{H_2}}{P_0}\right)^3 = 5.81 \times 10^5$$

in terms of partial pressures. In terms of mole fractions, it is given by

$$\frac{y_{NH_3}^2}{y_{N_2} y_{H_2}^3} = 2.32 \times 10^4.$$

At thermodynamic equilibrium at standard temperature and a pressure of 5 bar, the concentration of ammonia is larger than those for nitrogen and hydrogen. A mixture of just nitrogen and hydrogen gases is not stable thermodynamically relative to the formation of ammonia. §

However, this is not the complete picture. The kinetics for this reaction is very slow because of the strong bonds between the nitrogen atoms in N_2 molecules. Thus, even though the thermodynamic equilibrium state for this reaction is very much in the direction of the product, the reaction is so slow that we get essentially no ammonia if we mix nitrogen and hydrogen gas at room temperature. This kinetic barrier is overcome in the Haber process by making use of a catalyst, thus speeding up the kinetics but not changing the thermodynamics.

§ *Example 12.2 Combustion of hydrogen*

A mixture of one mole of hydrogen gas and half a mole of oxygen gas at room temperature and pressure reacts explosively when sparked to form water. This is a frequently-used dramatic demonstration that water is more stable than its elements at room temperature and pressure. Calculate the standard Gibbs free energy of reaction $\Delta_{rxn}G^{\emptyset}$ and the equilibrium constant at standard temperature and pressure. Is this reaction spontaneous at 298.15 K and 1 bar?

We have the reaction:

$$H_2(gas) + \frac{1}{2}O_2(gas) \rightarrow H_2O(liquid).$$

$$\Delta_{rxn}H^{\emptyset} = -285.83 \text{ kJ mol}^{-1}$$

$$\Delta_{rxn}S^{\emptyset} = -163.34 \text{ J K}^{-1} \text{ mol}^{-1}.$$

In Example 7.4, we calculated the standard Gibbs free energy of reaction, obtaining:

$$\Delta_{rxn}G^{\emptyset} = \Delta_{rxn}H^{\emptyset} - T_0\Delta_{rxn}S^{\emptyset} = -237.13 \text{ kJ mol}^{-1}.$$

Hence, the equilibrium constant is

$$K(T) = exp\left[-\frac{\Delta_{rxn}G^{\emptyset}}{298.15R}\right]$$
$$= 3.5 \times 10^{41}.$$

Thus, the product state is very much thermodynamically favoured over the reactants; this reaction is spontaneous around room temperature and pressure, although again kinetics may hold up the progress of the reaction as pointed out in Example 12.1. If we are careful not to expose the system to a spark, the reaction will not proceed even though a mixture of hydrogen and oxygen is explosively unstable relative to water. §

In both reactions in Examples 12.1 and 12.2, the number of moles of gases in the system decreases as the reaction proceeds. We can qualitative thus argue that the entropy of the system decreases as the reaction proceeds. Indeed, we have already seen in Example 7.4 that for the reaction of hydrogen and oxygen to form water, the entropy of reaction is equal to -163.34 J K^{-1} mol^{-1}. Despite that, these reactions are spontaneous. This is because in both cases the reactions are exothermic so that the enthalpy of the surroundings is increased by the reaction. This leads to an increase in the entropy of the surroundings that more than compensates for the decrease in the entropy of system due to its volume change.

We will discuss later how to calculate the mole fractions of the reactants and products from the equilibrium constant. We will also explain how to calculate the equilibrium constant at temperatures other than the standard temperature.

12.3 Equilibrium in reacting solutions

We obtain the equilibrium constant for reactions in liquid solutions in the same way. From Section 8.5, we have seen that the general form for the chemical potential of A in a solution is

$$\mu_A(liq, T, P, \{c_i\}) = \mu_A^\emptyset(liq, T, P) + RT \ln a_A$$
$$= \mu_A^\emptyset(liq, T, P) + RT \ln\left(\gamma_A \frac{c_A}{c_0}\right).$$

The standard state for which $\mu_A^\emptyset$ is defined is for a state with a_A equal to unity at a temperature T and pressure P. The activity generally depends upon temperature, pressure and composition, i.e. $a_A = a_A(T, P, \{c_i\})$. That is, all the composition dependence is through the logarithmic term in the activity a_A. A few possible choices of the standard state were discussed in Sections 9.5 for different systems.

With this form for the chemical potentials, the Gibbs free energy of reaction is generally written as

$$\Delta_{rxn}G(T, P) = \Delta_{rxn}G^\emptyset(T, P) + RT \ln Q$$

where the reaction quotient Q is

$$Q = \frac{\left(\gamma_C \frac{c_C}{c_0}\right)^{\nu_C} \left(\gamma_D \frac{c_D}{c_0}\right)^{\nu_D}}{\left(\gamma_A \frac{c_A}{c_0}\right)^{\nu_A} \left(\gamma_B \frac{c_B}{c_0}\right)^{\nu_B}}.$$

Therefore, when the reaction is at equilibrium at T and P, we have

$$K_c(T,P) = exp\left(-\frac{\Delta_{rxn}G^{\emptyset}(T,P)}{RT}\right) = \frac{\left(\gamma_C \frac{c_C}{c_0}\right)^{\nu_C}\left(\gamma_D \frac{c_D}{c_0}\right)^{\nu_D}}{\left(\gamma_A \frac{c_A}{c_0}\right)^{\nu_A}\left(\gamma_B \frac{c_B}{c_0}\right)^{\nu_B}}.$$

If the activity coefficients are approximately equal to unity, then this expression is simplified to the form you have probably encountered previously in school,

$$K_c(T,P) = \frac{\left(\frac{c_C}{c_0}\right)^{\nu_C}\left(\frac{c_D}{c_0}\right)^{\nu_D}}{\left(\frac{c_A}{c_0}\right)^{\nu_A}\left(\frac{c_B}{c_0}\right)^{\nu_B}}.$$

These expressions allow you to calculate the equilibrium concentrations in a reacting solution.

We have explicitly included the pressure as an argument for the equilibrium constant on the left-hand side even though this can be neglected for liquids because the compressibility of liquids is negligible. This is because solids and liquids have small molar volumes and thus their Gibbs free energy do not change appreciably compared to gases when the pressure is changed.

§   *Example 12.3 pH of vinegar*

The dissociation of aqueous ethanoic acid is

$$CH_3COOH(aq) \rightleftharpoons CH_3COO^-(aq) + H^+(aq).$$

Calculate its equilibrium constant. Vinegar contains 5% by mass of ethanoic acid in water, which is about 50 g of ethanoic acid per dm^3 of vinegar. Calculate the pH of vinegar.

The standard Gibbs free energy of formation are:

$$\Delta_f G^{\emptyset}(CH_3COOH(aq)) = -396.46 \text{ kJ mol}^{-1}$$
$$\Delta_f G^{\emptyset}(CH_3COO^-(aq)) = -369.31 \text{ kJ mol}^{-1}$$
$$\Delta_f G^{\emptyset}(H^+(aq)) = 0 \text{ kJ mol}^{-1}$$

Thus, the standard Gibbs free energy of reaction is

$$\Delta_{rxn}G^{\emptyset} = -369.31 + 396.46$$
$$= 27.15 \text{ kJ mol}^{-1},$$

giving an equilibrium constant equal to

$$K(298.15) = exp\left(-\frac{\Delta_{rxn}G^{\emptyset}}{298.15R}\right).$$

The reaction quotient at equilibrium is then equal to

$$\frac{a_{CH_3COO^-(aq)}a_{H^+(aq)}}{a_{CH_3COOH(aq)}} = 1.75 \times 10^{-5}.$$

This is approximately independent of the pressure since pressure changes do not change the chemical potential of solutions by much. If the activity coefficients of all the species involved can be taken to be one, then at equilibrium the reaction quotient in terms of molarity is

$$\frac{\left(\frac{m_{CH_3COO^-(aq)}}{m_0}\right)\left(\frac{m_{H^+(aq)}}{m_0}\right)}{\left(\frac{m_{CH_3COOH(aq)}}{m_0}\right)} = 1.75 \times 10^{-5}$$

where m_0 is the standard reference molarity of 1 mol dm^{-3}. With the ethanoic acid molar mass of 60.06, the total molarity of the dissociated and undissociated ethanoic acid is

$$m_{tot} = \frac{50}{60.06} = 0.833 \text{ mol dm}^{-3}.$$

Thus, we have

$$m_{CH_3COOH(aq)} = m_{tot} - m_{H^+(aq)} = 0.833 - m_{H^+(aq)}.$$

Therefore, chemical equilibrium gives

$$\frac{m_{H^+(aq)}^2}{0.833 - m_{H^+(aq)}} = 1.75 \times 10^{-5}$$

which can be solved to give $m_{H^+(aq)} = 3.809 \times 10^{-3}$ mol dm^{-3}. This is a pH of 2.419. §

12.4 Heterogeneous phase reacting systems

It is clear from the above two sections that the reaction quotient can be a product of pressures, mole fractions, activities or concentration units such as molarities or molalities depending upon the different conventions that are convenient for defining the chemical potential of the different species in the reacting system. This is particularly the case when the reaction system consists of more than one phase. Indeed, these different units may be mixed in the expression for the reaction quotient.

§   *Example 12.4 Equilibrium constant in a heterogeneous reaction*

Consider the example of the reaction of ammonia gas with water to form ammonium hydroxide solution. To simplify the discussion, we neglect ammonia molecules dissolved in the solution. In the gas phase we have ammonia and water vapour, and the liquid phase is an aqueous solution of ammonium hydroxide:

$$NH_3(gas) + H_2O(gas) \rightleftharpoons NH_4OH(aq).$$

Write expressions for the equilibrium constant and the reaction quotient, appropriately defining the chemical potentials of each species present in the reacting system.

In general, the chemical potentials, in terms of fugacities and activities are:

$$\mu_{NH_3}^*(gas, T, P) = \mu_{NH_3}^*(gas, T, P_0) + RT \ln\left(\frac{f_{NH_3}}{P_0}\right)$$

$$\mu_{H_2O}^*(gas, T, P) = \mu_{H_2O}^*(gas, T, P_0) + RT \ln\left(\frac{f_{H_2O}}{P_0}\right)$$

$$\mu_{NH_4OH}\left(aq, T, P, m_{NH_4OH}\right) = \mu_{NH_4OH}^{\emptyset}(liq, T, P, m_0) \\ + RT \ln\left(a_{NH_4OH}\right)$$

$\mu_{NH_3}^*(T, P_0)$ is the chemical potential of pure ammonia gas at standard pressure P_0 and f_{NH_3} is the fugacity of ammonia in the system; similarly, for water in the gas phase. $\mu_{NH_4OH}^{\emptyset}(aq, T, P, m_0)$ is the chemical potential of ammonium hydroxide in an aqueous solution for a reference state with molarity equal to m_0; m_{NH_4OH} is the molarity of ammonium hydroxide in the solution. Thus, chemical equilibrium at T and P requires

$$K = exp\left(-\frac{\Delta_{rxn}G^{\emptyset}}{RT}\right) = a_{NH_4OH(aq)} / \left(\frac{f_{NH_3(g)}}{P_0}\right)\left(\frac{f_{H_2O(g)}}{P_0}\right)$$

where the standard Gibbs free energy of reaction is given by:

$$\Delta_{rxn}G^{\emptyset} = \mu_{NH_4OH}^{\emptyset}(aq, T, P) - \mu_{NH_3}^*(gas, T, P_0) \\ - \mu_{H_2O}^*(gas, T, P_0).$$

The equilibrium constant K depends upon the standard chemical potentials of all reacting species, each of which may be defined using different conventions for the standard state. The expression for the reaction quotient includes activity and fugacity coefficients, which are unitless. If the gases and liquids can be assumed to behave ideally, then the equilibrium molarity and partial pressures are related by

$$K = exp\left(-\frac{\Delta_{rxn}G^{\emptyset}}{RT}\right)$$

$$= \left(\frac{m_{NH_4OH(aq)}}{m_0}\right) / \left(\frac{P_{NH_3(g)}}{P_0}\right)\left(\frac{P_{H_2O(g)}}{P_0}\right)^2 . \quad \S$$

In some systems, the reaction quotient may, to a good approximation, depend only upon the activity or concentration of some of the species. This is the case of a reacting system consisting of a mixture of gases, together with pure solids or liquids. Consider the reaction:

$$v_A A(solid) + v_B B(gas) \rightleftharpoons v_C C(liq) + v_D D(gas).$$

The chemical potential of a solid is

$$\mu_A^*(solid, T, P) = \mu_A^*(solid, T, P_0) + v_m(P - P_0)$$

where v_m is the molar volume of solid A, and P_0 is the standard pressure at which the chemical potential of pure solid A is equal to $\mu_A^*(solid, T, P_0)$ when the temperature is T. There is no composition dependence for the chemical potential of the solid because it is pure A; the activity of pure solids and liquids are equal to unity (Section 9.5). A similar expression holds for the chemical potential of the pure liquid C. In comparison to gases, the molar volumes of solids and liquids are generally much smaller. Thus, an approximation is to ignore the $v_m(P - P_0)$ term for solids and liquids in reaction mixtures with gases. See also Examples 9.4 and 9.5.

Keeping only the first terms we have

$$\mu_A^*(solid, T, P) \cong \mu_A^*(solid, T, P_0)$$

$$\mu_C^*(liquid, T, P) \cong \mu_C^*(liquid, T, P_0).$$

The chemical potentials for gases B and D are

$$\mu_B^*(T, P) = \mu_B^*(gas, T, P_0) + RT \ln\left(\frac{P_B}{P_0}\right)$$

$$\mu_D^*(T, P) = \mu_D^*(gas, T, P_0) + RT \ln\left(\frac{P_D}{P_0}\right)$$

where the respective mole fractions in the gas phase are $y_B = \frac{P_B}{P}$ and $y_D = \frac{P_D}{P}$.

Hence, equilibrium for this reaction is described by

$$K(T) = exp\left[-\frac{\Delta_{rxn}G^{\emptyset}(T)}{RT}\right] = \left(\frac{P}{P_0}\right)^{\nu_D-\nu_B}\frac{y_D^{\nu_D}}{y_B^{\nu_B}}$$

where the standard Gibbs free energy of reaction is

$$\Delta_{rxn}G^{\emptyset}(T,P_0) = \nu_C\mu_C^*(liq,T,P_0) + \nu_D\mu_D^*(gas,T,P_0)$$
$$-\nu_A\mu_A^*(solid,T,P_0) - \nu_B\mu_B^*(gas,T,P_0).$$

The reaction quotient includes only the concentrations of the species in the gas phase, but the standard Gibbs free energy of reaction accounts for the standard state chemical potentials of all the species present.

§   *Example 12.5 Equilibrium in a heterogeneous reaction*

Gaseous carbon dioxide reacts with solid quicklime to form solid limestone:

$$CaO(solid) + CO_2(gas) \rightleftharpoons CaCO_3(solid).$$

Write down an expression for its standard equilibrium constant assuming an ideal gas phase.

At equilibrium

$$K(T) = exp\left[-\frac{\Delta_{rxn}G^{\emptyset}(T)}{RT}\right] = \frac{P_0}{y_{CO_2}P}$$

and the standard Gibbs free energy of reaction is

$$\Delta_{rxn}G^{\emptyset}(T) = \mu_{CaCO_3}^*(solid,T,P_0)$$
$$-\mu_{CaO}^*(solid,T,P_0) - \mu_{CO_2}^*(gas,T,P_0)$$

where the first two terms on the right-hand side are the chemical potentials of the two pure solid phases in the reaction system; CaO and $CaCO_3$ do not form a solid solution. The third term is the chemical potential of pure carbon dioxide gas. Note that the reaction quotient includes only the concentration of carbon dioxide.

12.5 Pressure dependence

If a reaction mixture consists of only solid and liquid phases, changing the pressure does not significantly change the composition of the equilibrium mixture. This is because the overall volume of the solid or liquid reaction system does not change significantly with pressure. In reaction systems that contain a gas phase, however, the effect of pressure can be significant, as in Example 12.1.

In general, we have

$$K_P(T) = \left(\frac{P}{P_0}\right)^{\nu_C + \nu_D - \nu_A - \nu_B} \frac{y_C^{\nu_C} y_D^{\nu_D}}{y_A^{\nu_A} y_B^{\nu_B}}$$

from Section 12.2 for an ideal gas mixture. In a heterogeneous reaction system with a gas phase, the equilibrium product will contain one or more factors of the partial pressures of the gas phase species. Since $K_P(T)$ depends only upon temperature, if the stoichiometric coefficients in the exponent of the $\left(\frac{P}{P_0}\right)$ factor cancels, then the equilibrium composition is only a function of the temperature. This happens when the number of moles of the gaseous reactants is the same as the number of moles of the gaseous products, that is:

$$\nu_A + \nu_B = \nu_C + \nu_D.$$

However, if the reaction results in an increase in the number of moles of gases, i.e., $\nu_C + \nu_D > \nu_A + \nu_B$, then a higher pressure P would result in a smaller equilibrium product because

$$\frac{y_C^{\nu_C} y_D^{\nu_D}}{y_A^{\nu_A} y_B^{\nu_B}} = K_P(T) \left(\frac{P_0}{P}\right)^{\nu_C + \nu_D - \nu_A - \nu_B}.$$

As the pressure increases, the equilibrium is shifted to the reactant side of the reaction. On the other hand, if the reaction results in a decrease in the number of moles of gases, i.e., $v_C + v_D < v_A + v_B$, then a higher pressure would result in a larger equilibrium product; larger pressure then produces more products. This is summarized by the Le Chatelier principle (for pressure): an increase in pressure (the stress), shifts the equilibrium so as to decrease the volume of the system (the response) so that the increase in the stress on the system is reduced. We look at an example.

§   *Example 12.6 Effect of pressure on the dimerization of* NO_2

Nitrogen dioxide dimerizes in the gas phase reaction as follows. How does the equilibrium shift with pressure?

$$N_2O_4(g) \rightleftharpoons 2NO_2(g).$$

To be specific in our discussion, let us consider a reaction system that initially contains n moles of N_2O_4 and zero moles of NO_2. At equilibrium let there be $2m$ moles of NO_2. We want to calculate m given the equilibrium temperature and pressure. First, we count the number of moles of the reactants and the products given any extent of reaction by making use of the stoichiometry. If $2m$ moles of NO_2 are formed, then the number of moles of N_2O_4 consumed is equal to m. Hence, at equilibrium we have $n - m$ moles of N_2O_4. This gives a total number of moles in the system of $n + m$. We can therefore write the mole fractions of N_2O_4 and NO_2 in terms of n and m. If a fraction f of the initial amount of N_2O_4 is consumed by the reaction, then $f = \frac{m}{n}$. It is convenient to tabulate this mole counting as follows:

	$N_2O_4(g)$	$NO_2(g)$	Total
initial moles	n	0	n
equilibrium moles	$n - m$	$2m$	$n + m$
mole fraction at equilibrium	$\dfrac{n-m}{n+m}$ $= \dfrac{1-f}{1+f}$	$\dfrac{2m}{n+m}$ $= \dfrac{2f}{1+f}$	

We have done this counting starting with only N_2O_4 in the system. However, the method of accounting above can be applied to any initial composition of the reaction system.

With the above mole fractions at equilibrium, we can write

$$K_P(T) = exp\left(-\frac{\Delta_{rxn}\mu^\varnothing}{RT}\right) = \left(\frac{P}{P_0}\right)\left(\frac{y^2_{NO_2}}{y_{N_2O_4}}\right)$$

$$= \left(\frac{P}{P_0}\right)\left(\frac{4f^2}{1-f^2}\right).$$

Therefore, the fraction of N_2O_4 reacted at temperature T and pressure P is given by

$$f = \left[1 + \frac{4}{K_P}\left(\frac{P}{P_0}\right)\right]^{-\frac{1}{2}}.$$

If pressure increases, f decreases; the equilibrium shifts to the reactant side of the chemical reaction. This is as expected from Le Chatelier's principle since the number of moles of gases increases in the forward reaction direction. $\S$

12.6 Temperature dependence

We have seen that the equilibrium constant is $K(T) = exp\left(-\frac{\Delta_{rxn}G^{\emptyset}}{RT}\right)$. Taking the logarithm of this, we get

$$\ln K = -\frac{\Delta_{rxn}G^{\emptyset}}{RT}.$$

From the form of this equation, you realise quickly that the temperature dependence of $\ln K$ is most conveniently obtained by using the Gibbs-Helmholtz equation which we have encountered in Section 7.12:

$$\left[\frac{\partial}{\partial T}\left(\frac{\Delta G}{T}\right)\right]_P = -\frac{\Delta H}{T^2}$$

where H is the enthalpy. Therefore, we have

$$\frac{d(\ln K)}{dT} = \frac{\Delta_{rxn}H^{\emptyset}}{RT^2}$$

keeping the pressure constant. Here, the enthalpy of reaction, per mole of reaction is given by

$$\Delta_{rxn}H = -v_A h_A - v_B h_B + v_C h_C + v_D h_D$$

for the generic chemical reaction, where lowercase h are the molar enthalpies of the reactants and products, and v_i's are the stoichiometric coefficient for the reaction

$$v_A A + v_B B \rightleftharpoons v_C C + v_D D.$$

Integrating from T_1 to T_2, we get the van't Hoff equation

$$\ln K(T_2) - \ln K(T_1) = \int_{T_1}^{T_2} \frac{\Delta_{rxn}H^{\emptyset}(T)}{RT^2} dT.$$

In general, $\Delta_{rxn}H^{\emptyset}$ is a function of temperature because the molar enthalpies of the reacting substances are generally functions of temperature. However, a simple approximation can be obtained if we assume that it is a constant, independent of temperature. Then, we have the integrated form of the van't Hoff equation

$$\ln K(T_2) - \ln K(T_1) = -\frac{\Delta_{rxn} H^{\emptyset}}{R} \left(\frac{1}{T_2} - \frac{1}{T_1} \right),$$

which is essentially the Gibbs-Helmholtz equation. If the reaction is exothermic, $\Delta_{rxn} H^{\emptyset}$ is a negative quantity. Therefore, raising the temperature from T_1 to a higher temperature T_2 makes the right-hand side of this equation negative. Hence, the equilibrium constant $K(T_2)$ is smaller than $K(T_1)$: the equilibrium in an exothermic reaction shifts towards the reactants when temperature is increased. Vice versa, raising the temperature for an endothermic reaction drives the equilibrium towards the products. This is qualitatively expressed as the Le Chatelier's principle (for temperature): an increase in temperature (the stress), shifts the equilibrium so as to decrease the amount of heat in the system (the response) to reduce the stress on it.

§   *Example 12.7 Revisiting the dimerization of NO_2*

We look again at the dimerization reaction for nitrogen dioxide discussed in the previous Example:

$$N_2O_4(g) \rightleftharpoons 2NO_2(g).$$

Initially, you only have N_2O_4 in the reacting system.

Calculate the equilibrium constant and the fraction f of the initial N_2O_4 that is reacted at standard temperature and pressure.

Calculate the equilibrium constant and the fraction f of the initial N_2O_4 that is reacted when the temperature is $298.15\ K$ and pressure is $2\ bar$.

Calculate the equilibrium constant and the fraction f of the initial N_2O_4 that is reacted when the temperature is $320\ K$ and pressure is $2\ bar$.

The standard enthalpy of formation $\Delta_f H^\emptyset$ of formation and the standard (Third Law) entropy $S^\emptyset$ are:

$$\Delta_f H^\emptyset [N_2 O_4(g)] = 9.16 \text{ kJ mol}^{-1}$$

$$\Delta_f H^\emptyset [NO_2(g)] = 33.18 \text{ kJ mol}^{-1}$$

$$S^\emptyset [N_2 O_4(g)] = 304.29 \text{ J mol}^{-1}\text{K}^{-1}$$

$$S^\emptyset [NO_2(g)] = 240.06 \text{ J mol}^{-1}\text{K}^{-1}.$$

Thus, the standard enthalpy and entropy of reaction are:

$$\Delta_{rxn} H^\emptyset = 2(33.18) - 9.16 = 57.2 \text{ kJ mol}^{-1}$$

$$\Delta_{rxn} S^\emptyset = 2(240.06) - 304.29 = 175.83 \text{ J mol}^{-1}\text{K}^{-1}.$$

These are at a temperature of 298.15 K and the standard pressure of 1 bar. Since $G = H - TS$, these give the standard Gibbs free energy of reaction

$$\Delta_{rxn} G^\emptyset = \Delta_{rxn} H^\emptyset - T\Delta_{rxn} S^\emptyset = 4.78 \text{ kJ mol}^{-1}.$$

Therefore, the Gibbs free energy of reaction at 298.15 K is

$$K(298.15) = exp\left[-\frac{\Delta_{rxn} G^\phi}{298.15 R}\right] = 0.146.$$

Note that the superscript in $\Delta_{rxn} G^\phi$ indicates that its numerical value is dependent upon the pressure P_0 used to specify the standard state chemical potential for the reactants and products; the value of P_0 used here is 1 bar, as is typical. The equilibrium constant is independent of the pressure.

However, the fraction of the initial $N_2 O_4$ reacted depends upon the pressure. Starting with a system containing only the dimer $N_2 O_4$, we found that at equilibrium the fraction of $N_2 O_4$ reacted is $f = \left[1 + \frac{4}{K_P}\left(\frac{P}{P_0}\right)\right]^{-\frac{1}{2}}$ in Example 12.6. Hence, at a pressure of 1 bar and temperature of 298.15 K, the fraction of the initial N_2O_4 that is reacted is

$$f = \left[1 + \frac{4}{K_P}\left(\frac{P}{P_0}\right)\right]^{-\frac{1}{2}}$$

$$= \left[1 + \frac{4}{0.146}\left(\frac{1}{1}\right)\right]^{-\frac{1}{2}}$$

$$= 0.188.$$

Therefore, the mole fractions of N_2O_4 and NO_2 at equilibrium are

$$x_{N_2O_4} = \frac{1-f}{1+f} = 0.684$$

$$x_{NO_2} = \frac{2f}{1+f} = 0.316.$$

When the pressure is raised to 2 bar while keeping the temperature at 298.15 K,

$$f = \left[1 + \frac{4}{K_P}\left(\frac{P}{P_0}\right)\right]^{-\frac{1}{2}}$$

$$= \left[1 + \frac{4}{0.146}\left(\frac{2}{1}\right)\right]^{-\frac{1}{2}}$$

$$= 0.134.$$

This gives the equilibrium mole fractions

$$x_{N_2O_4} = \frac{1-f}{1+f} = 0.764$$

$$x_{NO_2} = \frac{2f}{1+f} = 0.236.$$

Thus, raising the pressure shifts the reaction towards the reactant. This is expected from Le Chatelier's principle.

Now we consider how the equilibrium shifts when the temperature is raised to 320 K, while keeping the pressure constant at 2 bar. We calculate the change in the equilibrium constant using the van't Hoff equation. Approximating the enthalpy of reaction as a constant equal to its value at 298.15 K, we have

$$\ln K(320) - \ln K(298.15) = -\frac{\Delta_{rxn}H^{\varnothing}}{R}\left(\frac{1}{320} - \frac{1}{298.15}\right)$$

$$= 1.576.$$

We have calculated above that $K(298.15) = 0.146$. Hence, the equilibrium constant at 320 K is raised to 0.706; this increase is expected because the reaction is endothermic.

With the equilibrium constant at 320 K, we can calculate the fraction of the initial N_2O_4 that is reacted at 320 K and 2 bar using

$$f = \left[1 + \frac{4}{K_P}\left(\frac{P}{P_0}\right)\right]^{-\frac{1}{2}}$$

$$= \left[1 + \frac{4}{0.706}\left(\frac{2}{1}\right)\right]^{-\frac{1}{2}}$$

$$= 0.285,$$

with equilibrium mole fractions of

$$x_{N_2O_4} = \frac{1-f}{1+f} = 0.557$$

$$x_{NO_2} = \frac{2f}{1+f} = 0.443.$$

Increasing the temperature shifts the equilibrium further to the product side of the reaction. We can understand, qualitatively, why this is so from Le Chatelier's principle since the reaction is endothermic. §

§ Example 12.8 The van't Hoff equation

Revisit the reaction of nitrogen and hydrogen to form ammonia in Example 12.1. Calculate the equilibrium constant at $310\ K$.

The Haber process uses the reaction:

$$N_2(g) + 3H_2(g) \rightleftharpoons 2NH_3(g)$$

for the synthesis of ammonia. At a temperature of 298.15 K, it has an equilibrium constant K_P equal to 5.8×10^5 as we have calculated in Example 12.1. The enthalpy of reaction is approximately constant at -92.22 kJ mol^{-1} for the temperature range from 298.15 K to 350 K. Thus, for any temperature T in this range, we can estimate the equilibrium constant using the van't Hoff equation:

$$K_P(T) = K_P(298.15)exp\left[-\frac{\Delta_{rxn}H^{\varnothing}}{R}\left(\frac{1}{T} - \frac{1}{298.15}\right)\right].$$

At 310 K, this gives an equilibrium constant of 1.4×10^5. Let us say we start with n moles of N_2 and $3n$ moles of H_2; this is the stoichiometric composition for the reactants. Then with m moles of nitrogen consumed by the reaction, the total number of moles at equilibrium is equal to $4n - 2m$. Hence, counting the number of moles at equilibrium, we have the following Table.

	$N_2(g)$	$H_2(g)$	$NH_3(g)$
initial moles	n	$3n$	0
equilibrium moles	$n - m$	$3(n - m)$	$2m$
mole fraction at equilibrium	$\dfrac{n - m}{4n - 2m}$ $= \dfrac{1 - f}{2(2 - f)}$	$\dfrac{3(n - m)}{4n - 2m}$ $= \dfrac{3(1 - f)}{2(2 - f)}$	$\dfrac{2m}{4n - 2m}$ $= \dfrac{2f}{2(2 - f)}$

From this, the equilibrium composition at 310 K, and pressure P can be calculated from:

$$K_P(T) = \left(\frac{P_0}{P}\right)^2 \frac{y_{NH_3}^2}{y_{N_2} y_{H_2}^3}$$

where P_0 is the standard pressure of 1 bar, so that

$$\left(\frac{P_0}{P}\right)^2 \frac{16f^2(2 - f)^2}{27(1 - f)^4} = 1.4 \times 10^5.$$

Taking the square root, you get a quadratic equation which you can solve to find that f is equal to 0.986 or 0.955 at pressures of 10 bar and 1 bar, respectively.   §

§   *Example 12.9 pH of water*

Calculate the pH of pure water assuming that the activities of all the species involved are equal to one.

Water dissociates as follows:

$$H_2O(l) \rightleftharpoons H^+(aq) + OH^-(aq).$$

The chemical potentials of the ions in aqueous solution are

$$\mu_{H^+(aq)} = \mu^{\emptyset}_{H^+(aq)} + RT \ln \left(\frac{\gamma_{H^+} m_{H^+}}{m_0} \right)$$

$$\mu_{OH^-(aq)} = \mu^{\emptyset}_{OH^-(aq)} + RT \ln \left(\frac{\gamma_{OH^-} m_{OH^-}}{m_0} \right)$$

$$\mu_{H_2O(liq)} = \mu^{\emptyset}_{H_2O(liq)} + RT \ln \left(\frac{\gamma_{H_2O} m_{H_2O}}{m_{0,H_2O}} \right)$$

where the superscript $\emptyset$ denotes the standard state. The usual "chemical" standard state used is 298.15 K, a pressure of 1 bar, and a standard concentration of $m_0 = 1$ mol dm^{-3} for all aqueous solutes except for liquid water itself. For the latter, the standard state is 298.15 K, 1 bar, and a concentration of $m_{0,H_2O} = 55.34$ mol dm^{-3}. This latter is the concentration of water molecules in pure water given its density at 298.15 K and 1 bar. See Section 9.5.4. Selecting these as standard states establishes the reference states relative to which chemical potentials are measured and gives the values of $\mu^{\emptyset}$ that we are using here.

With these reference states for the chemical potentials, the equilibrium constant is given by

$$K = exp\left[-\frac{\Delta_{rxn} G^{\emptyset}}{RT} \right] = exp\left[-\left(\frac{\mu^{\emptyset}_{H^+(aq)} + \mu^{\emptyset}_{OH^-(aq)} - \mu^{\emptyset}_{H_2O(liq)}}{RT} \right) \right]$$

where the standard chemical potentials, from Section 9.7.1 are:

$$\mu^{\emptyset}_{H^+(aq)} = 0$$

$$\mu^{\emptyset}_{OH^-(aq)} = -157.24 \text{ kJ mol}^{-1}$$

$$\mu^{\emptyset}_{H_2O(liq)} = -237.13 \text{ kJ mol}^{-1}.$$

Hence, the Gibbs free energy of reaction at 298.15 K is equal to 79.9 kJ mol^{-1}. This gives $K(298.15) = 1.007 \times 10^{-14}$.

Approximating all the activities to be equal to one, at equilibrium we have

$$\frac{\left(\frac{m_{H^+}}{m_0}\right)\left(\frac{m_{OH^-}}{m_0}\right)}{\frac{m_{H_2O}}{m_{0,H_2O}}} = 1.007 \times 10^{-14}$$

which immediately gives $m_{H^+} = m_{OH^-} = 1.004 \times 10^{-7}$ mol dm^{-3} if we set $m_{H_2O} \cong m_{0,H_2O}$, which is reasonable from the rather small extent of dissociation. Thus, the pH of neutral water at room temperature is equal to 7. We will revisit this below using the Debye-Hückel model to estimate the activities as a function of the concentration. §

For the same standard states, the standard enthalpy of reaction $\Delta_{rxn}H$ is found to be 55.8 kJ mol^{-1}. Given that this is approximately independent of temperature, we can use the van't Hoff equation to calculate the equilibrium constant for the dissociation of water at any temperature T:

$$K(T) = K(298.15)exp\left[-\frac{\Delta_{rxn}H^{\emptyset}}{R}\left(\frac{1}{T} - \frac{1}{298.15}\right)\right].$$

Therefore, for equilibrium at temperature T, the composition of pure water is given by:

$$K(T) = \frac{\left(\frac{m_{H^+}}{m_0}\right)\left(\frac{m_{OH^-}}{m_0}\right)}{\frac{m_{H_2O}}{m_{0,H_2O}}}.$$

For pure water, we have $m_{H^+} = m_{OH^-}$. Again, assuming that the extent of dissociation is very small, $m_{H_2O} \cong m_{0,H_2O}$. You can, more precisely, take m_{H_2O} to be equal to $(m_{0,H_2O} - m_{H^+})$, getting a quadratic equation in m_{H^+} to solve; but the numerical results will be much the same because of the rather small value of m_{H^+}.

Hence, the concentration of hydrogen ions in pure water varies with temperature according to:

$$m_{H^+} = m_0\sqrt{K(T)}$$

so that the pH of neutral water is dependent upon temperature as follows:

$$pH = -\log_{10} m_{H^+} = -\frac{1}{2}\log_{10} K(T).$$

It is only at 298.15 K that the pH of pure (neutral) water is equal to 7. As the temperature increases, the degree of dissociation of water changes and the pH varies as illustrated in Figure 12.1. At 60°C it decreases to about 6.5.

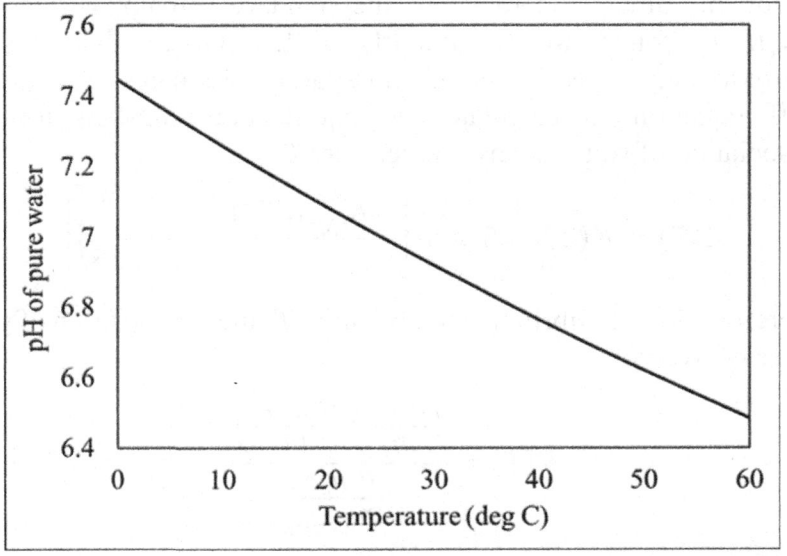

Figure 12.1 The dependence of the pH of pure water upon its temperature.

We have found in Example 9.9 that the activity coefficients of hydrogen and hydroxide ions in pure water at 298.15 K are equal to 0.99963. This is rather close to one. Therefore, we expect the calculations in Example 12.9 to be quite accurate. Nonetheless, we demonstrate how we approach chemical equilibrium problems if we do not assume ideal solution behaviour. For this, the Debye-Hückel model in Section 9.7.2 provides us with a reasonable estimate of the activities. We start with the equilibrium constant

$$K(T) = \frac{a_{H^+}a_{OH^-}}{a_{H_2O}}$$
$$= \frac{\gamma_{H^+}\gamma_{OH^-}}{\gamma_{H_2O}}\frac{(M_{H^+}/M_0)(M_{OH^-}/M_0)}{(M_{H_2O}/M_{0,H_2O})}.$$

Since the molality of the hydroxide ion is equal to that of the hydrogen ion, and some water molecules have dissociated, we can write

$$M_{H^+} = M_{OH^-}$$
$$M_{H_2O} = (M_{0,H_2O} - M_{H^+}).$$

From our discussions of non-ideal solutions in Section 9.5.4, we have the standard concentration of $M_0 = 1$ mol kg^{-1} for the hydrogen and hydroxide ions, and $M_{0,H_2O} = 55.51$ mol kg^{-1} for water; the activity coefficient of pure water is equal to one. From the definition of the mean activity coefficient in Section 9.7.1 we have

$$\gamma_\pm^2 = \gamma_{H^+}\gamma_{OH^-}.$$

Hence, we get

$$K(T) = \gamma_\pm^2 \frac{M_{0,H_2O}M_{H^+}^2}{(M_{0,H_2O} - M_{H^+})}.$$

Thus, the equilibrium condition can be written as

$$\frac{K(T)(M_{0,H_2O} - M_{H^+})}{M_{H^+}^2} = \gamma_\pm^2.$$

The left-hand side is in terms of the hydrogen ion molality. So is the right-hand side where the logarithm of the mean activity coefficient depends upon $M_{H^+}^{1/2}$ through the ionic strength of the solution as we have seen in Section 9.7.2. Using the Debye-Hückel model, the mean activity coefficient as a function of M_{H^+} is

$$\ln \gamma_{\pm} = -A_{\gamma}|z_+ z_-| \frac{I^{1/2}}{\left(1 + aB_{\gamma} I^{1/2}\right)}$$

where for pure water, the ionic strength is

$$I = \frac{1}{2}\left(M_{H^+} z_{H^+}^2 + M_{OH^-} z_{OH^-}^2\right)$$
$$= M_{H^+},$$

so that

$$\ln \gamma_{\pm} = -A_{\gamma}|z_+ z_-| \frac{M_{H^+}^{1/2}}{\left(1 + aB_{\gamma} M_{H^+}^{1/2}\right)}.$$

With this, both sides of the equilibrium equation are in terms of M_{H^+}. It is an elementary task to solve it numerically/graphically. Doing that we get obtain a hydrogen ion concentration in pure water at 298.15 K of 1.00192×10^{-7} mol kg^{-1}. We convert this to molarity units to get 0.99886 mol dm^{-1} which is very slightly different from our calculations in Example 12.9 assuming ideal solution. We will look at examples of equilibria in electrolytes where this difference is considerably larger.

12.7 Solubility of salts

In this section, we return to the example we started this chapter with, namely, the reaction that Berthollet observed when the highly saline water in a salt lake comes into contact with limestone ground:

$$CaCO_3(solid) + 2NaCl(aq) \rightleftharpoons CaCl_2(aq) + Na_2CO_3(solid).$$

We would like to quantify how sodium carbonate precipitates when a salt lake is on limestone ground. At the extreme end of saltiness for natural bodies of water, Lake Assal in Djibouti reputedly has about 300 g of dissolved $NaCl$ per litre. How does the high concentration of sodium and chloride ions affect the solubility of salts in such a solution?

We first discuss a saturated aqueous solution of calcium carbonate for which the following chemical reaction is in equilibrium

$$CaCO_3(solid) \rightleftharpoons Ca^{2+}(aq) + CO_3^{2-}(aq).$$

Solid calcium carbonate is in equilibrium with Ca^{2+} and CO_3^{2-} ions in solution. Approximating the activity coefficients of the ionic species to be unity, the chemical potentials are given by

$$\mu_{Ca^{2+}(aq)} = \mu_{Ca^{2+}(aq)}^{\emptyset} + RT \ln \left(\frac{m_{Ca^{2+}}}{m_0} \right)$$

$$\mu_{CO_3^{2-}(aq)} = \mu_{CO_3^{2-}(aq)}^{\emptyset} + RT \ln \left(\frac{m_{CO_3^{2-}}}{m_0} \right)$$

$$\mu_{CaCO_3(s)} = \mu_{CaCO_3(s)}^{\emptyset} + v_m (P - P_0)$$

where the standard states for the ions are at 298.15 K, 1 bar and $m_0 = 1$ mol dm^{-3}. Similarly, the standard state for the solid calcium carbonate is crystalline calcite at 298.15 K and $P_0 = 1$ bar. Equilibrium at temperature T and at pressure of 1 bar is given by

$$K(T) = exp \left(-\frac{\Delta_{soln} G^{\emptyset}}{RT} \right) = \left(\frac{m_{Ca^{2+}}}{m_0} \right) \left(\frac{m_{CO_3^{2-}}}{m_0} \right)$$

with the standard Gibbs free energy of solution being

$$\Delta_{soln} G^{\emptyset}(CaCO_3) = \mu_{Ca^{2+}(aq)}^{\emptyset} + \mu_{CO_3^{2-}(aq)}^{\emptyset} - \mu_{CaCO_3(s)}^{\emptyset}.$$

When discussing the solubility of salts, the equilibrium constant quantity $K(T)$ is referred to as the solubility product K_{sp}. The product of concentrations on the right-hand side is referred to as the ionic product. In a saturated solution when the solid is in equilibrium with the solution, these quantities are equal. In a solution that is not saturated, no solid is present, so that there is no equilibrium between the solid phase and the solution; the ionic product is then less than the solubility product. In general, for the dissolution of a salt $A_a B_b$,

$$A_a B_b(s) \rightleftharpoons a A^{b+}(aq) + b B^{a-}(aq)$$

we have

$$K_{sp}(T) = exp\left(-\frac{\Delta_{soln} G^{\emptyset}}{RT}\right) = \left(\frac{m_{A^{b+}}}{m_0}\right)^a \left(\frac{m_{B^{a-}}}{m_0}\right)^b$$

and

$$\Delta_{soln} G^{\emptyset} = a \mu^{\emptyset}_{A^{b+}(aq)} + b \mu^{\emptyset}_{B^{a-}(aq)} - \mu^{\emptyset}_{A_a B_b(s)}.$$

For calcium carbonate, we have the following thermochemical data for the standard states described above:

$$\mu^{\emptyset}_{Ca^{2+}(aq)} = -553.58 \text{ kJ mol}^{-1}$$

$$\mu^{\emptyset}_{CO_3^{2-}(aq)} = -527.81 \text{ kJ mol}^{-1}$$

$$\mu^{\emptyset}_{CaCO_3(s)} = -1128.8 \text{ kJ mol}^{-1}.$$

This gives $\Delta_{soln} G^{\emptyset}(CaCO_3) = 47.41 \text{ kJ mol}^{-1}$ and a solubility product equal to 4.94×10^{-9}. Therefore, at 298.15 K and 1 bar, the ionic product for a saturated solution of calcium carbonate in water is

$$\left(\frac{m_{Ca^{2+}}}{m_0}\right)\left(\frac{m_{CO_3^{2-}}}{m_0}\right) = K_{sp}(CaCO_3) = 4.94 \times 10^{-9}.$$

The standard state concentration is $m_0 = 1 \text{ mol dm}^{-3}$, so that the saturation concentration is rather small at

$$m_{CaCO_3(aq)} = m_0 \sqrt{K_{sp}} = 7.03 \times 10^{-5} \text{ mol dm}^{-3}.$$

There is a basic problem with this method of estimating solubility. By writing chemical potentials in terms of concentrations rather than activities we are assuming that the solution is ideal.

However, the standard chemical potentials are actually defined in terms of activities not the molarity.

We have seen in the previous section that for the dissociation of pure water into hydrogen and hydroxide ions at room temperature, a very good approximation to the pH can be obtained by doing your calculations with concentrations rather than activities. However, for many systems, this is not the case. We consider some examples to demonstrate how we use activities to solve chemical equilibrium problems. To illustrate this, we need to have an approximation to calculate activities; we will depend upon the Debye-Hückel model from Section 9.7.2 for this.

§   *Example 12.10 Solubility of calcium carbonate*

Estimate the solubility of calcium carbonate in pure water at 298.15 K using activity coefficients from the Debye-Hückel model.

You have a saturated solution of calcium carbonate with equilibrium as follows:

$$CaCO_3(solid) \rightleftharpoons Ca^{2+}(aq) + CO_3^{2-}(aq).$$

The equilibrium constant is

$$K_{sp}(CaCO_3) = \frac{a_{Ca^{2+}} a_{CO_3^{2-}}}{a_{CaCO_3}}$$

$$= \gamma_{Ca^{2+}} \gamma_{CO_3^{2-}} \left(\frac{M_{Ca^{2+}}}{M_0}\right)\left(\frac{M_{CO_3^{2-}}}{M_0}\right)$$

$$= \gamma_{Ca^{2+}} \gamma_{CO_3^{2-}} M_{Ca^{2+}}^2$$

where we have used the stoichiometry $M_{Ca^{2+}} = M_{CO_3^{2-}}$, M_0 equal to 1 mol kg^{-1}, and the activity of the pure solid is equal to unity. Therefore, the equilibrium condition is

$$\frac{K_{sp}}{M_{Ca^{2+}}^2} = \gamma_{Ca^{2+}} \gamma_{CO_3^{2-}}.$$

The activity coefficients can be estimated from $M_{Ca^{2+}}$ using the Debye-Hückel model. The left-hand side of this equation depends only upon the square of the molality. The right-hand side is a product of the activity coefficients, each of which is determined by the molality as follows:

$$\ln \gamma_{Ca^{2+}} = -A_\gamma z_{Ca^{2+}}^2 \frac{I^{1/2}}{\left(1 + a_{Ca^{2+}} B_\gamma I^{1/2}\right)}$$

$$\ln \gamma_{CO_3^{2-}} = -A_\gamma z_{CO_3^{2-}}^2 \frac{I^{1/2}}{\left(1 + a_{CO_3^{2-}} B_\gamma I^{1/2}\right)}$$

$$I = \frac{1}{2}\left[M_{Ca^{2+}}(2)^2 + M_{CO_3^{2-}}(-2)^2\right] = 4M_{Ca^{2+}}.$$

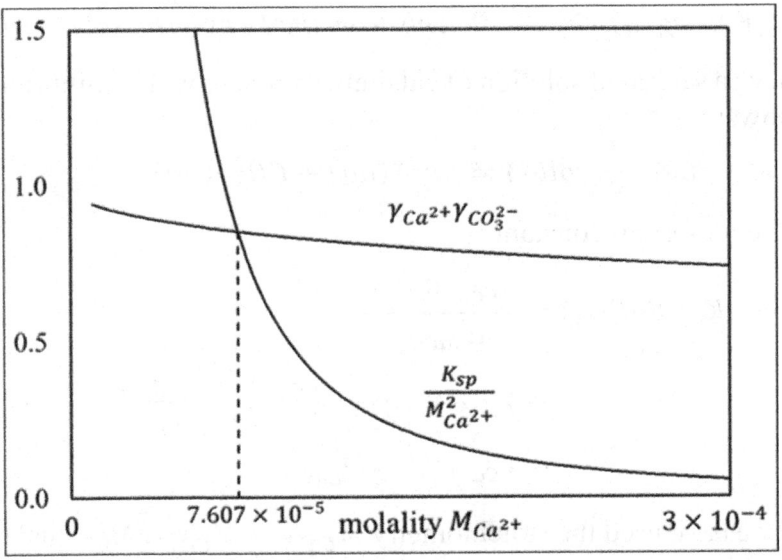

Figure 12.2 Graphical solution for Example 12.10. The product of activity coefficients is calculated using the Debye-Hückel model discussed in Section 9.7.2.

With the activity coefficients determined in terms of the molality, we can solve for $M_{Ca^{2+}}$ graphically. This is illustrated in Figure 12.2, giving $M_{Ca^{2+}} \cong 7.607 \times 10^{-5}$ mol kg^{-1}. This is the solubility at 298.15 K. At this saturation concentration, the activity coefficients for calcium and carbonate ions are 0.924. In units of molarity, the solubility is approximately

$$m_{Ca^{2+}} \cong 7.583 \times 10^{-5} \text{ mol dm}^{-3}.$$

Thus, our earlier estimate of 7.03×10^{-5} mol dm^{-3} assuming ideal solution behaviour is about 8% lower.   §

§   *Example 12.11 Dissociation of ethanoic acid in aqueous sodium chloride*

As a second example, consider adding ethanoic acid to an aqueous solution of sodium chloride at 298.15 K. The molality of the sodium chloride solution is 0.05 mol kg^{-1} and 0.15 mole of ethanoic acid is added per kilogram of water in the solution. What is the pH of the resulting solution?

The ethanoic acid dissociation equilibrium is

$$CH_3COOH(aq) \rightleftharpoons H^+(aq) + CH_3COO^{-1}(aq)$$

with standard state chemical potentials of

$$\mu^{\varnothing}_{H^+(aq)} = 0 \text{ kJ mol}^{-1}$$

$$\mu^{\varnothing}_{CH_3COO^-(aq)} = -369.31 \text{ kJ mol}^{-1}$$

$$\mu^{\varnothing}_{CH_3COOH(aq)} = -396.46 \text{ kJ mol}^{-1}.$$

The equilibrium (acid dissociation) constant is

$$K_a = \frac{a_{H^+}a_{CH_3COO^{-1}}}{a_{CH_3COOH}} = 1.75 \times 10^{-5}$$

$$= \gamma_{H^+}\gamma_{CH_3COO^-}\frac{M_{H^+}^2}{(M_{tot} - M_{H^+})}$$

with similar considerations as in Example 12.10. M_{tot} is the total molality of the dissociated and undissociated ethanoic acid, that is 0.15 moles per kilogram of water, and we have assumed the activity of ethanoic acid to be one. Thus, we have

$$\frac{K_a(0.15 - M_{H^+})}{M_{H^+}^2} = \gamma_{H^+}\gamma_{CH_3COO^-}.$$

But now, with sodium chloride dissolved in the solution, the activity coefficient depends upon an ionic strength which includes contributions from all the ionic species: Na^+, Cl^-, H^+ and CH_3COO^-:

$$I = \frac{1}{2}\left[0.05(1)^2 + 0.05(-1)^2 + M_{H^+}(1)^2 + M_{CH_3COO^-}(-1)^2\right]$$

$$= \frac{1}{2}(0.1 + 2M_{H^+}),$$

where we take sodium chloride to be completely dissociated. We can thus calculate $\frac{K_a(0.15-M_{H^+})}{M_{H^+}^2}$ and $\gamma_{H^+}\gamma_{CH_3COO^-}$ in terms of the molality of the hydrogen ion in the solution, and then graphically find when these two quantities are equal. This gives

$$M_{H^+} = 1.932 \times 10^{-3} \text{ mol kg}^{-1}$$

with activity coefficients equal to

$$\gamma_{H^+} = 0.853$$

$$\gamma_{CH_3COO^-} = 0.814.$$

The concentration in terms of molarity is 1.926×10^{-3} mol dm^{-3} and the pH is 2.72. If we treat the solution as ideal, we obtain at equilibrium

$$K_a = \frac{M_{H^+}^2}{(0.15 - M_{H^+})}$$

which gives

$$M_{H^+} = 1.611 \times 10^{-3} \text{ mol kg}^{-1}$$

and a pH of 2.79. The ideal solution estimate of the concentration is about 17% too low. §

§   **Example 12.12 The common ion effect**

Finally, let us consider a system with a common ion effect. You have an aqueous solution of magnesium sulphate of concentration 0.2 mol kg^{-1} ; $MgSO_4$ is very soluble in water. The solubility product of barium sulphate, which is rather sparingly soluble, is 1.084×10^{-10}. How much barium sulphate will dissolve in your magnesium sulphate solution?

The equilibrium for barium sulphate is described by

$$BaSO_4(s) \rightleftharpoons Ba^{2+}(aq) + SO_4^{2-}(aq)$$

$$K_{sp} = \frac{a_{Ba^{2+}} a_{SO_4^{2-}}}{a_{BaSO_4}} = 1.084 \times 10^{-10}$$

$$= \gamma_{Ba^{2+}} \gamma_{SO_4^{2-}} M_{Ba^{2+}} M_{SO_4^{2-}}$$

$$= \gamma_{Ba^{2+}} \gamma_{SO_4^{2-}} M_{Ba^{2+}} (0.2 + M_{Ba^{2+}})$$

where the contribution of 0.2 in the sulphate concentration comes from the magnesium sulphate.

The ionic strength of the solution is

$$I = \frac{1}{2} [0.2(2)^2 + 0.2(-2)^2 + M_{Ba^{2+}}(2)^2 + M_{Ba^{2+}}(-2)^2]$$

where the last term is the sulphate contribution from the dissolved barium sulphate. Thus, the product of the activity coefficients

$\gamma_{Ba^{2+}}\gamma_{SO_4^{2-}}$ can be calculated in terms of the molality of the barium ion. With this, we can numerically/graphically solve

$$\frac{K_{sp}}{M_{Ba^{2+}}(0.2 + M_{Ba^{2+}})} = \gamma_{Ba^{2+}}\gamma_{SO_4^{2-}}.$$

The results are:

$$M_{Ba^{2+}} = 1.62 \times 10^{-8} \text{ mol kg}^{-1}$$

$$\gamma_{Ba^{2+}} = 0.2026$$

$$\gamma_{SO_4^{2-}} = 0.1657.$$

In comparison, the ideal solution equilibrium condition is

$$\frac{K_{sp}}{M_{Ba^{2+}}(0.2 + M_{Ba^{2+}})} = 1$$

which gives at solubility of

$$M_{Ba^{2+}} = 5.421 \times 10^{-10} \text{ mol kg}^{-1}.$$

This underestimate of the dissolved barium ion concentration assuming an ideal solution might be a serious error in gauging the presence of impurities in a water supply for consumption or for use in demanding industries. §

12.8 Back to Berthollet

For the dissolution of sodium carbonate in aqueous solution,

$$Na_2CO_3(solid) \rightleftharpoons 2Na^+(aq) + CO_3^{2-}(aq)$$

we have the following thermochemical data:

$$\mu^{\varnothing}_{Na^+(aq)} = -261.91 \text{ kJ mol}^{-1}$$

$$\mu^{\varnothing}_{CO_3^{2-}(aq)} = -527.81 \text{ kJ mol}^{-1}$$

$$\mu^{\varnothing}_{Na_2CO_3(s)} = -1044.4 \text{ kJ mol}^{-1}.$$

Therefore, the standard Gibbs free energy for dissolution is

$$\Delta_{soln} G^{\emptyset}(Na_2CO_3) = 2\mu^{\emptyset}_{Na^+(aq)} + \mu^{\emptyset}_{CO_3^{2-}(aq)} - \mu^{\emptyset}_{Na_2CO_3(s)}$$

$$= -7.23 \text{ kJ mol}^{-1}$$

from which the solubility product is calculated to be 18.48, so that at saturation

$$\left(\frac{M_{Na^+}}{M_0}\right)^2 \left(\frac{M_{CO_3^{2-}}}{M_0}\right) = K_{sp}(Na_2CO_3) = 18.48.$$

Note the exponent of 2 for the sodium ion; this comes from the stoichiometric coefficient of the dissolution chemical reaction above. For a sodium carbonate solution, M_{Na^+} is equal to $2M_{CO_3^{2-}}$. Therefore, at 298.15 K and 1 bar, the concentration of a saturated sodium carbonate solution in water is:

$$M_{Na_2CO_3(aq)} = 1.67 \text{ mol kg}^{-1}.$$

Sodium carbonate is much more soluble in water than calcium carbonate. An estimate with the Debye-Hückel model gives an even higher solubility of about 6.35 mol kg^{-1}. At this high concentration, the estimates of activity coefficients are likely not good, but it is clear that sodium carbonate is highly soluble in water.

Using the same method, and the following additional information for the standard state chemical potentials for $Cl^-(aq)$ and solid sodium chloride

$$\mu^{\emptyset}_{Cl^-(aq)} = -131.23 \text{ kJ mol}^{-1}$$

$$\mu^{\emptyset}_{NaCl(s)} = -384.14 \text{ kJ mol}^{-1},$$

we can calculate the solubility product of sodium chloride to be

$$\left(\frac{M_{Na^+}}{M_0}\right)\left(\frac{M_{Cl^-}}{M_0}\right) = K_{sp}(NaCl) = 37.74.$$

Thus, the concentration of a saturated solution of sodium chloride is

$$M_{NaCl(aq)} = 6.14 \text{ mol kg}^{-1}.$$

This is an estimate of maximum common salt concentration in various salt lakes. Beyond this ionic product for $Na^+(aq)$ and $Cl^-(aq)$ solid sodium chloride would precipitate. At this concentration, there is about 359 g of sodium choride dissolved per litre of water, compared to the 300 g per litre at Lake Assal.

We can now go back to the reaction Berthollet observed at the limestone edges of an Egyptian salt lake:

$$CaCO_3(solid) + 2NaCl(aq) \rightleftharpoons CaCl_2(aq) + Na_2CO_3(solid).$$

Let us mimic this salt lake using a solution consisting of $Na^+(aq)$, $Ca^{2+}(aq)$, $CO_3^{2-}(aq)$, $Cl^-(aq)$ in equilibrium with $CaCO_3(s)$ and $Na_2CO_3(s)$. We assume that the solution behaviour is ideal, doing our calculations with molalities rather than activities since the ionic strength of the solution is rather high, as we shall see, and for these high concentrations the Debye-Hückel model is not accurate anyway.

Say, we know the salinity of a salt lake by measuring the concentration of the chloride ion M_{Cl^-}. We take this to be $100 \text{ g dm}^{-3} = 1.71 \text{ mol kg}^{-1}$, which is considerably lower than the amount of dissolved chloride ions in Lake Assal in Africa. We can then calculate the concentrations of the other ions. The following three equations relate these three concentrations:

$$\left(\frac{M_{Ca^{2+}}}{M_0}\right)\left(\frac{M_{CO_3^{2-}}}{M_0}\right) = K_{sp}(CaCO_3) = 4.94 \times 10^{-9}$$

$$\left(\frac{M_{Na^+}}{M_0}\right)^2\left(\frac{M_{CO_3^{2-}}}{M_0}\right) = K_{sp}(Na_2CO_3) = 18.48$$

$$2M_{Ca^{2+}} + M_{Na^+} = 2M_{CO_3^{2-}} + M_{Cl^-}.$$

The first two equations are from our equilibrium conditions above for calcium carbonate and sodium carbonate, while the third equation is for electrical neutrality for the solution. Solving these equations numerically give us

$$M_{CO_3^{2-}} = 1.15 \text{ mol kg}^{-1}$$

$$M_{Na^+} = 4.01 \text{ mol kg}^{-1}$$

$$M_{Ca^{2+}} = 4.30 \times 10^{-9} \text{ mol kg}^{-1}.$$

The ionic product $\left(\frac{M_{Na^+}}{M_0}\right)\left(\frac{M_{Cl^-}}{M_0}\right) = 6.86$. This is still below the solubility product for sodium chloride $K_{sp}(NaCl) = 37.74$, so that these concentrations are still considerably below concentrations when solid sodium chloride would precipitate out of solution, even though calcium carbonate and sodium carbonate are precipitating out from this salt lake. Any object, such as a dead animal, in this salt lake would be coated with a layer of carbonate crystals. Given the ions present, the other salt that can possibly precipitate out of this solution is calcium chloride. The solubility product for this is $K_{sp}(CaCl_2) = 8 \times 10^{11}$, which is many orders of magnitude larger than the ionic product $\left(\frac{M_{Ca^{2+}}}{M_0}\right)\left(\frac{M_{Cl^-}}{M_0}\right)^2 = 1.26 \times 10^{-8}$; calcium chloride is not anywhere close to precipitating out of this solution.

A point to note about the connection between the equilibrium constant for the reaction

$$CaCO_3(solid) + 2NaCl(aq) \rightleftharpoons CaCl_2(aq) + Na_2CO_3(solid)$$

and the solubility products for calcium and sodium carbonates. The equilibrium constant is

$$K(T) = exp\left(-\frac{\Delta_{rxn}G^{\emptyset}}{RT}\right) = \left(\frac{M_{Ca^{2+}}}{M_0}\right)\bigg/\left(\frac{M_{Na^+}}{M_0}\right)^2$$

with the standard Gibbs free energy of reaction being

$$\Delta_{rxn}G^{\emptyset} = \mu^{\emptyset}_{Na_2CO_3(s)} + \mu^{\emptyset}_{Ca^{2+}(aq)} - \mu^{\emptyset}_{CaCO_3(s)} - 2\mu^{\emptyset}_{Na^+(aq)}$$

$$= \Delta_{soln}G^{\emptyset}(CaCO_3) - \Delta_{soln}G^{\emptyset}(Na_2CO_3).$$

Comparing these expressions with those we have obtained above for the solubility of calcium carbonate and sodium carbonate, we see that

$$K(T) = \left(\frac{M_{Ca^{2+}}}{M_0}\right) \bigg/ \left(\frac{M_{Na^+}}{M_0}\right)^2 = \frac{K_{sp}(CaCO_3)}{K_{sp}(Na_2CO_3)}.$$

Using the solubility products for calcium carbonate and sodium carbonate, the equilibrium constant at 298.15 K and 1 bar is equal to 2.67×10^{-10}. Hence, we expect that at equilibrium the sodium ion concentration is much higher than the calcium ion concentration, as we have found above. And if the sodium ion concentration is sufficiently high, sodium carbonate precipitates out of the solution, as in the story about Berthollet that started this topic of chemical equilibrium. Using the approach we have discussed in these last two sections, we now consider a pressing contemporary situation due to the dissolution of solid calcium carbonate in the oceans.

12.9 The carbonate balance in the ocean

In the last two hundred years the global average pH of the oceans has decreased from 8.16 to 8.07. With this decrease, the hydrogen ion concentration of the oceans has increased by

$$\frac{10^{-8.07} - 10^{-8.16}}{10^{-8.16}} = 23\%.$$

This acidification of the oceans, which arises from the increased concentration of carbon dioxide in the atmosphere, is of concern because it is an important driving force for the dissolution of the calcium carbonate shells of many marine organisms and coral reefs. We close this chapter by examining the chemical equilibria which maintain the carbonate balance in the ocean. Specifically, let us calculate the pH of the ocean, an important problem given that the current concentration of carbon dioxide in the atmosphere has increased to a dismally high 420 ppm today from 315 ppm in 1958. This is also an illustrative problem on handling a few simultaneous chemical equilibria.

We first list the equilibria involved. Because of the increased amount of carbon dioxide in the atmosphere, the equilibrium

$$CO_2(g) \rightleftharpoons CO_2(aq)$$

has shifted further to the right since 1958. When dissolved in water, carbon dioxide reacts with water to form carbonic acid through the reaction

$$CO_2(aq) + H_2O(l) \rightleftharpoons H_2CO_3(aq).$$

Carbonic acid dissociates through two steps

$$H_2CO_3(aq) \rightleftharpoons H^+(aq) + HCO_3^-(aq)$$

$$HCO_3^-(aq) \rightleftharpoons H^+(aq) + CO_3^{2-}(aq)$$

producing hydrogen $H^+(aq)$, bicarbonate $HCO_3^-(aq)$ and carbonate $CO_3^{2-}(aq)$ ions in solution. The carbonate ion is directly in equilibrium with solid calcium carbonate in the ocean, which is present in the rocks and in the shells of many marine organisms:

$$CaCO_3(s) \rightleftharpoons Ca^{2+}(aq) + CO_3^{2-}(aq).$$

These equilibria determine the amount of carbonate in solution in the ocean. Because the hydrogen ion is involved, the dissociative equilibrium of water should also be considered:

$$H^+(aq) + OH^-(aq) \rightleftharpoons H_2O(l).$$

Thermodynamic equilibrium for each of these six reactions gives

$$K_s = \frac{a_{CO_2(aq)}}{a_{CO_2(g)}} = \frac{m_{CO_2(aq)}}{P_{CO_2(g)}}$$

$$K_0 = \frac{a_{H_2CO_3(aq)}}{a_{CO_2(aq)}a_{H_2O(l)}} = \frac{m_{H_2CO_3(aq)}}{m_{CO_2(aq)}}$$

$$K_1 = \frac{a_{H^+(aq)}a_{HCO_3^-(aq)}}{a_{H_2CO_3(aq)}} = \gamma_{H^+(aq)}\gamma_{HCO_3^-(aq)}\frac{m_{H^+(aq)}m_{HCO_3^-(aq)}}{m_{H_2CO_3(aq)}}$$

$$K_2 = \frac{a_{H^+(aq)}a_{CO_3^{2-}(aq)}}{a_{HCO_3^-(aq)}} = \frac{\gamma_{H^+(aq)}\gamma_{CO_3^{2-}(aq)}}{\gamma_{HCO_3^-(aq)}}\frac{m_{H^+(aq)}m_{CO_3^{2-}(aq)}}{m_{HCO_3^-(aq)}}$$

$$K_{sp} = \frac{a_{Ca^{2+}(aq)}a_{CO_3^{2-}(aq)}}{a_{CaCO_3(s)}} = \gamma_{Ca^{2+}(aq)}\gamma_{CO_3^{2-}(aq)}\,m_{Ca^{2+}(aq)}m_{CO_3^{2-}(aq)}$$

$$K_w = \frac{a_{H^+(aq)}a_{OH^-(aq)}}{a_{H_2O(l)}} = \gamma_{H^+(aq)}\gamma_{OH^-(aq)}\,m_{H^+(aq)}m_{OH^-(aq)}.$$

We have, to good approximation, taken the gas phase to be an ideal gas mixture. In the first two reactions, we have also approximated dissolved carbon dioxide $CO_2(aq)$ and undissociated carbonic acid $H_2CO_3(aq)$ as ideal dilute solutions. And the activity of water is approximated to be one in the Raoult's law limit. The activity of the pure solid calcium carbonate is one. To reduce equation clutter, we have also suppressed writing m_0 equal to one molar and P_0 equal to 1 bar, so that the concentration unit for these equations is mol dm^{-3} and for $P_{CO_2(g)}$ is bar.

Since all the activity coefficients are determined by the ionic strength of seawater, which can be independently measured as discussed in Section 9.7.1, these are all equations in terms of the concentrations of the reacting species. We will revisit the measurement of the activity coefficients in the context of electrochemistry in Section 13.9. Perhaps less accurately, we can also estimate the activity coefficients from an approximation such as the Debye-Hückel model. Instead of plugging experimentally measured values of the activity coefficients into the equations above, it may be slightly more instructive to follow this latter approximate approach.

You will notice that the first two equations, for the equilibration between gaseous and aqueous carbon dioxide, and the reaction of the latter with water to produce carbonic acid, can be solved separately to give us the concentrations $m_{CO_2(aq)}$ and $m_{H_2CO_3(aq)}$ from the pressure of the carbon dioxide $P_{CO_2(g)}$ in the gas phase. These give

$$m_{CO_2(aq)} = K_s P_{CO_2(g)}$$

$$m_{H_2CO_3(aq)} = K_0 m_{CO_2(aq)} = K_0 K_s P_{CO_2(g)}$$

in terms of the pressure of carbon dioxide in the atmosphere.

§   Example 12.13 Concentrations of dissolved carbon dioxide and carbonic acid in the oceans

The concentration of carbon dioxide in the atmosphere is 420 ppm today. Calculate the concentration of dissolved carbon dioxide and carbonic acids in the ocean. Calculate the same quantities for the 1958 atmospheric carbon dioxide concentration of 315 ppm.

$$\Delta_f G^{\varnothing}(CO_2(g)) = -394.36 \text{ kJ mol}^{-1}$$
$$\Delta_f G^{\varnothing}(CO_2(aq)) = -385.98 \text{ kJ mol}^{-1}.$$

Carbon dioxide dissolves in water through the reaction

$$CO_2(g) \rightleftharpoons CO_2(aq).$$

Thus, the standard Gibbs free energy of dissolution of carbon dioxide in water is

$$\Delta_{soln} G^{\varnothing} = -385.98 + 394.36 = 8.38 \text{ kJ mol}^{-1},$$

giving

$$K_s = exp\left(-\frac{8380}{8.314 \times 298.15}\right) = 3.403 \times 10^{-2}.$$

Therefore, the concentration of dissolved carbon dioxide is

$$m_{CO_2(aq)} = K_s P_{CO_2(g)} = K_s \frac{420}{10^6} \times 1.01325$$
$$= 1.448 \times 10^{-5} \text{ mol dm}^{-3}.$$

From this we can connect back to Example 9.7 by calculating the Henry's law constant K_{CO_2} for carbon dioxide in water. One dm^3 of seawater contains approximately $1000/18$ moles. Thus, the mole fraction of carbon dioxide is $x_{CO_2} \cong 2.61 \times 10^{-7}$. Given the pressure is $42\ Nm^{-2}$, we estimate a Henry's law constant of

$$K_{CO_2} = \frac{P_{CO_2}}{x_{CO_2}} \cong 1.61 \times 10^{-8}\ Nm^{-2}$$

which we used in Example 9.7.

The formation of carbonic acid is

$$CO_2(aq) + H_2O(l) \rightleftharpoons H_2CO_3(aq).$$

The Gibbs free energy of formation of carbonic acid in aqueous solution is $-623.08\ kJ\ mol^{-1}$ and for liquid water is $-237.13\ kJ\ mol^{-1}$. This gives

$$\Delta_{rxn} G^{\emptyset} = -623.08 + 237.13 + 385.98 = 0.030\ kJ\ mol^{-1}.$$

Thus, the equilibrium constant is

$$K_0 = exp\left(-\frac{30}{8.314 \times 298.15}\right) = 0.988.$$

This gives a carbonic acid concentration in the ocean of

$$m_{H_2CO_3(aq)} = K_0 K_s P_{CO_2(g)} = 1.431 \times 10^{-5}\ mol\ dm^{-3}.$$

For the 1958 atmospheric carbon dioxide concentration of 315 ppm, we obtain $m_{CO_2(aq)}$ equal to $1.086 \times 10^{-5}\ mol\ dm^{-3}$, and $m_{H_2CO_3(aq)}$ equal to $1.073 \times 10^{-5}\ mol\ dm^{-3}$. The amount of dissolved carbon dioxide in the oceans have increased by approximately 40%. §

With the concentration of dissolved carbonic acid, the remaining four chemical equilibria can be solved simultaneously for the five concentrations $m_{HCO_3^-(aq)}$, $m_{CO_3^{2-}(aq)}$, $m_{H^+(aq)}$, $m_{OH^-(aq)}$, and

$m_{Ca^{2+}(aq)}$. In addition to the equations describing these equilibria, the system is electrically neutral, so that

$$m_{H^+(aq)} + 2m_{Ca^{2+}(aq)} = m_{OH^-(aq)} + m_{HCO_3^-} + 2m_{CO_3^{2-}}$$

assuming just this set of chemical equilibria. Altogether, we have a set of five simultaneous equations with five unknown concentrations, consisting of the equations describing four chemical equilibria plus the ionic charge balance equation. Therefore, thermodynamics provides us with a well-defined problem statement as follows:

$$K_1' \equiv \frac{K_1}{\gamma_{H^+(aq)}\gamma_{HCO_3^-(aq)}} = \frac{m_{H^+(aq)}m_{HCO_3^-(aq)}}{m_{H_2CO_3(aq)}}$$

$$K_2' \equiv \frac{K_2\,\gamma_{HCO_3^-(aq)}}{\gamma_{H^+(aq)}\gamma_{CO_3^{2-}(aq)}} = \frac{m_{H^+(aq)}m_{CO_3^{2-}(aq)}}{m_{HCO_3^-(aq)}}$$

$$K_{sp}' \equiv \frac{K_{sp}}{\gamma_{Ca^{2+}(aq)}\gamma_{CO_3^{2-}(aq)}} = m_{Ca^{2+}(aq)}m_{CO_3^{2-}(aq)}$$

$$K_w' \equiv \frac{K_w}{\gamma_{H^+(aq)}\gamma_{OH^-(aq)}} = m_{H^+(aq)}m_{OH^-(aq)}.$$

§   **Example 12.14 Equilibrium constant for bicarbonate dissociation in seawater**

The standard chemical potential of the aqueous bicarbonate and carbonate ions are:

$$\mu^{\emptyset}_{HCO_3^-(aq)} = -586.77 \text{ kJ mol}^{-1}$$

$$\mu^{\emptyset}_{CO_3^{2-}(aq)} = -527.81 \text{ kJ mol}^{-1}.$$

Calculate the equilibrium constant K_2 for

$$HCO_3^-(aq) \rightleftharpoons H^+(aq) + CO_3^{2-}(aq).$$

Then calculate the equilibrium constant K_2' to take the activity coefficients of the ions into consideration, given that the ionic strength of seawater is 0.71 mol kg^{-1}. The Debye-Hückel ionic radii are 5 Å, 4 Å and 9 Å, respectively for the carbonate ion, the bicarbonate ion and the hydrogen ion.

The standard chemical potential of reaction is

$$\Delta_{rxn}\mu^{\varnothing} = \mu^{\varnothing}_{H^+(aq)} + \mu^{\varnothing}_{CO_3^{2-}(aq)} - \mu^{\varnothing}_{HCO_3^-(aq)}$$

$$= 0 - 527.81 + 586.77 = 58.96 \text{ kJ mol}^{-1}.$$

Hence, the equilibrium constant is

$$K_2 = \exp\left(-\frac{\Delta_{rxn}\mu^{\varnothing}}{RT}\right)$$

$$= 4.68 \times 10^{-11}$$

Using the Debye-Hückel model

$$\ln \gamma_i = -A_\gamma z_i^2 \frac{I^{1/2}}{\left(1 + aB_\gamma I^{1/2}\right)}$$

so that the activity coefficient of the carbonate ion in aqueous solution is

$$\ln \gamma_{CO_3^{2-}(aq)} = -4A_\gamma \frac{I^{1/2}}{\left(1 + (5 \times 10^{-10})B_\gamma I^{1/2}\right)}.$$

This gives an activity coefficient of 0.66 in seawater. Similarly, the activity coefficients for the bicarbonate and the hydrogen ions in seawater are 0.63 and 0.75. Therefore, we get

$$K_2' = \frac{K_2 \gamma_{HCO_3^-(aq)}}{\gamma_{H^+(aq)} \gamma_{CO_3^{2-}(aq)}} = 5.88 \times 10^{-11}. \qquad \S$$

Similarly, factoring in the activity coefficients of the respective ions, the equilibrium constants K_1', K_{sp}' and K_w' are equal to 9.21×10^{-7}, 1.08×10^{-8} and 2.27×10^{-14}. Using these equilibrium constants, the carbonate balance can be solved to obtain the pH of seawater.

§ **Example 12.15 pH of the oceans**

Solve the equations for chemical equilibrium for carbonate balance in seawater to estimate the pH of seawater for carbon dioxide concentrations of 315 ppm in 1958 and 420 ppm in 2022.

Using the equilibrium equations above, we can express the concentrations of the bicarbonate, carbonate, calcium and hydroxide ions in terms of the hydrogen ion concentration:

$$m_{HCO_3^-(aq)} = \frac{K_1' \, m_{H_2CO_3(aq)}}{m_{H^+(aq)}}$$

$$m_{CO_3^{2-}(aq)} = \frac{K_2' \, m_{HCO_3^-(aq)}}{m_{H^+(aq)}} = \frac{K_2'K_1' \, m_{H_2CO_3(aq)}}{m_{H^+(aq)}^2}$$

$$m_{Ca^{2+}(aq)} = \frac{K_{sp}'}{m_{CO_3^{2-}(aq)}} = \frac{K_{sp}' \, m_{H^+(aq)}^2}{K_1'K_2'm_{H_2CO_3(aq)}}$$

$$m_{OH^-(aq)} = \frac{K_w'}{m_{H^+(aq)}}.$$

Substituting these into the charge balance gives the following equation for the concentration of hydrogen ions which can be solved numerically:

$$m_{H^+(aq)} + \frac{2K_{sp}' \, m_{H^+(aq)}^2}{K_1'K_2'm_{H_2CO_3(aq)}}$$
$$= \frac{K_w'}{m_{H^+(aq)}} + \frac{K_1' \, m_{H_2CO_3(aq)}}{m_{H^+(aq)}} + \frac{K_2'K_1' \, m_{H_2CO_3(aq)}}{m_{H^+(aq)}^2}.$$

For the atmospheric concentration of carbon dioxide of 315 ppm in 1958 this gives the results

$$m_{H^+(aq)} = 6.470 \times 10^{-9} \text{ mol dm}^{-3}$$

$$pH = 8.19,$$

while for the current carbon dioxide concentration of 420 ppm we obtain

$$m_{H^+(aq)} = 7.828 \times 10^{-9} \text{ mol dm}^{-3}$$

$$pH = 8.11.$$

Our estimate gives a (large) increase of 21% in the hydrogen ion concentration in the ocean since 1958. As you might be aware, the increase in ocean acidity the past two hundred years has had a substantial direct detrimental effect on marine life. $\S$

Harnessing Chemical Energy

13.1 The electrochemical cell

Ions are mobile in electrolytes. Hence, when there is an electric potential gradient in an electrolyte, positively charged ions will move from the region of high (positive) potential to that of low (negative) potential, while negatively charged ions will move in the opposite direction. This occurs when electrodes connected to a battery are placed in the electrolyte. The battery thus sets up an electric field and its energy drives the ionic current in the electrolyte. When we have an external voltage driving the current through an electrochemical cell, we call it an electrolytic cell. We illustrate this with a simplest example in Figure 13.1 in which water is electrolysed into hydrogen gas and oxygen gas.

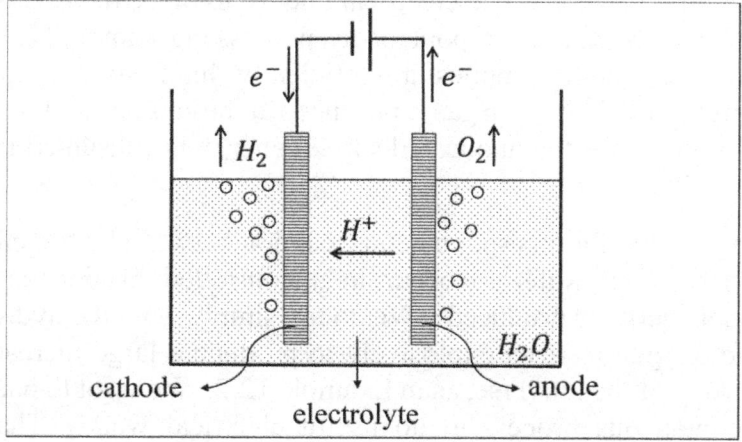

Figure 13.1 The electrolytic cell. The anode and cathode are carbon rods. The electrolyte is water with, for example, NaCl added to raise its electrical conductivity.

The reaction at the cathode is

$$H^+(aq) + e^- \to \frac{1}{2}H_2(gas)$$

and the reaction at the anode is

$$\frac{1}{2}H_2O(liq) \to \frac{1}{4}O_2(gas) + H^+(aq) + e^-.$$

We refer to each of these reactions as a half-cell reaction. Together they make up the net reaction:

$$\frac{1}{2}H_2O(liq) \to \frac{1}{2}H_2(gas) + \frac{1}{4}O_2(gas).$$

We have seen in Example 12.2 that the equilibrium under standard conditions for a gaseous mixture of hydrogen and oxygen, in the absence of an electric potential, lies far to the left of this identical overall electrolysis reaction. Thus, the application of an electrical potential can shift the equilibrium state towards $H_2(gas)$ and $O_2(gas)$. We can use electrical energy to drive the reverse of what we have seen to be a spontaneous process in Example 12.2. As long as the battery applies a sufficiently high voltage to the electrolytic cell, hydrogen gas is produced at the cathode and oxygen gas is produced at the anode. Electrical energy is consumed in the process.

Is it possible for this spontaneous reaction between hydrogen and oxygen forming water to occur in a controlled fashion so that electrical energy is produced? We do not simply want the hydrogen and the oxygen to react explosively to produce a large increase in the entropy of the universe, as in Example 12.2. We want to harness this spontaneous process to do useful electrical work. That is, starting with hydrogen and oxygen gases, can we react these to form water, and at the same time use the reaction to drive an electrical current in the reverse direction to what happens in electrolysis? This is precisely what is achieved in a fuel cell. We illustrate schematically the hydrogen fuel cell in Figure 13.2.

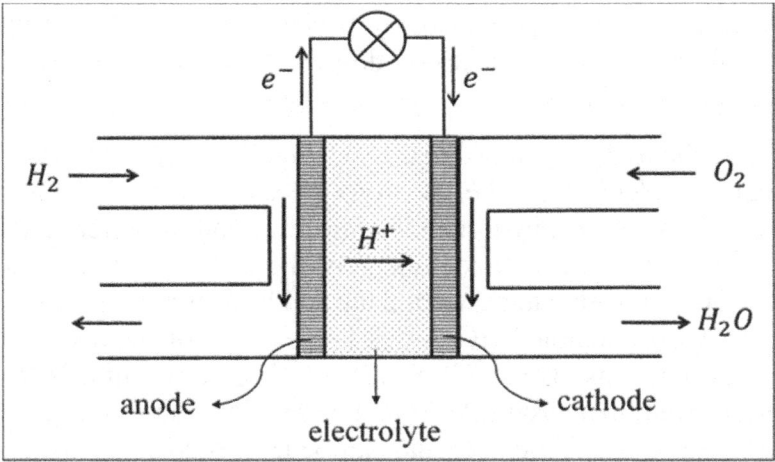

Figure 13.2 The hydrogen fuel cell as an example of a galvanic cell. The electrolyte is a proton-exchange membrane which allows protons to move from the anode to the cathode. An electric current is driven through the external resistive load which could be a lamp or a motor.

Such electrochemical systems, which generate electrical energy from chemical energy, are called galvanic cells, i.e., batteries. This galvanic cell consists of a steady supply of hydrogen gas and oxygen gas at separate electrodes. At the anode of the hydrogen fuel cell, hydrogen gas undergoes oxidation to form hydrogen ions and electrons:

$$\frac{1}{2}H_2(gas) \rightarrow H^+(aq) + e^-.$$

This is the reverse of what occurs at the cathode of the electrolytic cell. The hydrogen ions move through the electrolyte to the cathode where the corresponding reduction of oxygen gas to water occurs:

$$\frac{1}{4}O_2(gas) + H^+(aq) + e^- \rightarrow \frac{1}{2}H_2O(liq),$$

which is the reverse of the reaction at the anode of the electrolytic cell. Each of these two reactions is known as a half-cell reaction because each constitutes half of the electrochemical cell. The reaction in this galvanic cell occurs spontaneously because they are driven by the chemical potential of the excess hydrogen at the anode and the excess oxygen at the cathode.

While the hydrogen ions are driven from the anode to the cathode through the electrolyte, the electrons released by the oxidation of hydrogen gas to form hydrogen ions flow around the external circuit from the anode to the cathode; the galvanic cell is a battery with the cathode as its positive terminal and the anode as its negative terminal. The electrical potential driving this is the result of the spontaneous chemical reactions at the cell electrodes. The convention is that an electric current in a circuit flows in the direction opposite to the flow of electrons. Its electric potential can be harnessed to do work such as lighting a lamp or powering a motor as indicated by the symbol for an electrical load in Figure 13.2. If we replace the resistive load by a voltmeter, we can measure the electric potential of the galvanic cell φ. This is equal to the voltage *drop* from the cathode to the anode and is referred to as the cell potential. The electric current through the external resistive load is driven by this drop in electric potential, just as a mass in the Earth's gravitational field decreases its gravitational potential as it falls. Thus, if electric charge equal to $d\mathcal{Q}$ moves from the cathode to the anode, the electrical work done *by* the galvanic cell is $\varphi d\mathcal{Q}$. If we treat the galvanic cell as the system, then

$$dW_{elec} = -\varphi d\mathcal{Q}$$

where the negative sign arises with the convention that dW_{elec} is work done *on* the system; see Section 2.6.

To be specific, we consider a hydrogen fuel cell operating at a temperature of 298.15 K and 1 bar with both anode and cathode made of platinum. By adjusting the concentration of the electrolyte and the pressure of the gases, we can set the $H^+(aq)$ activity equal to unity and the fugacity of hydrogen and oxygen gas equal to 1 bar. As we have discussed in Sections 9.4 and 9.5 for non-ideal gases and solutions, these are standard conditions used to define reference states for chemical potentials. Operating under these conditions, the cell potential $\varphi^{\emptyset}$ for a hydrogen fuel cell is equal to 1.229 V. This is known as the standard cell potential, with the superscript $\emptyset$ indicating the standard conditions.

Now, consider replacing the resistive load in Figure 13.2 with a variable voltage power supply, connecting this external power source so that its electric potential φ_{ext} is in the opposite direction to the electric potential φ of the galvanic cell. If φ_{ext} is less than 1.229 V, then the galvanic cell potential dominates over φ_{ext} and the electric current flows in the direction driven by φ. Inside the galvanic cell, protons move from the anode to the cathode as illustrated in Figure 13.2, hydrogen and oxygen are consumed and water is produced. That is, the Gibbs free energy of the system decreases when the reaction goes in the direction

$$\frac{1}{2}H_2(gas) + \frac{1}{4}O_2(gas) \rightarrow \frac{1}{2}H_2O(liq).$$

On the other hand, if φ_{ext} is greater than 1.229 V, then the external voltage dominates over φ and the electric current now flows in the direction driven by φ_{ext}. Then, inside the galvanic cell protons move from the cathode to the anode; oxygen gas is produced at the cathode and hydrogen gas is produced at the anode. The galvanic cell is now operating as an electrolytic cell, and the Gibbs free energy of the system decreases when the reaction goes in the direction

$$\frac{1}{2}H_2O(liq) \rightarrow \frac{1}{2}H_2(gas) + \frac{1}{4}O_2(gas),$$

opposite to the case when φ_{ext} is less than 1.229 V.

When the external voltage is exactly equal to 1.229 V, no current flows and no chemical reactions occur at the electrodes; dG for the system is then equal to zero when φ_{ext} is equal to 1.229 V. From this we see that the difference $\left(\varphi^{\emptyset} - \varphi_{ext}\right)$ determines the direction in which the electrochemical reaction proceeds. If this difference is positive, the current flows in the direction driven by $\varphi^{\emptyset}$ and the reaction proceeds in the corresponding direction. And vice versa for a negative $\left(\varphi^\emptyset - \varphi_{ext}\right)$. Thus, the cell potential of a galvanic cell is related to the Gibbs free energy change of the reaction occurring in the cell.

13.2 From $\Delta_{rxn}G$ to φ

To connect the cell potential to the Gibbs free energy of reaction $\Delta_{rxn}G$ we go back to the basic definition of the Gibbs free energy for a system consisting of a galvanic cell at standard temperature and pressure connected to the "surroundings" consisting of just an external power supply. In general, we have

$$G \equiv H - TS \equiv U + PV - TS.$$

For a system in which only volume expansion work is done,

$$dU = TdS - PdV$$

where the volume expansion work done on the system is $-PdV$. When an electrochemical reaction occurs, electric charge is moved through the external circuit through an electric field and thus electrical work is done. Consider a cell operating under standard conditions. When $d\mathcal{Q}$ coulomb of electric charges moves through the constant electric potential of $\left(\varphi^\emptyset - \varphi_{ext}\right)$ volts, the electrical work done by the galvanic cell, is equal to $\left(\varphi^\emptyset - \varphi_{ext}\right)d\mathcal{Q}$ (see Section 2.6). Thus, including this electrical work done *on* the galvanic cell, we have

$$dU = TdS - PdV - \left(\varphi^\emptyset - \varphi_{ext}\right)d\mathcal{Q},$$

where $\varphi^\emptyset$ is the standard cell potential. Substituting this into the expression for the Gibbs free energy, we have,

$$d\mu = -SdT + VdP - \left(\varphi^\emptyset - \varphi_{ext}\right)d\mathcal{Q},$$

per mole of the reaction. For an electrochemical system at constant temperature, pressure and electric potential, the reaction chemical potential is thus

$$\Delta_{rxn}\mu = -\left(\varphi^\emptyset - \varphi_{ext}\right)\Delta\mathcal{Q}.$$

where ΔQ is the charge transferred through the external circuit per mole of reaction. If we have an electrochemical system at standard state without an additional voltage from an external power supply, we simply have

$$-\Delta_{rxn}\mu^{\emptyset} = \varphi^{\emptyset}\Delta Q.$$

We have used the notation for the standard chemical potential of reaction since the system is at standard state. Thus, measuring the cell potential of an electrochemical reaction gives us directly its Gibbs free energy of reaction. Notice that the cell potential is an intensive property of the system. The chemical potential change $\Delta_{rxn}\mu^{\emptyset}$ and the amount of charge transferred ΔQ depend upon the stoichiometry of the reaction as written. Since these are defined per mole of the reaction, if we scale the reaction up by multiplying all stoichiometric coefficients by a factor of two, for example, then both $\Delta_{rxn}\mu^{\emptyset}$ and ΔQ also increase by the same factor. Hence, the cell potential of an electrochemical cell is not dependent upon the stoichiometry. No matter how large a hydrogen fuel cell is or how we scale the stoichiometry of the reaction, at standard conditions the cell potential is 1.229 V.

When we discussed electrolytes in Section 9.7.1, we considered the problem of measuring the change in the standard chemical potential for an electrochemical reaction. There we simply stated that we can measure $\Delta_{rxn}\mu^{\emptyset}$. We now see that it is as easily done as connecting the electrochemical cell to a voltmeter to measure its standard cell potential. To be specific, we revisit out calculation in Example 7.4 for the water formation reaction

$$H_2(gas) + \frac{1}{2}O_2(gas) \rightleftharpoons H_2O(liq).$$

This has a standard reaction chemical potential of

$$\Delta_{rxn}\mu^{\emptyset} = -237.13 \text{ kJ}$$

per mole of the reaction as written here. The overall reaction can be written in terms of the half-cell reactions at the cathode and the anode as

$$\text{cathode: } H_2(gas) \rightarrow 2H^+(aq) + 2e^-$$
$$\text{anode: } \frac{1}{2}O_2(gas) + 2H^+(aq) + 2e^- \rightarrow H_2O(liq).$$

Therefore, for n moles of electrons transferred per mole of reaction, the charge transferred through the external circuit is

$$\Delta Q = neN_A = nF,$$

where $-e$ is the (negative) charge on an electron, N_A is the Avogadro number, and the quantity F, the charge on one mole of electrons, is called the Faraday constant. With the electronic charge equal to 1.6022×10^{-19} C the Faraday constant is equal to 96485 C. Thus,

$$-\Delta_{rxn}\mu^{\varnothing} = \varphi^{\varnothing}\Delta Q = \varphi^{\varnothing}nF.$$

For each mole of the overall reaction above, two moles of electrons flow through the external circuit even though the electrons are not written explicitly in the overall chemical reaction. With $\Delta_{rxn}\mu^{\varnothing}$ equal to -237.13 kJ per mole of reaction, then the standard cell potential for the hydrogen fuel cell is

$$\varphi^{\varnothing} = -\frac{\Delta_{rxn}\mu^{\varnothing}}{2F} = 1.229 \text{ V}$$

as we have seen above. This is the same cell potential whether we write the overall reaction as

$$H_2(gas) + \frac{1}{2}O_2(gas) \rightleftharpoons H_2O(liquid)$$

with two moles of electrons transferred through the external circuit per mole of reaction, or as

$$\frac{1}{2}H_2(gas) + \frac{1}{4}O_2(gas) \rightarrow \frac{1}{2}H_2O(liquid)$$

with one mole transferred per mole of reaction. As another example of the connection between the standard chemical potential of reaction and the standard cell potential, we consider the reaction of zinc with an acid.

§ **Example 13.1 Cell potential of the zinc-hydrogen galvanic cell**

A galvanic cell can be constructed as illustrated in Figure 13.3 with the hydrogen electrode as the cathode. The overall reaction in the cell is:

$$Zn(s) + 2H^+(aq) \rightleftharpoons H_2(g) + Zn^{2+}(aq).$$

Calculate its cell potential. The standard chemical potentials are:

$$\mu^{\emptyset}[Ag(s)] = 0 \text{ kJ mol}^{-1}$$

$$\mu^{\emptyset}[H^+(aq)] = 0 \text{ kJ mol}^{-1}$$

$$\mu^{\emptyset}[H_2(g)] = 0 \text{ kJ mol}^{-1}$$

$$\mu^{\emptyset}[Zn^{2+}(aq)] = -147.06 \text{ kJ mol}^{-1}.$$

To clearly see that two moles of electrons are transferred per mole of reaction, we write the overall reaction in two steps as follows:

$$Zn(s) \rightarrow Zn^{2+}(aq) + 2e^-$$

$$2H^+(aq) + 2e^- \rightarrow H_2(g).$$

The first half-reaction occurs at the anode consisting of the zinc electrode while the second half-reaction occurs at the platinum cathode. The standard chemical potential of the reaction is -147.060 kJ mol^{-1}. Hence the standard cell potential is

$$\varphi^{\emptyset} = -\frac{\Delta_{rxn}\mu^{\emptyset}}{nF}$$

$$= \frac{147060}{2 \times 96485} \text{ V}$$

$$= 0.762 \text{ V}.$$

The hydrogen electrode is at a higher electric potential than the zinc electrode by 0.762 V. Thus, the former is the cathode in this galvanic cell. §

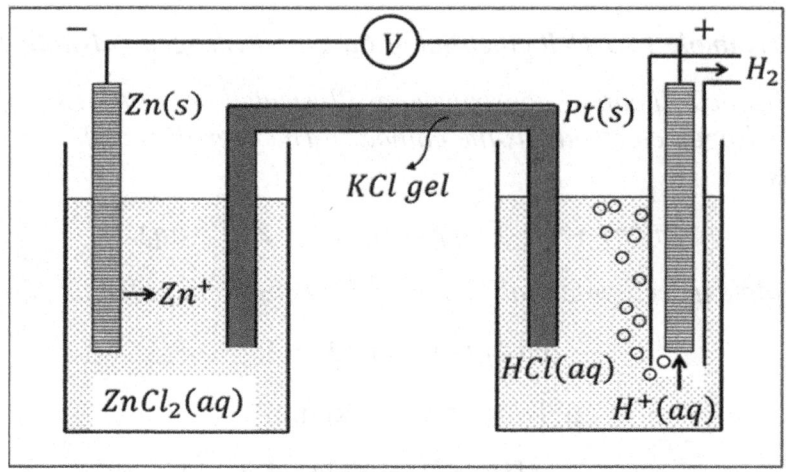

Figure 13.3 Galvanic cell with zinc and hydrogen electrodes.

The explosion of a mixture of hydrogen and oxygen gas to form water is clearly a very different process from what happens in the hydrogen fuel cell. However, the Gibbs free energy of reaction for the gas phase reaction can indeed be measured through the hydrogen fuel cell potential because the electrochemical process produces the same change of state. This illustrates the power of thermodynamics in relating rather different macroscopic properties.

13.3 The half-cell potential

In Section 13.1 we have defined the cell potential φ. The two half-cell reactions, one at the anode and the other at the cathode, together provide the conversion of chemical energy into electrical energy. The change in state from reactants to products for each half-cell reaction is accompanied by a change in the electric potential of the system. This is an intensive thermodynamic variable so that it does not depend upon the number of moles of the reaction. Where does this electrical energy of a galvanic cell come from?

If we write the overall reaction occurring in the hydrogen fuel cell in steps, we have

$$\frac{1}{2}H_2(gas) + \frac{1}{4}O_2(gas) \xrightarrow{\varphi_1} H^+(aq) + e^- + \frac{1}{4}O_2(gas)$$
$$\xrightarrow{\varphi_2} \frac{1}{2}H_2O(liquid),$$

where φ_1 and φ_2 are the electric potential changes for the reactions taking place at the anode and the cathode, respectively. At the anode of the hydrogen fuel cell, we have electrons transferring from electronic states in the hydrogen molecule into conduction electronic states in the electrode. We associate this process with the electric potential of φ_1 so that the electrical work that can be done by this half-cell reaction is $\varphi_1 F$ per mole of electron transferred. Similarly, at the cathode of the hydrogen fuel cell, we have electrons transferring from the conducting electronic states of the electrode to electronic states of hydrogen ions and oxygen molecules to form water. The electrical work that can be done by the half-cell reaction is $\varphi_2 F$ per mole of electron transferred. Thus, the total electrical work that can be extracted from the overall reaction in the hydrogen fuel cell is

$$\varphi F = (\varphi_1 + \varphi_2)F$$

per mole of electron flowing through the external circuit where φ is the cell potential, as the equation for the chemical potential of the reaction indicates.

We have defined in Section 13.1 the standard conditions for a hydrogen fuel cell. These are a temperature of 298.15 K, an electrolyte with the $H^+(aq)$ activity equal to unity. The anode is made of platinum and in contact with hydrogen gas at a fugacity of 1 bar. These conditions define the standard hydrogen electrode. Similarly, if the cathode is made of platinum and is in contact with an oxygen gas fugacity of 1 bar, it is a standard electrode for oxygen.

A hydrogen fuel cell operating with these standard electrodes has a cell potential equal to 1.229 V. Thus, we have a standard cell potential of

$$\varphi^{\emptyset} = \left(\varphi_1^{\emptyset} + \varphi_2^{\emptyset}\right) = 1.229 \text{ V}.$$

We conventionally write the half-cell reactions and their corresponding half-cell potentials as

$$H^+(aq) + e^- \rightarrow \frac{1}{2}H_2(gas): \qquad -\varphi_1^{\emptyset}$$

$$\frac{1}{4}O_2(gas) + H^+(aq) + e^- \rightarrow \frac{1}{2}H_2O(liquid): \quad \varphi_2^{\emptyset}.$$

Note the negative sign for the anode reaction; this is because the half-cell reaction is written in the opposite direction to how it occurs in the overall reaction above. At this point, we note the convention of setting the half-cell potential for the standard hydrogen electrode to be equal to zero. That is, by convention, $\varphi_1^{\emptyset} = 0$. With this reference point for electric potential, the standard half-cell potential for the oxygen electrode is $\varphi_2^{\emptyset} = 1.229$ V in order that the standard cell potential φ^0 is equal to 1.229 V as measured. With the standard hydrogen electrode half-cell potential set to zero, the standard half-cell potential of any other half-cell reaction can be defined and measured relative to it.

With this convention, the standard half-reactions and potentials for the hydrogen fuel cell are as follows:

$$H^+(aq) + e^- \rightarrow \frac{1}{2}H_2(gas): \qquad\qquad \varphi_A^{\emptyset} = 0 \text{ V}$$

$$\frac{1}{4}O_2(gas) + H^+(aq) + e^- \rightarrow \frac{1}{2}H_2O(liq): \quad \varphi_B^{\emptyset} = 1.229 \text{ V}$$

with $\varphi_A^{\emptyset} = -\varphi_1^{\emptyset}$ and $\varphi_B^{\emptyset} = \varphi_2^{\emptyset}$ from above. We note here the convention of writing the half-cell reactions as reduction reactions where the chemical species on the left-hand side are reduced to the products on the right-hand side. Each half-cell reaction involves a redox couple. At the hydrogen electrode, the redox couple is the *reduction* of the aqueous proton to hydrogen gas; the proton is the oxidised state and the hydrogen gas is the reduced state for that half-reaction. At the oxygen electrode, it is the *reduction* of oxygen gas and the aqueous proton to water. Thus, combining the half-cell reactions for the hydrogen-fuel cell gives

$$\frac{1}{4}O_2(gas) + \frac{1}{2}H_2(gas) \rightleftharpoons \frac{1}{2}H_2O(liq)$$

where we have subtracted reaction A from reaction B. The standard cell potential for the overall reaction is

$$\varphi_B^{\emptyset} - \varphi_A^{\emptyset} = 1.229 \text{ V}.$$

Since this is positive, and therefore the change in the Gibbs free energy $\Delta_{rxn}\mu^0$ is negative, the spontaneous direction of the overall reaction is from left to right under standard conditions. The cathode is formed by the left-hand side of electrode B. That is, the oxygen electrode is the positive electrical terminal of the hydrogen-fuel cell.

§ *Example 13.2 Half-cell potential for the zinc electrode*

The galvanic cell in Example 13.3 has a cell potential of 0.762 V. What is the standard half-cell potential for the zinc electrode with the following half-cell reaction?

$$Zn^{2+}(aq) + 2e^- \rightarrow Zn(s).$$

The two half-cell reactions occurring in the galvanic cell are:

$$2H^+(aq) + 2e^- \rightarrow H_2(g): \quad \varphi_H^{\emptyset}$$
$$Zn^{2+}(aq) + 2e^- \rightarrow Zn(s): \quad \varphi_{Zn}^{\emptyset}.$$

In terms of these half-cell reactions, the overall reaction

$$Zn(s) + 2H^+(aq) \rightarrow Zn^{2+}(aq) + H_2(g)$$

can be written in two steps as:

$$2H^+(aq) + Zn(s) \xrightarrow{\varphi_H^\emptyset} H_2(g) + 2e^- + Zn(s)$$

$$\xrightarrow{-\varphi_{Zn}^\emptyset} H_2(g) + Zn^{2+}(aq),$$

so that the overall cell potential is equal to

$$\varphi^\emptyset = \varphi_H^\emptyset - \varphi_{Zn}^\emptyset.$$

We have seen in Example 13.1 that the standard cell potential is 0.762 V. We have also seen that the standard hydrogen potential is set to zero as a reference potential. Hence, the half-cell potential for the standard zinc electrode is

$$\varphi_{Zn}^\emptyset = -0.762 \text{ V}.$$

That is,

$$Zn^{2+}(aq) + 2e^- \rightleftharpoons Zn(s): \qquad \varphi_{Zn}^\emptyset = -0.762 \text{ V}.$$

Since the cell potential is opposite in sign to the reaction chemical potential, the zinc half-cell reaction going in the reverse direction to how it is written here contributes a decrease in the Gibbs free energy for the overall reaction. In a tussle for electrons between $H^+(aq)$ and $Zn^{2+}(aq)$, the former wins because the total free energy decreases when metallic zinc loses electrons to form zinc ions while the hydrogen ions pick up the electrons that are released to form hydrogen gas. §

We further connect the half-cell potential for an aqueous ion to its standard chemical potential. The overall cell potential for the zinc-hydrogen cell is 0.762 V and the overall reaction is

$$Zn(s) + 2H^+(aq) \rightarrow Zn^{2+}(aq) + H_2(g).$$

Thus, the standard reaction chemical potential per mole of the reaction is

$$\Delta_{rxn}\mu^\emptyset = \mu_{H_2(g)}^\emptyset + \mu_{Zn^{2+}(aq)}^\emptyset - 2\mu_{H^+(aq)}^\emptyset - \mu_{Zn(s)}^\emptyset$$

$$= -n\varphi^{\emptyset} F$$

$$= -2 \times 0.762 \times 96485$$

$$= -147.06 \text{ kJ mol}^{-1}.$$

With the standard chemical potentials $\mu_{H_2(g)}^{\emptyset}$, $\mu_{H^+(aq)}^{\emptyset}$ and $\mu_{Zn(s)}^{\emptyset}$ being zero, the standard chemical potential of the aqueous zinc ion is therefore

$$\mu_{Zn^{2+}(aq)}^{\emptyset} = -147.06 \text{ kJ mol}^{-1}.$$

Note that this is per mole of aqueous zinc ion.

We repeat what we stated in Section 9.7.1. By selecting the appropriate electrochemical reactions to measure the cell potentials, the standard chemical potentials of aqueous ions can be determined relative to the arbitrary reference value of $\mu_{H^+(aq)}^{\emptyset} = 0$. For example, given the half-cell potential of the zinc electrode, and measuring the Daniell cell potential, we can determine the half-cell potential of the copper electrode and the standard chemical potential of $Cu^{2+}(aq)$, as we will see in Example 13.3. With such a tabulation of standard chemical potentials, we can calculate the standard Gibbs free energy of reaction for any reaction involving electrolyte species.

§   ***Example 13.3 The Daniell cell***

A Daniell cell is formed by placing a copper electrode into a solution of copper sulphate and a zinc electrode into a solution of zinc sulphate. The two electrolytes are electrically connected through a salt bridge which consists of a concentrated electrolyte such as KCl in a jelly. The salt bridge keeps the electrolytes for each electrode from mixing. The standard cell potential of the Daniell cell is 1.104 V *with the copper electrode serving as the cathode. Calculate the standard half-cell potential for the copper electrode and the standard chemical potential for* $Cu^{2+}(aq)$.

The half-cell reactions and standard half-cell potentials are:

$$\frac{1}{2}Cu^{2+}(aq) + e^- \rightarrow \frac{1}{2}Cu(s) \qquad \varphi^\emptyset_{Cu} = ?$$

$$\frac{1}{2}Zn^{2+}(aq) + e^- \rightarrow \frac{1}{2}Zn(s) \qquad \varphi^\emptyset_{Zn} = -0.762 \text{ V.}$$

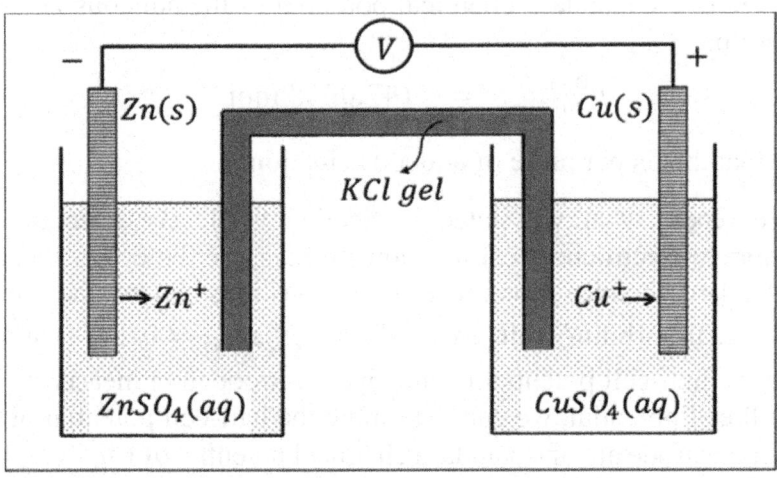

Figure 13.4 The Daniell cell.

Since we observe that the copper electrode is the cathode, electrons flow from the zinc electrode to the copper electrode through the external circuit. Thus, the overall reaction is from left to right:

$$Cu^{2+}(aq) + Zn(s) \rightleftharpoons Cu(s) + Zn^{2+}(aq),$$

giving a standard cell potential for the Daniell cell equal to

$$\varphi^\emptyset = \varphi^\emptyset_{Cu} - \varphi^\emptyset_{Zn} = 1.104 \text{ V.}$$

From measurements of a galvanic cell such as in Example 13.2, we already have the zinc half-cell potential $\varphi^\emptyset_{Zn} = -0.762$ V. Hence, the copper half-cell potential is

$$\varphi^\emptyset_{Cu} = \varphi^\emptyset + \varphi^\emptyset_{Zn} = 1.104 - 0.762 = 0.342 \text{ V.}$$

From the cell potential, the standard chemical potential per mole of the overall reaction written with the stoichiometry above is

$$\Delta_{rxn}\mu^{\emptyset} = -\varphi^{\emptyset} nF$$

$$= -1.104 \times 1 \times 96485$$

$$= -106.52 \text{ kJ mol}^{-1}.$$

But, this is also equal to

$$\Delta_{rxn}\mu^{\emptyset} = \frac{1}{2}\mu^{\emptyset}_{Cu(s)} + \frac{1}{2}\mu^{\emptyset}_{Zn^{2+}(aq)} - \frac{1}{2}\mu^{\emptyset}_{Cu^{2+}(aq)} - \frac{1}{2}\mu^{\emptyset}_{Zn(s)}$$

$$= \frac{1}{2}\mu^{\emptyset}_{Zn^{2+}(aq)} - \frac{1}{2}\mu^{\emptyset}_{Cu^{2+}(aq)}$$

where the standard state chemical potentials of solid crystalline copper and zinc are set equal to zero since these are the most stable states of the elements at standard temperature and pressure. We have seen above that the standard chemical potential for the aqueous zinc ion $\mu^{\emptyset}_{Zn^{2+}(aq)}$ is equal to $-147.06 \text{ kJ mol}^{-1}$. Therefore, measuring the cell potential of the Daniell cell allows us to obtain the standard chemical potential of $Cu^{2+}(aq)$:

$$\mu^{\emptyset}_{Cu^{2+}(aq)} = \mu^{\emptyset}_{Zn^{2+}(aq)} - 2\Delta_{rxn}\mu^{\emptyset}$$

$$= -147.06 - 2(-106.52) \text{ kJ mol}^{-1}$$

$$= 65.98 \text{ kJ mol}^{-1}.$$

In this way, we can establish the standard chemical potential of any ionic species starting from the reference value of $\mu^{\emptyset}_{H^+} = 0$. §

The standard chemical potential for the Daniell cell reaction is $-2 \times 106.52 = -213.04 \text{ kJ mol}^{-1}$ with two electrons transferred per mole of $Zn(s)$ converted to $Zn^{2+}(aq)$. From the discussion in Section 7.3 and, especially, Example 7.4, the maximum amount of electrical energy that can be extracted from a Daniell cell operating at standard conditions is thus $213.04 \text{ kJ mol}^{-1}$ per mole of zinc atoms ionised. The thermodynamics we have discussed in this chapter is fundamental to the operation of fuel cells which are essentially systems that allow electrical energy to be extracted from chemical reactions, that is electrochemical cells. We have confined our discussions to aqueous electrolytes because the thermodynamic potentials for these are readily available for illustration. Even though fuel cells use rather different liquids and even solids as electrolytes, the fundamental thermodynamic principles are the same. Indeed, the fundamental thermodynamics of fuel cells are the same for the steam engines examined by Carnot even though the physical systems are completely different.

Real pistons and mechanical devices have friction so that the effective pressure of a cylinder of gas available for doing work by pushing against a piston bearing an external force is less than the actual pressure of the gas. Similarly, real electrochemical cells have what is called an overpotential. For a real galvanic cell, the electrical potential that is available to perform electrical work is less than its cell potential so that a galvanic cell does less work than might be expected. Conversely, for an electrolytic cell the electrical potential required to drive the electrolytic process is more than its cell potential so that more work than expected is needed to drive the electrolytic process in the cell. Thus, overpotentials play the same role in real electrochemical cells as friction does in real pistons. There exists no truly reversible macroscopic chemical process because of these overpotentials.

There are various sources of overpotentials. For example, there is always a resistance to charge transfer at the junction of two dissimilar substances. These include the contact points between the electrodes and the external electrical circuit, the interface between each electrode and its electrolyte, and the junctions between any connecting salt-bridge or porous membrane between the electrolytes. Another source of overpotentials is the activation barrier that chemical reactions generally have for the transfer of electrons or the transport of ions through an electrolyte. Additionally, the concentration in a real electrochemical cell is not uniform so that the concentration gradient poses another source of dissipation of energy. Thus, the maximum amount of work that can be extracted from a fuel cell is somewhat less than what is expected from the theoretical cell potential. In analysing reversible gas expansions, we ignore friction in pistons. Similarly, we ignore overpotentials when we discuss reversible chemical reactions.

13.4 The electrochemical series

Putting the results from the zinc-hydrogen electrochemical cell together with the results from the Daniell cell, we can arrange the three half-cell reactions in order of increasing standard half-cell potentials:

$$\frac{1}{2}Zn^{2+}(aq) + e^- \rightarrow \frac{1}{2}Zn(s): \qquad \varphi^{\emptyset} = -0.762 \text{ V}$$

$$H^+(aq) + e^- \rightarrow \frac{1}{2}H_2(g): \qquad \varphi^{\emptyset} = 0 \text{ V}$$

$$\frac{1}{2}Cu^{2+}(aq) + e^- \rightarrow \frac{1}{2}Cu(s): \qquad \varphi^{\emptyset} = 0.342 \text{ V}$$

These half-cell reactions are written in terms of the gain of one electron, but since the cell potential is an intensive quantity, the half-cell potential is independent of the scale of each half-cell reaction. The electrochemical series is a list of such half-cell reactions arranged in the order of their potentials. A short electrochemical series for a selected number of ionic reactions is illustrated in the table below. At the top of the series are the highly reactive metals, the atoms of which are readily oxidised to their ions. Using such an electrochemical series, the standard cell potential and thus $\Delta_{rxn}\mu^{\emptyset}$ for any electrolytic reaction can be calculated. Or, vice versa, the cell potential measurement of an appropriate electrochemical reaction can be used to add data to any existing table of half-cell potentials.

Electrode reaction	$\varphi^{\varnothing}/V$
$Ca^{2+}(aq) + 2e^- \rightarrow Ca(s)$	-2.87
$Na^+(aq) + e^- \rightarrow Na(s)$	-2.71
$H_2O(l) + e^- \rightarrow \frac{1}{2}H_2(g) + OH^-(aq)$	-0.83
$Zn^{2+}(aq) + 2e^- \rightarrow Zn(s)$	-0.76
$Fe^{2+}(aq) + 2e^- \rightarrow Fe(s)$	-0.44
$H^+(aq) + e^- \rightarrow \frac{1}{2}H_2(g)$	0
$AgCl(s) + e^- \rightarrow Ag(s) + Cl^-(aq)$	0.22
$Cu^{2+}(aq) + 2e^- \rightarrow Cu(s)$	0.34
$\frac{1}{2}O_2(g) + H_2O(l) + 2e^- \rightarrow 2OH^-(aq)$	0.40
$Ag^+(aq) + e^- \rightarrow Ag(s)$	0.80
$\frac{1}{4}O_2(g) + H^+(aq) + e^- \rightarrow \frac{1}{2}H_2O(l)$	1.23
$\frac{1}{2}Cl_2(g) + e^- \rightarrow Cl^-(aq)$	1.36
$Au^+(aq) + e^- \rightarrow Au(s)$	1.69

§   *Example 13.4 Sodium reacts explosively with water*

From the table above, what is $\Delta_{rxn}\mu^{\varnothing}$ *for the reaction of sodium metal with water?*

$$Na(s) + H_2O(l) \rightleftharpoons Na^+(aq) + OH^-(aq) + \frac{1}{2}H_2(g).$$

The overall reaction consists of the sum of the following two half-cell reactions

$$Na(s) \rightarrow Na^+(aq) + e^-$$

$$H_2O(l) + e^- \rightarrow \frac{1}{2}H_2(g) + OH^-(aq)$$

with standard potentials of 2.71 V and –0.83 V, respectively. Hence, the cell potential at standard conditions is 1.88 V. For the reaction stoichiometry written with one mole of sodium atoms, one mole of electrons is consumed. Thus, per mole of reaction

$$\Delta_{rxn}\mu^{\emptyset} = -1.88 \times 96.485 = -181.39 \text{ kJ mol}^{-1}.$$

This reaction is spontaneous at standard conditions. Equilibrium lies far to the product side of this reaction, which is explosive because the heat generated can ignite the hydrogen gas produced. §

§   *Example 13.5 Zinc displaces silver*

If a piece of metallic zinc is place in an aqueous solution of silver nitrate, metallic silver is displaced from the solution. The reaction is

$$2Ag^+(aq) + Zn(s) \rightleftharpoons 2Ag(s) + Zn^{2+}(aq).$$

Calculate the Gibbs free energy of reaction at standard state using the electrochemical series.

The overall reaction consists of the half-cell reactions

$$2Ag^+(aq) + 2e^- \rightarrow 2Ag(s)$$

$$Zn^{2+}(aq) + 2e^- \rightarrow Zn(s)$$

with standard half-cell potentials of 0.80 V and -0.76 V, respectively. Hence, the net potential is 1.56 V. Two moles of electrons are transferred per mole of the reaction as written above. Thus,

$$\Delta_{rxn}\mu^{\emptyset} = -1.56 \times 2 \times 96.485 = -301.03 \text{ kJ mol}^{-1}.$$

At standard state this is a spontaneous reaction, with the equilibrium state being far to the product side of the reaction. The equilibrium constant is

$$K = exp\left(-\frac{\Delta_{rxn}\mu^{\emptyset}}{RT}\right) = 5.5 \times 10^{52}.$$

Metallic zinc can be used to almost completely displace aqueous silver ions from solution. §

In Example 13.5, we see that a metal higher up in the electrochemical series displaces the ions of a metal lower in the series, precipitating this latter metal from a solution of its ions. You realise that this provides a means of protecting against corrosion. For example, many buried pipelines are made of iron. When exposed to water and oxygen, these pipes corrode. Two of the relevant half-cell reactions are:

$$Fe^{2+}(aq) + 2e^- \rightarrow Fe(s)$$

$$\frac{1}{2}O_2(g) + H_2O(l) + 2e^- \rightarrow 2OH^-(aq)$$

with standard half-cell potentials of -0.44 V and 0.40 V, respectively. Thus, the overall reaction goes from left to right as follows:

$$\frac{1}{2}O_2(g) + H_2O(l) + Fe(s) \rightarrow 2OH^-(aq) + Fe^{2+}(aq)$$

with a cell potential of 0.84 V and $\Delta_{rxn}\mu$ equal to -162.10 kJ mol^{-1} at standard conditions. However, if the iron pipe is in electrical contact with a piece of zinc, then electrochemical equilibrium between zinc and iron will be established. This consists of the half-cell reactions

$$Fe^{2+}(aq) + 2e^- \rightarrow Fe(s)$$

$$Zn^{2+}(aq) + 2e^- \rightarrow Zn(s)$$

with standard half-cell potentials of -0.44 V and -0.76 V, respectively, so that the overall reaction is

$$Fe^{2+}(aq) + Zn(s) \rightarrow Fe(s) + Zn^{2+}(aq)$$

going from left to right with a cell potential of $+0.32$ V and $\Delta_{rxn}\mu$ equal to -61.75 kJ mol^{-1} at standard conditions. The equilibrium constant for this reaction is

$$K = exp\left(\frac{61750}{8.314 \times 298.15}\right) = 6.6 \times 10^{10}.$$

Taking the activity coefficients to be one, the ratio of the ion molarities is

$$\frac{m_{Zn^{2+}(aq)}}{m_{Fe^{2+}(aq)}} \cong 6.6 \times 10^{10}.$$

Electrically connecting a piece of zinc to the iron pipe results in the zinc corroding instead of the iron. This is simply a consequence of the thermodynamics. This method of preventing corrosion is known as cathodic protection, because in its electrochemical reaction with iron, the piece of zinc is the cathode. The zinc is sacrificially corroded.

13.5 The electrochemical cell: notation

To complete the description of electrochemical cells, we briefly describe the standard notation used to represent an electrochemical cell. The hydrogen fuel cell with, say platinum electrodes for both the anode and the cathode is written as:

$$Pt(s)|H_2(g)|H^+(aq), H_2O|O_2(g)|Pt(s).$$

This is in the format:

left electrode|reductant|electrolyte|oxidant|right electrode.

The Daniell cell in Example 13.3 is represented as:

$$Zn(s)|ZnSO_4(aq)||CuSO_4(aq)|Cu(s).$$

The cell potential is then equal to the half-cell potential at the right electrode minus the half-cell potential at the left electrode:

$$\varphi = \varphi_{right} - \varphi_{left}.$$

The double-vertical lines represent the salt-bridge that electrically connects the electrolytes but prevents them from mixing. The zinc-hydrogen cell in Figure 13.3 is

$$Zn(s)|ZnCl_2(aq)||H^+(aq)|H_2(g)|Pt(s).$$

In Section 9.7.1, we discussed how the chemical potentials of electrolytes are determined, using as an example the reaction

$$AgCl(s) + \frac{1}{2}H_2(gas) \rightleftharpoons Ag(s) + H^+(aq) + Cl^-(aq).$$

For this reaction, the standard chemical potential of reaction is

$$\Delta_{rxn}\mu^{\emptyset} = \mu^{\emptyset}_{Ag(s)} + \mu^{\emptyset}_{H^+(aq)} + \mu^{\emptyset}_{Cl^-(aq)} - \mu^{\emptyset}_{AgCl(s)} - \frac{1}{2}\mu^{\emptyset}_{H_2(g)}$$

which can be determined electrochemically to be -21.45 kJ mol^{-1}. Of course, the determination of this quantity depends upon the measurement of the cell potential for the appropriate electrochemical cell. For this reaction the electrochemical cell is

$$Pt(s)|H_2(g)|AgCl(aq)|AgCl(s)|Ag(s)$$

with a hydrogen electrode paired with a silver/silver chloride electrode. The latter electrode consists of solid silver coated with a solid silver chloride layer. This electrochemical cell is known as the Harned cell and has a standard cell potential equal to

$$\varphi^{\emptyset} = -\frac{\Delta_{rxn}\mu^{\emptyset}}{nF} = 0.22 \text{ V}$$

from the electrochemical series.

§   ***Example 13.6 The silver-hydrogen electrochemical cell***

You have the electrochemical cell

$$Pt(s)|H_2(g)|H^+(aq)|Ag^+(aq)|Ag(s)$$

where the anode is a hydrogen electrode and the cathode is a silver electrode. The following reaction occurs in it:

$$Ag^+(aq) + \frac{1}{2}H_2(g) \rightleftharpoons Ag(s) + H^+(aq).$$

The standard cell potential is equal to 0.80 V. Determine the standard chemical potential for $Ag^+(aq)$.

We have

$$\Delta_{rxn}\mu^{\emptyset} = -\varphi^{\emptyset}nF = 0.80 \times 1 \times 96485 = -77.19 \text{ kJ mol}^{-1}.$$

But we also have

$$\Delta_{rxn}\mu^{\emptyset} = \mu^{\emptyset}_{Ag(s)} + \mu^{\emptyset}_{H^+(aq)} - \mu^{\emptyset}_{Ag^+(aq)} - \frac{1}{2}\mu^{\emptyset}_{H_2(g)}$$

$$= -\mu^{\emptyset}_{Ag^+(aq)}.$$

Hence,

$$\mu^{\emptyset}_{Ag^+(aq)} = -\Delta_{rxn}\mu^{\emptyset} = F\varphi^{\emptyset} = 77.19 \text{ kJ mol}^{-1}. \qquad \S$$

13.6 Nernst equation

An important lesson from Section 7.2 is that at constant temperature and pressure, a system reaches equilibrium when its Gibbs free energy is minimized. This means that the change in the Gibbs free energy of reaction $\Delta_{rxn}\mu$ is zero at equilibrium. But since the cell potential φ of an electrochemical system is proportional to its Gibbs free energy of reaction, φ measures how far the system is from equilibrium. Thus, the equilibrium concentration of the various chemical species in an electrolytic cell must be related to its, directly and relatively easily measurable, cell potential. In this section we discuss how φ changes with the concentrations of the reacting species.

In Section 9.7.1 we have seen that the chemical potential of an ionic species A^{+z_A} with a molarity of m_A in an electrolytic solution is

$$\mu_A(liq, T, P, m_A) = \mu^{\emptyset}_A(liq, T, P, m_0) + RT \ln a_A.$$

For brevity of notation, the ionic charge is not included in the subscripts. Hence, for the electrochemical reaction

$$\nu_A A^{+z_A} + \nu_B B^{-z_B} \rightleftharpoons \nu_C C^{+z_C} + \nu_D D^{-z_D}$$

$$\Delta_{rxn}\mu = \Delta_{rxn}\mu^{\emptyset} + RT \ln Q$$

where the Gibbs free energy of reaction

$$\Delta_{rxn}\mu = (\nu_C\mu_C + \nu_D\mu_D) - (\nu_A\mu_A + \nu_B\mu_B)$$

and the reaction quotient

$$Q = \frac{a_C^{\nu_C} a_D^{\nu_D}}{a_A^{\nu_A} a_B^{\nu_B}}$$

as in Section 12.1. But we have also seen in Section 13.2 that the relationship between the cell potential and the Gibbs free energy change per mole of reaction is

$$\Delta_{rxn}\mu = -\varphi nF$$

where n is the number of moles of electrons transferred in the external circuit per mole of reaction, that is

$$n = \nu_C z_C - \nu_D z_D = \nu_A z_A - \nu_B z_B$$

recognizing that the reacting system remains electrically neutral. Therefore, we have

$$\varphi = \varphi^\emptyset - \frac{RT}{nF}\ln Q.$$

This is the Nernst equation which relates the cell potential to the reaction quotient Q. The exponents in Q are the stoichiometric coefficients of the reactants and products. The stoichiometric coefficients of the reaction also determine the number of moles of electrons transferred per mole of the reaction so that $\frac{1}{n}\ln Q$ is independent of the stoichiometry of the reaction used.

When the reaction reaches equilibrium, we have $\Delta_{rxn}\mu = \varphi = 0$, so that

$$K(T) = exp\left(-\frac{\Delta_{rxn}\mu^\emptyset}{RT}\right) = exp\left(\frac{nF\varphi^\emptyset}{RT}\right) = \left(\frac{a_C^{\nu_C} a_D^{\nu_D}}{a_A^{\nu_A} a_B^{\nu_B}}\right)_{eq} = Q_{eq}.$$

Measuring the standard cell potential $\varphi^{\varnothing}$ thus allows you to figure out the reaction quotient at equilibrium Q_{eq}. Conversely, varying the concentration of the reactants and products will change the cell potential of the system.

§   ***Example 13.7 The cell potential is an intensive variable***

For any generic reaction

$$\nu_A A^{z_A+} + \nu_B B \rightleftharpoons \nu_C C^{z_C+} + \nu_D D$$

where A and C are ionic species, the number of moles of electrons transferred per mole of reaction is

$$n = \nu_A z_A = \nu_C z_C.$$

Show that scaling the reaction stoichiometry by any factor does not change the quantity $\frac{1}{n} \ln Q$.

The reaction quotient is given by

$$Q = \frac{a_C^{\nu_C} a_D^{\nu_D}}{a_A^{\nu_A} a_B^{\nu_B}}.$$

If the reaction is scaled by a factor λ, then it becomes

$$\lambda \nu_A A^{z_A+} + \lambda \nu_B B \rightleftharpoons \lambda \nu_C C^{z_C+} + \lambda \nu_D D.$$

The number of moles of electrons transferred becomes λn, and the reaction quotient becomes

$$Q = \frac{a_C^{\lambda \nu_C} a_D^{\lambda \nu_D}}{a_A^{\lambda \nu_A} a_B^{\lambda \nu_B}} = \left(\frac{a_C^{\nu_C} a_D^{\nu_D}}{a_A^{\nu_A} a_B^{\nu_B}} \right)^{\lambda}.$$

Thus, we have

$$\frac{1}{\lambda n} \ln \left(\frac{a_C^{\nu_C} a_D^{\nu_D}}{a_A^{\nu_A} a_B^{\nu_B}} \right)^{\lambda} = \frac{1}{n} \ln \left(\frac{a_C^{\nu_C} a_D^{\nu_D}}{a_A^{\nu_A} a_B^{\nu_B}} \right)$$

so that the term $\frac{1}{n}\ln Q$ in the Nernst equation is not dependent upon the scale of the reaction; it is an intensive quantity just as φ is. §

The Nernst equation

$$\varphi = \varphi^{\emptyset} - \frac{RT}{nF}\ln Q$$

gives the dependence of an electrochemical cell potential upon the concentrations of the electrolytes. How does the cell potential of the zinc-hydrogen cell in Figure 13.3 depend upon the concentration of $Zn^{2+}(aq)$, the pH of the hydrochloric acid and the hydrogen gas pressure? The overall reaction in the electrochemical cell is

$$2H^+(aq) + Zn(s) \rightleftharpoons H_2(g) + Zn^{2+}(aq).$$

Approximating the activity coefficients of $Zn^{2+}(aq)$, $H^+(aq)$ and the hydrogen gas to be equal to one, we have

$$\varphi = \varphi^{\emptyset} - \frac{RT}{nF}\ln\frac{m_{Zn^{2+}(aq)}P_{H_2(g)}}{m_{H^+(aq)}^2},$$

where the concentrations are in units of mol dm^{-3} and pressure is in units of bar. We remind ourselves that this approximation for the reaction quotient means

$$a_{Zn^{2+}(aq)} = \gamma_{Zn^{2+}(aq)}\frac{m_{Zn^{2+}(aq)}}{m_0} \cong \frac{m_{Zn^{2+}(aq)}}{m_0}$$

$$a_{H^+(aq)} = \gamma_{H^+(aq)}\frac{m_{H^+(aq)}}{m_0} \cong \frac{m_{H^+(aq)}}{m_0}$$

$$a_{H_2(g)} = \gamma_{H_2(g)}\frac{f_{H_2(g)}}{P_0} \cong \frac{P_{H_2(g)}}{P_0}.$$

The standard molarity is equal to 1 mol dm^{-3} and the standard pressure is 1 bar. With the standard cell potential equal to 0.762 V, the voltage of the zinc-hydrogen cell at any electrolyte concentration and hydrogen gas pressure is thus given by

$$\varphi = 0.762 - \frac{RT}{nF} \ln \frac{m_{Zn^{2+}(aq)} P_{H_2(g)}}{m_{H^+(aq)}^2}$$

where φ is the electrical potential difference between the hydrogen electrode and the zinc electrode.

At standard concentrations and pressure, the hydrogen electrode is the cathode, and the cell potential is equal to 0.762 V. When the concentration of the zinc chloride electrolyte is reduced the cell potential is increased. Similarly, when the hydrogen pressure is decreased, the cell potential increases. It is interesting to consider the effect of changing the *pH* of the hydrochloric acid electrolyte for the hydrogen electrode. Decreasing the hydrogen ion concentration decreases the cell potential.

§   *Example 13.8 pH dependence of the zinc-hydrogen cell potential*

We add some sodium hydroxide solution to increase the pH of the hydrogen electrode in the zinc-hydrogen cell illustrated in Figure 13.3. What would the cell potential be if the pH of the solution is increased to 6, the zinc ion concentration is equal to 0.01 mol dm^{-3} and the hydrogen gas pressure equal to 1 bar?

For a *pH* of 6 the hydrogen ion concentration is 10^{-6} mol dm^{-3}. Thus, the cell potential is decreased to

$$\varphi = 0.762 - \frac{RT}{nF} \ln \frac{0.01}{(10^{-6})^2}$$

$$= 0.466 \text{ V}.$$

The *pH* of the electrolyte has a significant effect on the cell potential of electrochemical cells. Conversely, from this we see that the cell potential can be readily used to measure the *pH* of a solution. §

We need to be a little more careful in this example. Multiple equilibria can occur in an electrolyte system. In any aqueous solution, the dissociative equilibrium of water occurs, so that with a pH equal to 6, the concentration of OH^- is 10^{-8} mol dm^{-3}. Since both zinc and hydroxide ions present in solution, it is necessary to consider whether $Zn(OH)_2(s)$, which is sparingly soluble, precipitates out of solution. This has a solubility product is

$$K_{sp} = 2 \times 10^{-17}$$

while the ionic product is

$$\frac{m_{Zn^{2+}(aq)} m_{OH^-(aq)}^2}{m_0^3} = 1 \times 10^{-18}$$

which is smaller than K_{sp}, so that the zinc hydroxide concentration has not yet reached saturation. Further increase in the pH could increase the ionic product beyond K_{sp}, upon which the equilibrium

$$Zn(OH)_2(s) \rightleftharpoons Zn^{2+}(aq) + 2OH^-(aq)$$

would need to be considered.

We now revisit the reaction in the Harned cell to further examine how cell potentials are measured. The reaction is

$$AgCl(s) + \frac{1}{2}H_2(gas) \rightleftharpoons Ag(s) + H^+(aq) + Cl^-(aq).$$

From the discussion in Section 9.7.1 and using the Nernst equation,

$$\varphi = \varphi^\emptyset - \frac{RT}{nF} \ln\left(\frac{a_{Ag(s)} a_{H^+(aq)} a_{Cl^-(aq)}}{a_{AgCl(s)} a_{H_2(g)}^{\frac{1}{2}}} \right)$$

$$= \varphi^\emptyset - \frac{RT}{nF} \ln\left(\frac{a_{H^+(aq)} a_{Cl^-(aq)}}{a_{H_2(g)}^{\frac{1}{2}}} \right)$$

$$= \varphi^{\emptyset} - \frac{RT}{F} \ln\left(\frac{a_{\pm}^2}{a_{H_2(g)}^{\frac{1}{2}}} \right),$$

with n equal to one mole of electrons transferred per mole of reaction. In the second line, we have set the standard state activities of the pure solid silver and silver chloride to unity. In the third line, we have written the activities of $H^+(aq)$ and $Cl^-(aq)$ in terms of the mean activity

$$a_{\pm} = \gamma_{\pm} \frac{m_{\pm}}{m_0}.$$

The mean molarity is equal to the concentration of the hydrochloric acid solution

$$m_{\pm} = \left(m_{H^+(aq)} \, m_{Cl^-(aq)} \right)^{1/2} = m_{HCl}$$

assuming that it is fully dissociated. The activity of hydrogen gas can be separately measured or estimated as we have seen in Section 9.4. For convenience, let us approximate hydrogen as an ideal gas so that we simply have

$$a_{H_2(g)} = \frac{P_{H_2}}{P_0}.$$

Then the cell potential is given by

$$\varphi = \varphi^{\emptyset} + \frac{RT}{2F} \ln\left(\frac{P_{H_2}}{P_0} \right) - \frac{2RT}{F} \ln(a_{\pm})$$

$$= \varphi^{\emptyset} + \frac{RT}{2F} \ln\left(\frac{P_{H_2}}{P_0} \right) - \frac{2RT}{F} \ln\left(\frac{m_{HCl}}{m_0} \right) - \frac{2RT}{F} \ln(\gamma_{\pm}).$$

As discussed in Section 9.7.1, the value of φ_0 is obtained from measuring the value of φ in the limit of low concentration m_{HCl} when the activity coefficient $\gamma_\pm$ is unity. That is:

$$\varphi^\emptyset = \lim_{m_{HCl} \to 0} \left(\varphi - \frac{RT}{2F} \ln \left(\frac{P_{H_2}}{P_0} \right) + \frac{2RT}{F} \ln \left(\frac{m_{HCl}}{m_0} \right) \right).$$

Such measurements give the value of $\varphi^\emptyset = 0.22$ V we have listed in the table on the electrochemical series. Thus, the cell potential for the Harned cell is

$$\varphi = 0.22 + \frac{RT}{2F} \ln \left(\frac{P_{H_2}}{P_0} \right) - \frac{2RT}{F} \ln \left(\frac{m_{HCl}}{m_0} \right) - \frac{2RT}{F} \ln(\gamma_\pm).$$

Both the hydrogen pressure and the concentration of the hydrochloric acid can be separately measured. Hence, measurements of the cell potential give us the values of the activity coefficient at any concentration of the solution. An application is illustrated in the following example.

§   *Example 13.9 Measuring the activity coefficient*

In an experiment at a temperature of 298.15 *K and a hydrogen gas pressure of* 1 *bar, the cell potential of a Harned cell is measured to be* 0.188 *V when the concentration of the hydrochloric acid is* 1.93 *mol dm*$^{-3}$. *Assuming that the hydrogen gas activity coefficient is equal to unity, estimate the mean activity of a* 1.93 *mol dm*$^{-3}$ *solution of hydrochloric acid from the measured cell potential. Compare this to the value estimated from the Debye-Hückel model with the ion size parameter equal to* 9 Å *and* 3 Å, *respectively, for* $H^+(aq)$ *and* $Cl^-(aq)$. *Comment on the model estimate.*

From the experimental data, we have

$$\varphi = 0.188 \text{ V}$$

$$\varphi_0 = 0.22 \text{ V}.$$

Hence,

$$\ln(\gamma_\pm) = (\varphi_0 - \varphi)\frac{F}{2RT} - \ln\left(\frac{m_{HCl}}{m_0}\right)$$

$$= -0.0347.$$

This gives an activity coefficient of $\gamma_\pm = 0.966$.

The solution contains only hydrochloric acid, assumed completely dissociated, so that the ionic strength is

$$I = \frac{1000}{2\rho}\left(m_{H^+(aq)} + m_{Cl^-(aq)}\right)$$

$$= 1.936 \text{ mol kg}^{-1}.$$

Using the Debye-Hückel model,

$$\ln\gamma_{H^+} = -1.170\left[\frac{1.936^{1/2}}{1 + (9 \times 10^{-10} \times 3.281 \times 10^9 \times 1.936^{1/2})}\right]$$

$$= -0.319.$$

$$\ln\gamma_{Cl^-} = -1.170\left[\frac{1.936^{1/2}}{1 + (3 \times 10^{-10} \times 3.281 \times 10^9 \times 1.936^{1/2})}\right]$$

$$= -0.687.$$

Hence, the activity coefficient are

$$\gamma_{H^+} = 0.727$$

$$\gamma_{Cl^-} = 0.503.$$

These are substantially lower than the experimental value. This is not too surprising because the concentration of 1.93 mol dm^{-3} is considerably higher than the range for which the Debye-Hückel model is expected to give satisfactory estimates.   §

§   *Example 13.10 Pressure dependence of the cell potential*

The hydrogen gas pressure in the Harned cell in Example 13.9 is increased to 5 bar. How does its cell potential change?

The hydrogen electrode is the anode of the Harned cell. At the anode, the half-cell reaction is

$$\frac{1}{2} H_2(g) \rightarrow H^+(aq) + e^-$$

so that we expect increasing the hydrogen gas pressure to push this reaction to the right. Then, the anode of the cell becomes more effective at pushing electrons into an external circuit; the electrical potential of the anode becomes more negative. Thus, the cell potential of the Harned cell is expected to increase when the hydrogen gas pressure increases. Taking the mean activity of the electrolyte to be 0.966 as calculated in Example 13.9, we estimate the cell potential at an increased hydrogen pressure using

$$\ln(\gamma_\pm) = \ln(0.966) = (\varphi_0 - \varphi)\frac{F}{2RT} - \ln\left(\frac{m_{HCl}}{m_0}\right) + \frac{1}{4}\ln\left(\frac{P_{H_2}}{P_0}\right).$$

This gives $\varphi = 0.209$ V, which is an increase as we have expected qualitatively. §

§   *Example 13.11 Equilibrium constant from the cell potential*

The Daniell cell reaction

$$Cu^{2+}(aq) + Zn(s) \rightleftharpoons Cu(s) + Zn^{2+}(aq)$$

is also the same reaction that occurs when you simply toss a piece of zinc into an aqueous solution of copper sulphate. This is, of course, not an electrochemical cell, but the generality of thermodynamics means that the Daniell cell potential provides information that is applicable to it. Using the standard cell potential for the Daniell cell, calculate the equilibrium constant and the equilibrium concentrations of $Zn^{2+}(aq)$ and $Cu^{2+}(aq)$ when you place a sufficiently large piece of zinc into a 0.01 mol dm^{-3} solution of copper sulphate.

At equilibrium, the Nernst equation gives

$$0 = \varphi^{\emptyset} - \frac{RT}{nF} \ln \left(\frac{m_{Zn^{2+}(aq)}}{m_{Cu^{2+}(aq)}} \right).$$

Each mole of $Cu^{2+}(aq)$ displaced from solution produces one mole of $Zn^{2+}(aq)$ so that

$$m_{Zn^{2+}(aq)} = 0.01 - m_{Cu^{2+}(aq)}.$$

Hence, at equilibrium

$$exp \left(\frac{2F \times 1.104}{RT} \right) = 2.1 \times 10^{37} = \frac{0.01 - m_{Cu^{2+}(aq)}}{m_{Cu^{2+}(aq)}}.$$

This gives the equilibrium concentrations of

$$m_{Cu^{2+}(aq)} = 4.73 \times 10^{-40} \cong 0 \text{ mol dm}^{-3}$$

$$m_{Zn^{2+}(aq)} = 0.01 \text{ mol dm}^{-3}.$$

We have explicitly written a numerical value for $m_{Cu^{2+}(aq)}$ just to indicate its rather small value. Qualitatively, you know from the large value 2.1×10^{37} of the equilibrium constant that $m_{Cu^{2+}(aq)}$ is practically zero. The point here is that the near zero equilibrium concentration, and hence the large equilibrium constant, is not practical to measure directly. However, it is relatively easy to measure the cell potential for the same reaction in an electrochemical cell. The cell voltage of $\varphi^{\emptyset} = 1.104$ V can be easily measured using a simple voltmeter. It is frequently the case that multiple electrochemical equilibria occur simultaneously. In all aqueous solutions, the dissociation of water occurs, and could be important. We illustrate this with the following example. §

§   *Example 13.12 Measuring the dissociation constant of water*

You have the following electrochemical cell

$$Pt(s)|H_2(g)|NaCl(aq), NaOH(aq)|AgCl(s)|Ag(s).$$

with a hydrogen anode and a silver-silver chloride cathode. The electrolyte is a solution of sodium chloride and sodium hydroxide.

The reaction at the anode is the usual oxidation of hydrogen gas

$$\frac{1}{2}H_2(g) \rightleftharpoons H^+(aq) + e^-,$$

as a result of which protons go into solution. The reaction at the cathode is

$$AgCl(s) + e^- \rightleftharpoons Cl^-(aq) + Ag(s).$$

Thus, chloride ions go into solution from the cathode, converting the silver chloride in the cathode to silver.

Show how this Harned cell can be used to measure the dissociation constant of water.

The overall reaction is

$$AgCl(s) + \frac{1}{2}H_2(g) \rightleftharpoons Cl^-(aq) + Ag(s) + H^+(aq).$$

Thus, the cell potential is related to the activities by

$$\varphi = \varphi^\varnothing - \frac{RT}{F}\ln\left(\frac{a_{Cl^-(aq)}\,a_{Ag(s)}\,a_{H^+(aq)}}{a_{AgCl(s)}\,a_{H_2(g)}^{1/2}}\right).$$

Now, in any aqueous solution, we have the equilibrium

$$H_2O(liq) \rightleftharpoons H^+(aq) + OH^-(aq)$$

which has the equilibrium constant

$$K_w = \frac{a_{H^+(aq)}\,a_{OH^-(aq)}}{a_{H_2O}(liq)}.$$

Substituting for $a_{H^+(aq)}$ in the expression for the Harned cell potential, we get

$$\varphi = \varphi^\varnothing - \frac{RT}{F}\ln\left(K_w \frac{a_{Cl^-(aq)}\, a_{Ag(s)}\, a_{H_2O}}{a_{AgCl(s)}\, a_{H_2(g)}^{1/2}\, a_{OH^-(aq)}}\right).$$

In the limit of a dilute solution so that both

$$m_{Cl^-(aq)} \to 0$$

$$m_{OH^-(aq)} \to 0$$

whereupon the activity of liquid water also goes to unity because the solution tends towards pure water. Furthermore, we can set the hydrogen gas pressure at the anode such that the activity of hydrogen gas is equal to one. Using the usual unit activities for the pure solids, we then have

$$\ln K_w = \lim_{\substack{m_{Cl^-(aq)} \to 0 \\ m_{OH^-(aq)} \to 0}} \left[\frac{F}{RT}(\varphi^\varnothing - \varphi) - \ln\left(\frac{m_{Cl^-(aq)}}{m_{OH^-(aq)}}\right)\right].$$

If the hydroxide and chloride ion concentrations are adjusted to be equal, then

$$\ln K_w = \lim_{\substack{m_{Cl^-(aq)} \to 0 \\ m_{OH^-(aq)} \to 0}} \left[\frac{F}{RT}(\varphi^\varnothing - \varphi)\right].$$

Therefore, the dissociation constant of water can be relatively easily measured using the cell potential of this Harned cell with hydrogen-silver chloride/silver electrodes. The silver/silver-chloride electrode is commonly used in pH meters. §

You should note that while the cell potential is relatively easy to measure, it can be considerably more difficult to determine K_w by directly measuring the typically rather small concentrations of the aqueous hydrogen and hydroxide ions.

13.7 The temperature dependence of $\varphi^{\varnothing}$

We have seen in Chapter 9 that the chemical potential of a substance A depends upon its activity a_A through:

$$\mu_A = \mu_A^{\varnothing} + RT \ln a_A$$

where $\mu_A^{\varnothing}$ is its chemical potential for the state with unit activity at a standard temperature, typically 298.15 K and a standard pressure, typically 1 bar. The logarithmic term in the activity describes the dependence upon the composition of the system. For a gas phase species, this term gives the pressure dependence of the chemical potential μ_A, and hence the cell potential, as we have seen in Example 13.10. For liquid and solid phase species, the pressure dependence is small and generally neglected. For any generic reaction

$$\nu_A A^{+z_A} + \nu_B B^{-z_B} \rightleftharpoons \nu_C C^{+z_C} + \nu_D D^{-z_D}$$

the standard cell potential is

$$\varphi^{\varnothing} nF = -\Delta_{rxn}\mu^{\varnothing} = -\nu_C \mu_C^{\varnothing} - \nu_D \mu_D^{\varnothing} + \nu_A \mu_A^{\varnothing} + \nu_B \mu_B^{\varnothing}$$

where the standard chemical potential for each of the reacting species is at a specific standard temperature and standard pressure. Thus, the dependence of the standard cell potential upon the temperature is given by

$$nF \frac{d\varphi^{\varnothing}}{dT} = -\left[\frac{\partial\left(\Delta_{rxn}\mu^{\varnothing}\right)}{\partial T}\right]_P$$

where the derivatives of the chemical potentials with respect to temperature are evaluated at constant pressure equal to the standard pressure P_0. But, we have seen from Section 7.6 that

$$\left(\frac{\partial G}{\partial T}\right)_P = -S.$$

Therefore,

$$nF\frac{d\varphi^{\emptyset}}{dT} = \left[\frac{\partial\left(\nu_C\mu_C^{\emptyset} + \nu_D\mu_D^{\emptyset} - \nu_A\mu_A^{\emptyset} - \nu_B\mu_B^{\emptyset}\right)}{\partial T}\right]_P$$

$$= \nu_C s_C^{\emptyset} + \nu_D s_D^{\emptyset} - \nu_A s_A^{\emptyset} - \nu_B s_B^{\emptyset}$$

$$= \Delta_{rxn}S^{\emptyset}$$

where $s_i^{\emptyset}$ is the molar entropy of species i at the standard temperature and pressure, and $\Delta_{rxn}S^{\emptyset}$ is the standard change in the entropy per mole of the reaction. Measuring the standard cell potential and its rate of change with respect to temperature then gives the standard Gibbs free energy and the standard entropy of reaction.

$$\Delta_{rxn}\mu^{\emptyset} = -nF\varphi^{\emptyset}$$

$$\Delta_{rxn}S^{\emptyset} = nF\frac{d\varphi^{\emptyset}}{dT}.$$

Since G is equal to $H - TS$, we can determine the standard enthalpy of reaction from these two quantities

$$\Delta_{rxn}H^{\emptyset} = -nF\left(\varphi^{\emptyset} - T\frac{d\varphi^{\emptyset}}{dT}\right).$$

Here, the standard enthalpy of reaction is for one mole of the reaction

$$\Delta_{rxn}H^{\emptyset} = \nu_C h_C^{\emptyset} + \nu_D h_D^{\emptyset} - \nu_A h_A^{\emptyset} - \nu_B h_B^{\emptyset}$$

where $h_i^{\emptyset}$ is the molar enthalpy of species i at the standard temperature and pressure.

The change in the standard cell potential with temperature should not be puzzling. This is because it is simply a measure of the standard chemical potential of reaction $\Delta_{rxn}\mu^{\emptyset}$, which is the stoichiometry-weighted change in the Gibbs free energy per mole of the reaction at standard temperature. Since G is equal to $H - TS$ and both enthalpy and entropy of the reactants and products are dependent upon temperature, $\Delta_{rxn}\mu^{\emptyset}$ generally depends upon temperature. The equilibrium constant generally depends upon temperature.

§   *Example 13.13 Temperature dependence of the cell potential*

For temperatures from zero to 95 degrees Celsius the standard cell potential of the Harned cell is experimentally found to depend upon temperature as follows:

$$\varphi^{\emptyset} = 0.23659 - 4.8564 \times 10^{-4}(T - 273)$$

$$-3.4205 \times 10^{-6}(T - 273)^2$$

$$+5.869 \times 10^{-9}(T - 273)^3.$$

Calculate the Gibbs free energy, entropy and enthalpy of reaction for

$$AgCl(s) + \frac{1}{2}H_2(g) \rightleftharpoons Cl^-(aq) + Ag(s) + H^+(aq).$$

At 298.15 K, we get

$$\varphi^{\emptyset} = 0.2223 \text{ V}$$

$$\frac{d\varphi^{\emptyset}}{dT} = -0.6465 \text{ mV K}^{-1}.$$

For the reaction as written, the number of moles of electrons transferred per mole of reaction is one. Thus, at standard state

$$\Delta_{rxn}\mu^{\emptyset} = -nF\varphi^{\emptyset} = -21.45 \text{ kJ mol}^{-1}$$

$$\Delta_{rxn}S^{\emptyset} = nF\frac{d\varphi^{\emptyset}}{dT} = -62.38 \text{ J mol}^{-1} \text{ K}^{-1}$$

$$\Delta_{rxn}H^{\emptyset} = -nF\left(\varphi^{\emptyset} - T\frac{d\varphi^{\emptyset}}{dT}\right) = -40.05 \text{ kJ mol}^{-1}. \quad §$$

Electrochemistry is a means of converting chemical energy to electrical energy. This is the basis of much current work on fuel cell technology to make devices that extract energy from fuels as efficiently as possible. Regardless of the actual chemical reaction used in a fuel cell or the construction of the cell, its thermodynamics determines the maximum amount of useful work we can extract. Consider the Harned cell in the final example above. Since the entropy of the Harned cell decreases by 62.38 J mol^{-1} K^{-1}, the minimum amount of energy that needs to be *lost as heat* to the environment in order for the Harned cell reaction to proceed spontaneously is

$$-T\Delta_{rxn}S^{\emptyset} = 298.15 \times \frac{62.38}{1000} = 18.6 \text{ kJ mol}^{-1}.$$

Thus, the maximum electrical work that can be extracted using the Harned cell is

$$-\Delta_{rxn}H^{\emptyset} + -T\Delta_{rxn}S^{\emptyset} = 40.05 - 18.6 = 21.45 \text{ kJ mol}^{-1},$$

which is simply the decrease in the Gibbs free energy of the system per mole of reaction $-\Delta_{rxn}\mu^{\emptyset}$.

This brings us back to the fundamentals in Chapter 7 where we learned to account for heat and work exchange given the laws of Thermodynamics: the maximum useful work is extracted when we minimize the energy dissipated to the surroundings as heat. From the Second Law, we know that heat is a form of energy like no other; the more heat is produced in *any* process, the more the Universe winds down. This is the fundamental lesson from Carnot in his *Reflections on the Motive Power of Fire*.

Index

Made in the USA
Monee, IL
27 December 2025

40422753R00272